Bioeconomy of *Streptomyces*

This book provides a comprehensive understanding of streptomycetes, highlighting their various habitats, diversity, genetic structure, and metabolic engineering techniques to enhance the production of secondary metabolites. It also presents techniques for the isolation, cultivation, biochemical identification, chemotaxonomy, and phylogenetic analysis of streptomycetes. The book explores the production of bioactive compounds from streptomycetes, including novel natural products, antimicrobial agents, pharmaceuticals, medicinal compounds, and bioactive enzymes. Furthermore, the book examines the diverse applications of streptomycetes in agriculture, medicine, industry, and bioremediation. The chapters emphasize the latest technological advancements such as bioprocess optimization, genetic engineering, metabolic engineering, genome mining, and pathway engineering for enhancing the production of secondary metabolites from *Streptomyces*. As such, this book is of interest to academicians, researchers, and professionals working in the fields of bioprocessing, microbiology, industrial microbiology, medical microbiology, fermentation technology, and biotechnology.

Bioeconomy of *Streptomyces*

Edited by
Kashyap Kumar Dubey
Punit Kumar

CRC Press
Taylor & Francis Group
Boca Raton London New York

CRC Press is an imprint of the
Taylor & Francis Group, an **informa** business

Designed cover image: Shutterstock

First edition published 2025
by CRC Press
2385 NW Executive Center Drive, Suite 320, Boca Raton FL 33431

and by CRC Press
4 Park Square, Milton Park, Abingdon, Oxon, OX14 4RN

CRC Press is an imprint of Taylor & Francis Group, LLC

ISBN: 978-1-032-69457-3 (hbk)
ISBN: 978-1-032-69459-7 (pbk)
ISBN: 978-1-032-69463-4 (ebk)

DOI: 10.1201/9781032694634

Typeset in Sabon
by SPi Technologies India Pvt Ltd (Straive)

Contents

About the Editors

Kashyap Kumar Dubey is currently Professor and Dean, School of Biotechnology at Jawaharlal Nehru University, New Delhi (India). He is working in the areas of sustainability, process development, and removal of antineoplastic compounds from aqueous waste. He is running two research projects sanctioned from ICMR-DHR, CRG-DST and has completed six projects funded by DBT (Horizon 2020) on Clean Water for Health, BIRAC-DBT (PACE), BRICS-DST, UGC, and SERB on microbial process development. He has been significantly published in the areas of biomanufacturing and hospital aqueous wastewater treatment. His motivation is towards development of low-cost decentralised technology for onsite treatment of hospital aqueous water.

Dr. Punit Kumar is currently working in the Department of Morphology at Karaganda Medical University, Kazakhstan. He is an accomplished researcher and educator in the field of life science and biotechnology, with a focus on health science, microbial bioprocessing and bioengineering. He holds a Ph.D. in Biotechnology. His research work during his Ph.D. revolved around enhancing lipstatin production from *Streptomyces toxytricini* under submerged fermentation. With a keen interest in teaching and imparting knowledge, Punit Kumar has several teaching experiences at various prestigious institutes. He has mentored both undergraduate and postgraduate students, contributing to their academic growth and fostering their interest in life science. He has made significant contributions to the field through his research articles and book chapters published in reputable international journals and books. His research interest focuses on cytotoxicity study, histology, secondary metabolites, microbial biodegradation, bioengineering, and bioprocessing. Dr. Punit Kumar is an active member of esteemed research societies, including the Indian Science Congress, the Association of Microbiologists of India, and the Biotech Research Society of India.

Contributors

Aarzoo
University College Dublin
Ireland

Neeraj Agarwal
Department of Life Sciences
Sharda University, Knowledge Park III
Greater Noida, India

Rabab Anjum
Department of Biotechnology
IMS Engineering College
Ghaziabad, India

Archana
Department of Electronic and Communication
Kashi Institute of Technology
Varanasi, India
and
Department of Electronics and Communication
Motilal Nehru National Institute of Technology Allahabad,
Uttar Pradesh, India

Kanchan Chatterjee
Department of Zoology
Doon (PG) College of Agriculture Science and Technology Sealqui
Dehradun, India

Preeti Chauhan
Department of Biotechnology
Chandigarh Colleges of Technology
Mohali, India

Kashyap Kumar Dubey
School of Biotechnology
Jawaharlal Nehru University
New Delhi, India

Ekta
Department of Biotechnology
IMS Engineering College
Ghaziabad, India

Oluwabusola Omolola Fatade-Ogunseye
Department of Microbiology
University of Ibadan
Ibadan, Nigeria

Asim Kumar Jana
Department of Biotechnology
National Institute of Technology
Jalandhar, India

Palki Sahib Kaur
Department of Biotechnology
Chandigarh Colleges of Technology
Mohali, India

Vivek Keshri
Department of Pharmacology
Kashi Institute of Pharmacy
Varanasi, India

Khushboo
Department of Botany
Banaras Hindu University
Varanasi, India

Deepak Kumar
Department of Botany
Banaras Hindu University
Varanasi, India

Mohit Kumar
Department of Life Science, SSBSR
Sharda University
Greater Noida, India

Naveen Kumar
Kurukshetra University
Haryana, India

Punit Kumar
Department of Morphology
Karaganda Medical University
Karaganda, Kazakhstan

Yatin Kumar
Department of Microbiology
Central University of Haryana
Mahendergarh, India

Harmanpreet Meehnian
Department of Biotechnology Engineering and Food Technology
Chandigarh University
Mohali, India

Santosh Kumar Mishra
Department of Life Science, SSBSR
Sharda University
Greater Noida, India

Antonina Nikolayeva
Medical University of Karaganda
Kazakhstan

Nupur
Department of Life Sciences
Sharda University, Knowledge Park III
Greater Noida, India

Kehinde Caleb Omidoyin
Department of Agricultural Chemistry
Chonnam National University
Gwangju, South Korea

Darshna Pandey
Department of Microbiology
Central University of Haryana
Mahendergarh, India

Anushree Pant
Department of Biotechnology
National Institute of Technology
Jalandhar, India

Tilak Raj
Department of Zoology
Pandit Neki Ram Sharma Government College
Rohtak, India

Deepanshi Rajput
School of Biotechnology
Jawaharlal Nehru University
New Delhi, India

Ranu A.
Department of Biotechnology
National Institute of Technology
Jalandhar, India

Sandeep Rathor
Department of Pharmaceutics
MM College of Pharmacy
Maharishi Markandeshwar Mullana
Ambala, India

Saheli Sabnam
Department of Biotechnology
National Institute of Technology
Jalandhar, India

Nisha Sharma
Baba Haridas College of Pharmacy
Nazafgarh, India

Tanisha Sharma
Department of Biotechnology
IMS Engineering College
Ghaziabad, India

Pritam Singh
Chandigarh College of Pharmacy
Chandigarh Group of College
Landran
Greater Mohali, India

Ananya Tanwar
Department of Pharmaceutics
MM College of Pharmacy
Maharishi Markandeshwar Mullana
Ambala, India

Mony Thakur
Department of Microbiology
Central University of Haryana
Mahendergarh, India

D. Vasanth
Department of Biotechnology
National Institute of Technology
Raipur, India

Sakshi Verma
Department of Pharmaceutics, MM College of Pharmacy
Maharishi Markandeshwar Mullana
Ambala, India

Vinod Yadav
Department of Microbiology
Central University of Haryana
Mahendergarh, India

Kashish Zoya
Department of Biotechnology
IMS Engineering College
Ghaziabad, India

Chapter 1

Exploring the ecological and molecular frontiers of *Streptomyces*

From soil to sea

Mony Thakur
Central University of Haryana, Mahendergarh, India

Khushboo
Banaras Hindu University, Varanasi, India

Yatin Kumar, Darshna Pandey and Vinod Yadav
Central University of Haryana, Mahendergarh, India

Deepak Kumar
Banaras Hindu University, Varanasi, India

Kashyap Kumar Dubey
Jawaharlal Nehru University, New Delhi, India

1.1 INTRODUCTION

Natural products have long served as indispensable therapeutic agents and resources for humanity, owing to their historical utilization and manifold benefits (Thakur et al. 2024). They are very important in pharmaceutical research because of their large biological resources and ability to make new molecules with altered and often beneficial properties (K. Alam et al. 2022). However, the emergence of pathogenic and multidrug-resistant microorganisms poses a formidable challenge (Kumar et al. 2024; Thakur et al. 2023). The term "actinomycetes" derives from the Greek words "atkis" (ray) and "mykes" (fungus), encompassing both bacterial and fungal entities. Recently, actinobacteriologists have pushed for the name "Actinobacteria" instead of "actinomycetes" to be used to describe these soil-dwelling microbes (Gajalakshmi, Bi, and Udhayasudhan 2022). They are similar to bacteria and fungi but also have enough differences to be considered a separate group. Taxonomically, Actinobacteria align with bacteria within the Schizomycetes class but within the Actinomycetales order (Azman et al. 2015). Hence, this chapter discusses these borderline microorganisms under

DOI: 10.1201/9781032694634-1

the nomenclature of Actinobacteria. Actinobacteria, primarily free-living saprophytes inhabiting soil, water, and plant surfaces, represent a significant component of soil microbial communities, their abundance varying with soil type (Yadav et al. 2018). This group makes many different secondary metabolites, such as enzymes, antibiotics, antitumor drugs, and immunosuppressants (Khushboo and Dubey 2023). Actinobacteria present a wide range of morphologies, including rod-coccoid forms like *Arthrobacter* and coccoid forms like *Micrococcus*. Certain species, such as those found in *Nocardia*, exhibit fragmenting hyphal structures, while other species, such as those found in *Streptomyces* and *Frankia*, grow complex, highly distinct branched mycelia. On substrates, rhodococci lengthen into filaments but do not produce genuine mycelium. Conversely, corynebacteria completely avoid mycelial development. But unlike other Actinobacteria, filamentous growth in corynebacteria mostly happens at the apex instead of through the extension of the lateral wall. Members of the *Oerskovia* genus are identified by their branched substrate hyphae, which break down into motile components with flagella. Notably, with few exceptions, mycobacteria and rhodococci usually do not generate aerial hyphae (Barka et al. 2016). Actinomycetes have been extensively studied since the discovery of streptomycin from *Streptomyces griseus* by Selman A. Waksman (1953). Streptomycetes, Gram-positive bacteria belonging to the Actinomycetales order, produce filamentous spores and trace their evolutionary lineage back approximately 450 million years to plant residue consumption. Renowned for their prolific production of secondary metabolites, streptomycetes contribute significantly to commercial industries (Ferrer et al. 2018). Their unique filamentous growth pattern and the fact that they often release colored bioactive compounds show how important they are to the decomposition of both organic and inorganic substrates in soil ecosystems. Their capacity to synthesize numerous enzymes aids in organic matter breakdown, rendering them pivotal contributors to nutrient cycling processes (Mitra et al. 2022). *Streptomyces* species are producers of different identified antibiotics to date, representing only a fraction of their bioactive chemical repertoire. Different screening methods have proven to be effective in discovering a diverse range of bioactive compounds with potential therapeutic applications from these microorganisms, including antiviral, anticancer, and immunomodulatory effects. Genomic analyses have revealed the remarkable biosynthetic potential of *Streptomyces* strains (Singhvi et al. 2023). Metagenomic studies reveal a substantial abundance of biosynthetic gene clusters (BGCs) inside these bacteria. Consequently, exploring these BGCs has become a promising strategy for identifying novel compounds with therapeutic potential to fight against drug resistance. Streptomycetes, exhibiting symbiotic relationships with fungi and plants, contribute to cell wall recycling processes, although certain strains display pathogenic traits (Shepherdson, Baglio, and Elliot 2023). The complex growth and developmental processes of model organisms such as *S. coelicolor* and *S. venezuelae*, offer insights into their ecological roles and evolutionary history (Seipke, Kaltenpoth, and Hutchings 2012). Genome analyses underscore the conservation of their ecological niche and fundamental roles in primordial soil formation (Timmis and Ramos 2021). Overall, the bioactive compounds produced by *Streptomyces* strains have great potential for drug discovery and development. By exploring the genetic makeup of these microorganisms, researchers are uncovering new possibilities for combating antibiotic resistance and other health challenges.

1.1.1 Distribution pattern of *Streptomyces* in different habitats

Actinomycetes encompass 120 genera, including *Streptomyces* (E. M. H. Wellington, personal communication). At first, actinomycetes were thought to be a bridge between bacteria and fungi. Actinomycetes represent a major branch of Gram +ve bacteria, and their DNA has a lot more G+C than that of other Gram +ve bacteria like *Staphylococcus*, *Streptococcus*, and *Bacillus* (Singhvi et al. 2023). Soil-dwelling actinomycetes are producers of thousands of known antibiotics to date, surpassing figures for all other bacteria and fungi combined (Khushboo and Dubey 2023). Since its inception, *Streptomyces* genetics has undergone significant evolution as a result of the discovery of mitosis and meiosis-like chromosomal activities. *Streptomyces* chromosomes are made up of a "core" with important housekeeping genes and "arms" that are full of conditionally adaptable genes that were probably gained through horizontal transfer as the species evolved (Thakur et al. 2023). Nearly 4,110 genomes of the *Streptomyces* genus are cataloged in the NCBI database. Most streptomycetes produce spores and emit a characteristic "earthy" odor due to geosmin synthesis, primarily found in soil and decaying plant matter (Cheng et al. 2020). Resting spores undergo development where one or more germ tubes emerge, extending apically to form new hyphal tips featuring side branches and cross walls. This process culminates in the formation of sporulation compartments upon depletion of nutrients. The genome sequence (whole) of *S. coelicolor* was unveiled in 2002, boasting the highest gene count among bacteria at the time. *S. coelicolor* was initially considered a subspecies of *S. violaceoruber*, with subsequent transcriptome studies conducted on strain A3(2) in 2016 (Komaki and Tamura 2021). The first whole genome sequence of *S. avermitilis* was elucidated in 2003 (Ikeda et al. 2003). Unlike other bacterial species with circular chromosomes, each *Streptomyces* chromosome is linear in structure, with *S. scabies* harboring the largest genome, attributed to a vast pathogenicity island causing potato scab disease.

1.1.1.1 Soil ecosystem

Streptomyces spp. naturally occur in soil, and their distribution is significantly high when isolated from different ecosystems as well as from extreme environments (Donald et al. 2022; Das and Mangwani 2015). Recent studies from Calamba City, Laguna, in the Philippines showed that the number of *Streptomyces* isolates varied with the site of collection (Bontemps et al. 2013). There are several factors, like soil nutrients, soil type, temperature, precipitation, humidity, and pH, that influence the structure and distribution pattern of the *Streptomyces* community (Antido and Climacosa 2022; Davelos, Kinkel, and Samac 2004). On the basis of the pH of the soil, there are acid-resistant groups that grow in acidic soil and are most abundant in nature. The *Streptomyces* genus is also responsible for producing more than 70% of known antibiotics. Interestingly, *Streptomyces* has the capability to thrive in extreme environmental conditions like high salinity, high exposure to UV rays, high pH, high temperature, low nutrient and oxygen supply, and so on, which is detrimental to other bacteria. Some unique features such as antibiosis, the synthesis of specific enzymes, and switching among the metabolic modes allow them to survive in extreme conditions (Tanaka and Omura 1990). The correlation between nutrient acquisition and utilization plays a significant role in shaping the antibiotic production and pathogen-inhibitory capacity of *Streptomyces* spp. in soil (Shivlata and Tulasi 2015). Another group reported the abundance of the *Streptomyces* community in areas where rainfall is low and the soil type is Yermosol (Hamid et al. 2020). Several other groups studied the forest ecosystem and demonstrated that strains of *Streptomyces* play an effective role in the decomposition of wood constituents, leading to the breakdown of cellulose and other organic matter. *Streptomyces* also aids in hydrogen uptake and regulates ecosystem dynamics (Bontemps et al. 2013).

1.1.1.2 Marine ecosystem

The isolation and culturing of *Streptomyces* species from marine environments often require specific growth conditions (Dharmaraj, Ashokkumar, and Dhevendaran 2010). Maldonado et al. have demonstrated that the isolation medium should be supplemented with salts to adjust the osmotic levels and mimic the conditions of the seawater (Maldonado et al. 2005). Certain isolation media are fortified with supplements such as 50% (v/v) seawater, 3% NaCl, and, in some cases, even up to 70% seawater to facilitate the growth of the bacteria (Katz et al. 2017; Huang et al. 2016). Moran et al. demonstrated that the marine habitat is very dynamic, and there are several species that dwell at the bottom of the marine ecosystem, and they face extreme situations like nutrient deficiency, high salinity, and high pressure (Moran, Rutherford, and Hodson 1995). Hence, they adapt according to the environment, evolve over time, and become genetically distinct from the terrestrial ecosystem. There are several reports that have explained the unexplored marine sediments for microbial diversities (Tiwari and Gupta 2013). The term "sediments" refers to sandy beaches, shallow to deep-water sediments, and mangrove sediments. Marine sediment samples were utilized for the isolation of *Streptomyces* employing various isolation media. The

International *Streptomyces* Project (ISP2) media is one of the most commonly used media (Law et al. 2017). But these *Streptomyces* are not confined to the seawater and sediments; they are also discovered in a diverse array of biological sources, such as mangroves, algae, jellyfish, corals, sponges, and fish. They are found mainly in mangrove sediments, and they play effective role in regulating the biogeochemical cycles and maintaining the food chain (Das and Mangwani 2015).

1.1.2 Traditional methods for the identification of *Streptomyces*

Different approaches, such as macroscopic and microscopic morphology, growth characteristics, and biochemical tests, were used for the identification of *Streptomyces* (Akshatha and Kalyani 2022). *Streptomyces* strains exhibit diverse growth characteristics, including growth rate, colony size, texture, and morphology. *Streptomyces* colonies are typically observed as raised, wrinkled, or powdery with a fuzzy texture. They often exhibit various colors ranging from white, gray, yellow, red, or even blue-green (Goredema et al. 2020). The production of pigmented compounds can affect coloration. Colonies may also produce an earthy odor due to the synthesis of geosmin, a volatile compound. Environmental factors like temperature, pH, and nutrient availability can have an impact on these characteristics. Observing the growth patterns of *Streptomyces* cultures on agar plates or in liquid media provides valuable information for strain characterization. *Streptomyces* form long, filamentous structures called hyphae, which branch extensively, forming a network of mycelium (Boukaew et al. 2022). Aerial hyphae rise above the substrate mycelium and differentiate into chains of spores (conidia). The arrangement of spores and the morphology of aerial hyphae can vary significantly between species and even strains.

Streptomyces species produce the catalase enzyme associated with the breakdown of hydrogen peroxide into water and oxygen, producing bubbles (Yuan et al. 2021). Some *Streptomyces* species have the ability to reduce nitrate to nitrite or nitrogen gas, which can be observed in nitrate-containing mediums (Haihan Zhang et al. 2021). *Streptomyces* strains have the ability to hydrolyze substrates such as starch, gelatin, or casein (Akintunde 2021). These are the traditional methods through which researchers can comprehensively characterize and identify *Streptomyces* strains, gaining insights into their taxonomic diversity, metabolic capabilities, and potential applications in various fields such as medicine, agriculture, and biotechnology.

1.1.3 Modern approaches for molecular characterization

1.1.3.1 Evolution of taxonomic classification

Accurate taxonomic classification is paramount for effective scientific discourse, facilitating a deeper understanding of bacterial species and their evolutionary dynamics (Goris et al. 2007). Haeckel's fundamental phylogeny classification system underwent revisions by Copeland, who proposed a separate kingdom for bacteria (Herron, Conlin, and Ratcliff 2022). Eventually, Whittaker introduced a five-kingdom classification, which gained widespread acceptance following the inclusion of fungi (Singh 2021). Bacterial identification poses challenges due to the unique properties of bacteria, which necessitate time-consuming biochemical, nutritional, and molecular analyses. Initially, bacterial classification relied on observable traits such as color, colony morphology, and growth conditions (temperature, pH). Subsequently, physiological and biochemical characteristics became the basis for taxonomic delineation of bacteria (Komaki 2023). Further chemo-taxonomic markers such as alkaloids, steroids, amino acids, and sugar metabolism were also considered for bacterial characterization as morphological characteristics were still unable to establish a true phylogeny of the bacterium (Stadler and Hellwig 2004). In the late 1970s, R. R. Colwell provided a well-defined system for bacterial taxonomy on the basis of genotypic, phenotypic, and chemotaxonomic markers to help strengthen bacterial classification. The name "polyphasic method" was used to describe this classification technique, which was based on data from phenotypic and chemotaxonomic analyses, DNA-DNA hybridization (DDH), and 16S rRNA gene phylogenetic analysis, as shown in "Polyphasic approaches used for the taxonomic demarcation of microbes" (Figure 1.1) (Law et al. 2018). In addition, other genomic

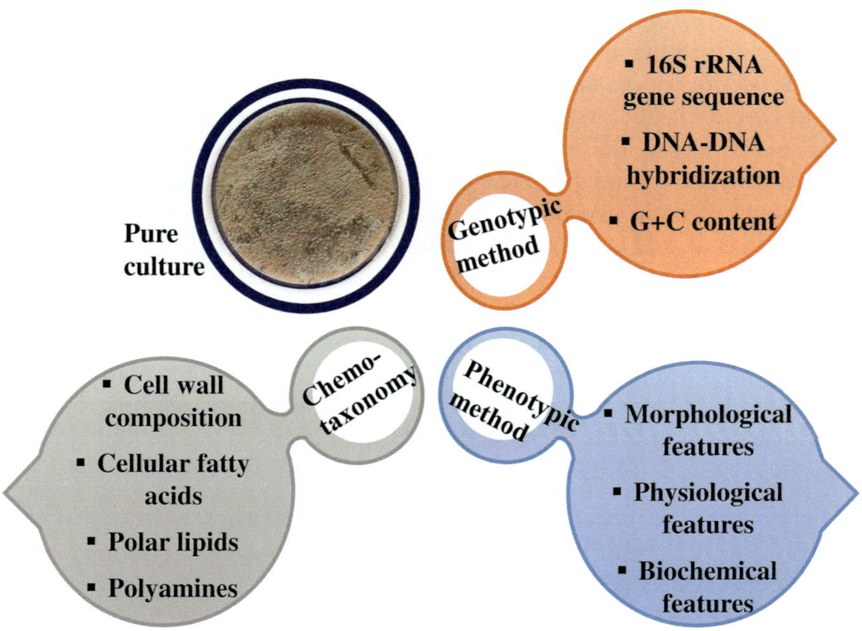

Figure 1.1 Polyphasic approaches used for the taxonomic demarcation of microbes.

regions such as 23S rRNA gene sequence, citrate synthetase gene, and so on have been used as marker genes for taxonomic identification (Qadir 2017). Genotypic markers evaluate GC content, phylogeny creation using rRNA, and DDH method. Morphological, physiological, biochemical features are considered as phenotypic markers whereas polyamines, fatty acids, whole cell protein, cell wall analyses, and isoprenoids are some of the chemotaxonomic markers used to identify bacterial species. Evolutionary research elucidates *Streptomyces* adaptation to soil environments, offering insights into comparative genome reannotation, visualization, and sequence comparison (Y. Li and Huang 2022). Phylogenetic analyses aid in reconstructing metabolic models crucial for strain engineering in biotechnology. The field of molecular systematics uses genomic DNA hybridization and modern sequencing tools to study how species evolved using 16S rDNA sequences as a starting point; see "Steps involved to determine the genetic similarity between two organisms" (Figure 1.2) (Croce et al. 2021). This means that molecular morphometrics needs to be studied more to find evolutionary connections. The ecological and economic significance of *Streptomyces* underscores ongoing research to explore actinobacterial diversity in diverse environmental settings.

1.1.3.2 16S rRNA sequencing

16S rRNA gene sequence has served as a molecular clock for identifying microorganisms. Its presence in all prokaryotes, high degree of conservation, functional stability, and low susceptibility to horizontal gene transfer made it ideal for the classification system. The molecular and phylogenetic

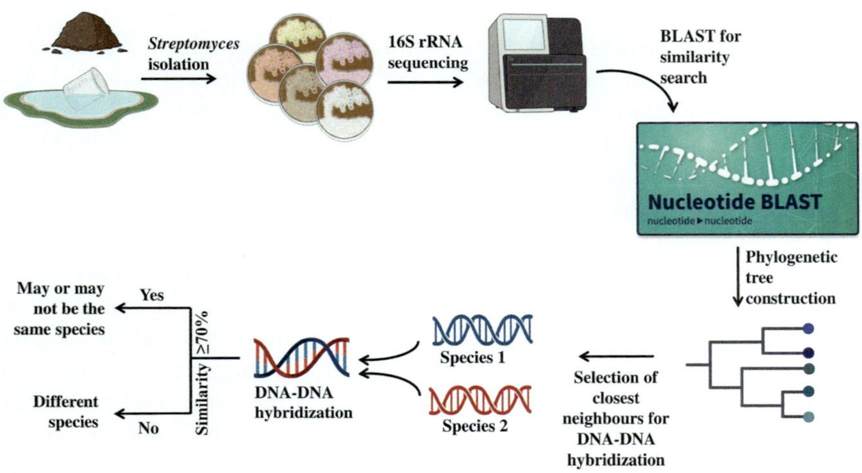

Figure 1.2 Steps involved to determine the genetic similarity between two organisms.

characterization improved with the development of polymerase chain reaction (PCR), DDH, and DNA sequencing techniques to examine the gene sequence in which the linear 16S rRNA gene sequence is targeted. 16S rRNA gene sequencing is one of the best standard methods for taxonomic identification and phylogenetic analysis (Law et al. 2018). The sequence match algorithm was considered to compare the 16S rRNA sequences to sequences within the Ribosomal Database Project. According to the standard genetic distances, molecular evolutionary genetics analysis (Mega) 5.1 was used to evaluate the genetic relationship of strains. The unweighted pair group method with arithmetic mean (UPGMA) was used for the results to construct a dendrogram with bootstrap repetitions of 1000 (Escalante-Réndiz et al. 2019). Being related closely and with a high number of members, the 16S rRNA gene gives low resolution to distinguish among species of *Streptomyces* (Law et al. 2018). Recent studies have used some basic standard methods with 16S rRNA for bacterial identification and linked them with the Biolog system (Algarni 2022). The 16S rRNA gene phylogeny analysis by Bontemps et al. indicated that there was a significantly high diversity of *Streptomyces* within the soil samples with at least nine distinct species, some of which can be previously unidentified or potentially new species (Bontemps et al. 2013). Also, several copies of 16S rRNA can be present in the whole genome sequence, but all 16S rRNA need not be molecularly sequenced; instead, metabolite profiling, DDH, and analysis of fatty acid methyl ester are preferable (Khadayat et al. 2020). Because of this, we need to find more phylogenetic markers besides the 16S rRNA gene that can be used to tell the difference between evolutionary relationships in the *Streptomyces* family (Law et al. 2018).

1.1.3.3 Phylogenetic tree reconstruction

A phylogenetic tree illustrates the evolutionary descent paths of distinct species from a common ancestor, while the history of descent from a common ancestor is known as phylogeny. The distance between one group of organisms from the groups indicating the degree of relationship is displayed by the phylogenetic tree, also known as a dendrogram (Abaza 2020). The visual depiction of the relationships between various species for their evolutionary chronology unites the diversity of groups (McCullough et al. 2020). In addition to being essential for deducing the origin of novel genes, identifying molecular adaptation, comprehending the growth of morphological characteristics, and reconstructing demographic alterations in recently diverged species, a precise phylogenetic tree aids in our comprehension of the key evolutionary transitions, including the appearance of novel body designs and metabolic processes (Kapli, Yang, and Telford 2020). Scientists utilize phylogenetics to comprehend the mechanisms governing an organism's evolution and dispersal across space and time (Hassler et al. 2023).

The neighbor-joining, maximum-likelihood (ML), and maximum-parsimony (MP) methods are used to construct phylogenetic trees. Using the auto Maximum-Relative-Error (MRE) criterion and rapid bootstrapping, an ML tree can be constructed from alignments using RAxML. Alternatively, the Tree Analysis New Technology program can be used to construct an MP tree from the alignments using 1000 bootstraps, tree-bisection-and-reconnection branch swapping, and ten random sequence replicates. The PAUP (Phylogenetic Analysis Using Parsimony) X2 test can be used to assess the sequences for computational bias (Kusuma, Nouioui, and Goodfellow 2021).

A collection of gene sequences (usually translated proteins) obtained from transcriptome sequencing or predicted from a genome sequence serves as the pipeline for phylogenetic tree construction. It is necessary to detect and eliminate contamination originating from commensals, parasites, animal gastrointestinal contents, environmental sources, or experimental mistakes, particularly in multiplexed transcriptome sequencing. Based on the GC content, taxonomy of sequence similarity matches, and read coverage, contaminants can be found and eliminated. The homologous sequences across all species of interest are found using programs like BLAST or others of a similar nature (Tizabi, Bachvaroff, and Hill 2022). To find putative orthologous genes whose associations should match the species phylogeny, clustering algorithms are applied. Multiple sequence alignments (MSAs) are produced by aligning the sequences of putative orthologues. For the putative orthologues, the MSA can be analyzed to construct an evolutionary tree to identify contaminants, other problematic sequences, and remaining paralogues. Unusually lengthy branches in the tree indicate the presence of these contaminants. Usually, the MSA is filtered to eliminate areas with inconsistent alignment. The super matrix created by concatenating the orthologues is analyzed to determine the species phylogeny. For distinct MSA partitions, various models may be applied (Kapli, Yang, and Telford 2020).

In a study, concatenated sequences of 231 single-copy core genes from the pathogenic isolates, their closest relatives *S. pratensis* and *S. xiamenensis* were used for the phylogenetic analysis. After RAST annotation, the gene identifiers were compared to choose the set of single-copy genes. MUSCLE was used to align the homologous gene sequences, and G-block implementation was used to clean the alignments and enhance phylogenetic reconstruction. Then, using the Maximum Likelihood approach and the RAxML programme on the CIPRES Science Gateway, evolutionary tree topology was created on the basis of aligned sequences. For bootstrap analysis, the data was resampled 1000 times, and use of GRTGAMMA model was done as the substitution model (Henao et al. 2023).

For the first phylogenetic study, it was safe and helpful to use ERIC-, Rep-, and BOX-PCR fingerprinting molecular typing methods to describe the scab-causing *Streptomyces* species. Paleontological Statistics Software (PAST) was used to study the banding patterns of the preceding additional methods. The UPGMA was used to carry out related dendrograms. The analysis of phylogenetic trees of data by ERIC-, Rep-, and BOX-PCR was according to reference guidelines (Karagoz et al. 2024). Different primers were used, among which better-amplified fragments were given by primers of BOX-A1R, being different from the patterns of amplification of the REP 1R and 2R primers. It was observed that amplification by REP-PCR was more diverse, whereas the amplified fragments by the BOX-A1R primer were higher in number (Borba et al. 2020).

The phylogenies may be poor or inaccurate if based on single-gene analysis, and to overcome the limitation whole-genome analysis can be used (M. T. Alam et al. 2010). ClustalX 2.1 or PhyloPhlAn 3.0 can be used to construct whole-genome phylogeny based on the results of average amino acid identity, an optimized subset of conserved universal proteins and percentage of conserved proteins analyses (Madhaiyan et al. 2022; Komaki 2023). Multilocus sequence analysis is evolutionarily sensible and can be a good option to check conclusions derived from the overall genome relatedness index (Chun and Rainey 2014).

1.1.3.4 Whole-genome sequencing

Whole-genome sequencing (WGS) typically identifies about 3 million single nucleotide variations, enabling nearly full genome sequencing (Kremer, Wortmann, and Prokisch 2018). This approach allows for a more detailed analysis of genetic variations compared to other sequencing methods, making it a valuable tool in research and clinical settings. Additionally, WGS can provide insights into disease risk, ancestry, and potential treatment options based on an individual's unique genetic makeup. Because they can find genetic links between species of interest, WGS–based techniques are likely to replace older methods like multilocus sequence typing and pulsed-field gel electrophoresis as the best way to type bacterial strains. Also, progress in long-read sequencing techniques would make it much easier to study the whole genome and find useful extra genomic elements, such as mobile genetic elements that carry antibiotic resistance markers in real time (Simar, Hanson, and Arias 2021). WGS can provide valuable insights into the evolution and spread of antibiotic resistance genes within bacterial populations. By identifying genetic variations at the whole-genome level, researchers can better understand how resistance mechanisms develop and adapt over time. WGS lets scientists look closely at parts of the genome that don't code for proteins, like repetitive sequences and regulatory elements. Scientists can identify mutations that could alter gene regulation and potentially cause illness, as well as genetic diseases with multiple underlying causes (Brlek et al. 2024).

WGS allows correct taxonomic classification, as in the case of *S. cyaneus* B-2296T, which was published as *Actinospica acidiphila* B-2296 in a study due to 95% similarity in sequencing and taxonomic analysis by GDTBD-Tk

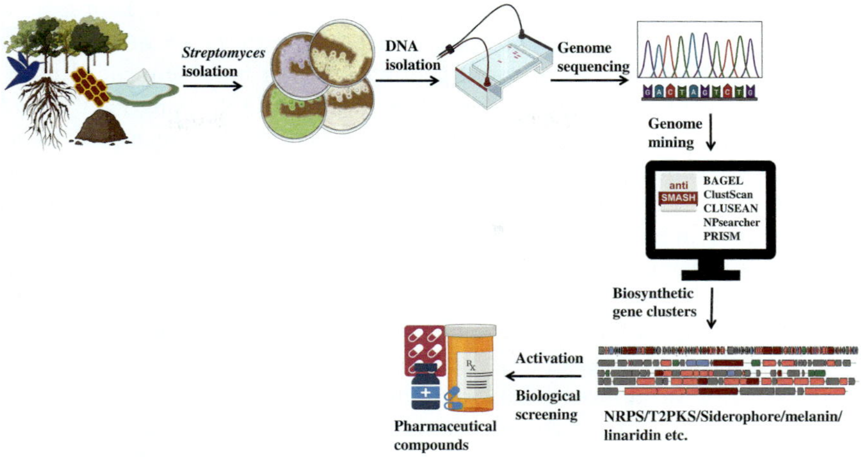

Figure 1.3 Isolation of *Streptomyces* from different ecological habitats for the identification of different secondary metabolites with pharmaceutical and industrial applications.

(Majer et al. 2021). Using the full genome sequence for genome mining helps find the biosynthetic gene clusters (BGCs) in *Streptomyces*; see "Isolation of *Streptomyces* from different ecological habitats for the identification of different secondary metabolites with pharmaceutical and industrial applications" (Figure 1.3). The biosynthesis of diazepinomicin, as predicted by antiSMASH, aligns with a seven-gene pathway in the terpene gene clusters of *Streptomyces*, but not in the sponge-associated *Micromonospora* sp. The DPJ12 genome was the only one in which the BCG of diazepinomicin was identified, which showed antibacterial effect against Gram +ve strains (Gosse et al. 2019). WGS's BGC analysis deepens our understanding of gene families, shedding light on the mechanisms involved in pathogenesis, the adaptive responses to environmental challenges, and the production of novel antibiotics among strain-specific phenotypes (Z. Zhou et al. 2012).

1.1.3.5 Next-generation sequencing

The cost of sequencing the whole genome deeply is effectively brought down by next-generation sequencing (NGS), which includes RNA-seq Illumina and WGS (R. Li et al. 2010). Contigs are being formed by compiling millions of parallel reads that are generated by *de novo* sequencing in WGS if the reference file is absent and assembling them depending on the affordable depth of coverage provided by NGS platforms. In the presence of a reference file, the multiple smaller reads are aligned to it generated by deep sequencing (Brown and Nybo 2020). 454 (Roche), Solexa (Illumina), and SOLiD (Applied Biosystems) are commercial instruments for parallel sequencing that give better data than Sanger's method by providing a high amount of data at a low cost when compared (Law et al. 2018). As of February 6, 2020, the RefSeq database showed the deposition of 1,749 *Streptomyces* genomes, with over 73% of those genomes having been sequenced using NGS methods like Illumina, PacBio, 454, and MinION. Using PacBio and Illumina sequencing techniques, a high-quality genome sequence of *S. clavuligerus* was recently produced, which revealed that 2,184 out of 7,163 total genes were missed in the former genome sequences. Among these, 47 genes were encoded in smBGCs (Lee, Kim, et al. 2020b).

With the introduction of NGS, genetics research, genetic illness diagnosis, and treatment were all significantly enhanced. From library preparations to final results, its technologies have allowed billions of sequencing reactions to be resolved in a matter of days. Managing this kind of large-scale data led to the creation of numerous NGS bioinformatic tools, each with a distinct function. The majority of computational tools, including Burrows–Wheeler Aligner (BWA), Bowtie2, Strelka2, and Mutect2 are freely accessible for the processing of NGS data including base calling, variant calling, and sequence mapping, among other purposes (Vestergaard et al. 2021). It has several uses, including RNA-Sequencing, WGS, WES (whole-exome sequencing), and the investigation of modifications in DNA and histone (ChIP-Seq, Methyl-Seq, ATAC-Seq). Illumina currently holds a dominant position in the commercial sequencing technologies market (Goodwin, McPherson, and McCombie 2016). 2 X 100 bp paired-end protocol was used in the Illumina HiSeq 2500 NGS, and WGS of potato scab–causing *Streptomyces* sp. IBSBF 2867T was done (Corrêa et al. 2021). Hybrid strategies for high-quality genome sequence and methods for both short-read and long-read genome sequencing can be applied, as done in eight strains of *Streptomyces venezuelae* (Lee, Hwang, et al. 2020a).

1.1.3.6 Metabolomics

To encourage and aid the identification of this significant genus of bacteria, some studies advise creating databases of metabolic fingerprints for each of their diverse forms and connecting them to fundamental molecular fingerprints (Algarni 2022). The biologically active compounds among *Streptomyces* are produced by biosynthetic systems, which are non-ribosomal peptide synthetases (NRPS) and type I polyketide synthases (PKS-I). For the specific amplification of NRPS and PKS-I gene sequences, new primers are designed, and these biosynthetic systems are assessed in actinomycetes.

Literature and genomic sequencing show that *Streptomyces coelicolor* A3(2) and two new strains of *Streptomyces*, named F1 and F7, can break down cellulosic materials (de Melo et al. 2018). pUWL201PW, a *Streptomyces/Escherichia coli* shuttle vector, was used to clone sco6548, one of the eight genes in *S. coelicolor* A3(2) that might code for cellulases

(Lim et al. 2016). 329-C1 and 329-C3 were the two herbicidal compounds of *Streptomyces* strain 329 isolated from the liquid culture broth by fractionation, chromatography (Sephadex LH-20), and extraction. It was seen that strain 329 caused the death of *D. sanguinalis* through leaf spots and blight on the foliage (Bo et al. 2019). pyrrolo[1,2-a]pyrazine-1,4-dione, hexahydro-3, a compound with anti-cancer properties against MCF-7 human breast cancer cells, was isolated from the marine *S. akiyoshiensis* GRG 6 (KY457710) (Rajivgandhi et al. 2020). The probability of discovering natural products can be raised through genomics-metabolomics profiling by the combination of databases like antiSMASH with the bioinformatic analysis of metabolomics (Osama et al. 2022; Paulus et al. 2017).

1.1.3.7 Genome mining

Following the development of WGS technology, numerous silent or cryptic BGCs were discovered in *Streptomyces*, among which only a small number of natural compounds were found. Activating these pathways to find novel secondary metabolites from *Streptomyces* is still a very difficult task. The culture conditions used in laboratories and industries for the growth of *Streptomyces* are very different from those found in nature, due to which many BGCs are either unexpressed or only minimally expressed. Furthermore, the majority of *Streptomyces* (>90%) are difficult or impossible to cultivate in laboratory settings. One possible solution to this issue is to apply the functional metagenomics method (K. Alam et al. 2022).

The process of "genome mining" aids in understanding the biosynthesis route used to produce natural substances, which depends on BGCs (L.-F. Zhou et al. 2021). According to genome mining techniques used *in silico*, *Streptomyces* genomes include 25–70 BGCs, although a very small portion of these bioactive chemicals are produced in laboratories using culture techniques (Belknap et al. 2020). With the development of computer programs like BLAST, Diamond, and HMMer that can align amino acid or nucleotide sequences, scientists can now search databases and genome sequences for new smBGCs using conserved sequences instead of the time-consuming Southern blot method (Lee, Hwang, et al. 2020a). By mining genomes, scientists can find biosynthetic pathways from wild *Streptomyces* that are hidden in a variety of unexplored environments. They can also find biosynthetic processes that have not been identified in the genomes of *Streptomyces* that have been cultured (Niu et al. 2013). Versions as recent as PRISM v3 and antiSMASH v5 can anticipate up to 52 distinct forms of smBGCs and determine 22 different types of smBGCs, respectively (Hannigan et al. 2019). Researchers can identify the BGCs of known secondary metabolites using the reverse approach and the products of unknown smBGCs using the forward approach. See "Streptomyces strains smBGC genome mining by reverse approach" (Table 1.1).

Table 1.1 *Streptomyces* strains smBGC genome mining by reverse approach

S. no	*Strains*	*Reference*
1	*Streptomyces chromofuscus* ATCC 49982	(Shao et al. 2012)
2	*Streptomyces clavuligerus*	(Lee, Kim, et al. 2020b)
3	*Streptomyces venezuelae* ATCC 10712	(Lee, Hwang, et al. 2020a)
4	*Streptomyces cacaoi* subsp. cacaoi NBRC 12748 T	(Kaweewan et al. 2020)
5	*Streptomyces caniferus* CA-271066	(Pérez-Victoria et al. 2019)
6	*Streptomyces olivaceus* SCSIO T05	(C. Zhang et al. 2019)
7	*Streptomyces coelicolor* M145	(Lautru et al. 2005)
8	*Streptomyces* sp. W007	(Hongyu Zhang et al. 2012)
9	*Streptomyces chartreusis* AN1542	(Park et al. 2016)
10	*Streptomyces* sp. DUT11	(Xu et al. 2018)

1.1.4 Future prospectives

Streptomycetes, a biotechnologically important bacteria, produces a significant amount of secondary metabolites. We can precisely and repidly identify different bacterial strains and differences in bioactivities using high-throughput techniques like NGS, WGS, genome mining, and metagenomics. Genome mining is done progressively, which leads to the fact that there are many unexplored biosynthetic gene clusters (Majer et al. 2021). Many *Streptomyces* strains' genomic assemblies have revealed the role of genes for traits, but we should confirm the results by creating knockouts or analyzing gene expression. This presents both a challenge for the study and a chance for future research. Modern approaches like metagenomics and meta-transcriptomics can also study the diversity of viable but non-culturable bacteria. Furthermore, the studies' progress may pave the way for target gene cloning, heterologous expression, and genome mining, particularly for biosynthetic gene clusters from *Streptomyces* and other microorganisms (Subramaniam et al. 2020). *Streptomyces* genomes are useful for finding new drugs because they have a lot of silent gene clusters (BGCs) that are involved in making natural products. Using Omics studies, one can also investigate the roles of various genes, focusing on the target gene to generate secondary metabolites that aid in the creation of antibiotics, enzymes, anticancer agents, immunomodulators, and more. Moreover, there are several other tools used in bioinformatics that help in the identification of phylogenetic relationships and adaptations and establish the ecological balance of bacteria with other soil-dwelling organisms. By identifying the biodiversity among the different strains of *Streptomyces*, researchers can discover a plethora of natural products.

1.1.5 Conclusion

In soil ecosystems, *Streptomyces* species are important because they aid in the breakdown of organic matter, the cycling of nutrients, and the control of microbial communities. *Streptomyces* are still widely distributed and may be unknown in a variety of uncharted areas, including freshwater, marine, desert, and mangrove habitats. Organisms in these locations have great survival difficulties, which frequently lead to the synthesis of novel bioactive compounds. They demonstrate their versatility by flourishing in harsh situations as well. For scientific discourse and comprehension, *Streptomyces* must be accurately classified according to its taxonomy. The enormous metabolic potential of *Streptomyces* strains is revealed by genomic investigations, and the identification of physiologically active chemicals is aided by metabolomic research. The identification of new secondary metabolites and insights into mysterious BGCs can be gained using genome mining approaches. Bacterial typing and characterization are revolutionized by NGS technologies, which also make it possible to analyze genomic and metabolic profiles in great depth. Our knowledge of the variety and evolution of *Streptomyces* could be improved by these methods. Overall, the chapter highlights the diverse significance of *Streptomyces* in ecological balance, evolutionary studies, and pharmaceutical research, as well as the continuous efforts to use their potential for a range of uses, such as environmental cleanup and drug development.

ACKNOWLEDGEMENT

Khushboo acknowledges DBT-RA program (fellowship) of Department of Biotechnology, Govt. of India.

REFERENCES

Abaza, Sherif. 2020. "What Is and Why Do We Have to Know the Phylogenetic Tree?" *Parasitologists United Journal* 13 (2). Egyptian Parasitologists United Society (EPU): 68–71.

Akintunde, Olaitan. 2021. "Detecting and Analyzing Hydrolytic Enzymes of Industrial Significance in Two Streptomyces Strains Isolated from the Soil." *European Journal of Biology and Biotechnology* 2 (3): 85–88.

Akshatha, S. J., and Manjula Ishwara Kalyani. 2022. "Evaluation of Mangrove Soil Streptomyces Spp. Exhibiting Culture and Biochemical Variation for Determination of Antibacterial Activity." *Journal of Pure and Applied Microbiology* 16 (4): 2458–76.

Alam, Khorshed, Suranjana Sikdar, Yi-Ming Zhao, Jinfang Hao, Youming Zhang, and Aiying Li. 2022. "Streptomyces: The Biofactory of Secondary Metabolites." *Frontiers in Microbiology* 13. Frontiers: 968053.

Alam, Mohammad Tauqeer, Maria Elena Merlo, Eriko Takano, and Rainer Breitling. 2010. "Genome-Based Phylogenetic Analysis of Streptomyces and Its Relatives." *Molecular Phylogenetics and Evolution* 54 (3). Elsevier: 763–72.

Algarni, Awad A. 2022. "Combining of Molecular 16S RRNA Gene and Metabolic Fingerprinting through Biolog System for the Identification of Streptomycetes in Saudi Arabia." *Journal of King Saud University-Science* 34 (3). Elsevier: 101889.

Antido, Jhon Wilson A., and Fresthel Monica M. Climacosa. 2022. "Enhanced Isolation of Streptomyces from Different Soil Habitats in Calamba City, Laguna, Philippines Using a Modified Integrated Approach." *International Journal of Microbiology*. Hindawi.

Azman, Adzzie-Shazleen, Iekhsan Othman, Saraswati S. Velu, Kok-Gan Chan, and Learn-Han Lee. 2015. "Mangrove Rare Actinobacteria: Taxonomy, Natural Compound, and Discovery of Bioactivity." *Frontiers in Microbiology* 6. Frontiers: 134793.

Barka, Essaid Ait, Parul Vatsa, Lisa Sanchez, Nathalie Gaveau-Vaillant, Cedric Jacquard, Hans-Peter Klenk, Christophe Clément, Yder Ouhdouch, and Gilles P. van Wezel. 2016. "Taxonomy, Physiology, and Natural Products of Actinobacteria." *Microbiology and Molecular Biology Reviews* 80 (1). Am Soc Microbiol: 1–43.

Belknap, Kaitlyn C., Cooper J. Park, Brian M. Barth, and Cheryl P. Andam. 2020. "Genome Mining of Biosynthetic and Chemotherapeutic Gene Clusters in Streptomyces Bacteria." *Scientific Reports* 10 (1). Nature Publishing Group UK London: 2003.

Bo, Aung B., Jae D. Kim, Young S. Kim, Hun T. Sin, Hye J. Kim, Botir Khaitov, Young K. Ko, Kee W. Park, and Jung S. Choi. 2019. "Isolation, Identification and Characterization of Streptomyces Metabolites as a Potential Bioherbicide." *PLoS One* 14 (9). Public Library of Science San Francisco, CA USA: e0222933.

Bontemps, Cyril, Maxime Toussaint, Pierre-Vincent Revol, Laurence Hotel, Mathilde Jeanbille, Stéphane Uroz, Marie-Pierre Turpault, Damien Blaudez, and Pierre Leblond. 2013. "Taxonomic and Functional Diversity of Streptomyces in a Forest Soil." *FEMS Microbiology Letters* 342 (2). Blackwell Publishing Ltd Oxford, UK: 157–67.

Borba, Marcela Proença, Ana Elisa Ballarini, Joao Paulo Duarte Witusk, Paris Lavin, and Sueli Van Der Sand. 2020. "Evaluation of BOX-PCR and REP-PCR as Molecular Typing Tools for Antarctic Streptomyces." *Current Microbiology* 77 (11). Springer: 3573–81.

Boukaew, Sawai, Siriporn Yossan, Benjamas Cheirsilp, and Poonsuk Prasertsan. 2022. "Impact of Environmental Factors on Streptomyces Spp. Metabolites against Botrytis Cinerea." *Journal of Basic Microbiology* 62 (5). Wiley Online Library: 611–22.

Brlek, Petar, Luka Bulić, Matea Bračić, Petar Projić, Vedrana Škaro, Nidhi Shah, Parth Shah, and Dragan Primorac. 2024. "Implementing Whole Genome Sequencing (WGS) in Clinical Practice: Advantages, Challenges, and Future Perspectives." *Cells* 13 (6). Multidisciplinary Digital Publishing Institute: 504.

Brown, Katelyn, and S. Eric Nybo. 2020. "Next-Generation Sequencing of PIJ101-Based Streptomyces-E. Coli Shuttle Vectors." *BioRxiv*. Cold Spring Harbor Laboratory, 2002–20.

Cheng, Zhenlong, Sean McCann, Nicoletta Faraone, Jody-Ann Clarke, E. Abbie Hudson, Kevin Cloonan, N Kirk Hillier, and Kapil Tahlan. 2020. "Production of Plant-Associated Volatiles by Select Model and Industrially Important Streptomyces Spp." *Microorganisms* 8 (11). MDPI: 1767.

Chun, Jongsik, and Fred A. Rainey. 2014. "Integrating Genomics into the Taxonomy and Systematics of the Bacteria and Archaea." *International Journal of Systematic and Evolutionary Microbiology* 64 (Pt_2). Microbiology Society: 316–24.

Corrêa, Daniele Bussioli Alves, Danilo Trabuco do Amaral, Márcio José da Silva, and Suzete Aparecida Lanza Destéfano. 2021. "Streptomyces Brasiliscabiei, a New Species Causing Potato Scab in South Brazil." *Antonie Van Leeuwenhoek* 114. Springer: 913–31.

Croce, Valentina, Andrés López-Radcenco, María Inés Lapaz, María Julia Pianzzola, Guillermo Moyna, and María Inés Siri. 2021. "An Integrative Approach for the Characterization of Plant-Pathogenic Streptomyces Spp. Strains Based on Metabolomic, Bioactivity, and Phylogenetic Analysis." *Frontiers in Microbiology* 12. Frontiers: 643792.

Das, Surajit, and Neelam Mangwani. 2015. "Ocean Acidification and Marine Microorganisms: Responses and Consequences." *Oceanologia* 57 (4). Elsevier: 349–61.

Davelos, Anita L., Linda L. Kinkel, and Deborah A. Samac. 2004. "Spatial Variation in Frequency and Intensity of Antibiotic Interactions among Streptomycetes from Prairie Soil." *Applied and Environmental Microbiology* 70 (2). Am Soc Microbiol: 1051–58.

Dharmaraj, Selvakumar, B. Ashokkumar, and K. Dhevendaran. 2010. "Isolation of Marine Streptomyces and the Evaluation of Its Bioactive Potential." *African Journal of Microbiology Research* 4 (4): 240–48.

Donald, Lavinia, Atanas Pipite, Ramesh Subramani, Jeremy Owen, Robert A. Keyzers, and Taitusi Taufa. 2022. "Streptomyces: Still the Biggest Producer of New Natural Secondary Metabolites, a Current Perspective." *Microbiology Research* 13 (3). MDPI: 418–65.

Escalante-Réndiz, Diana, Susana De-La-Rosa-García, Raúl Tapia-Tussell, Jesús Martín, Fernando Reyes, Francisca Vicente, and Marcela Gamboa-Angulo. 2019. "Molecular Identification of Selected Streptomyces Strains Isolated from Mexican Tropical Soils and Their Anti-Candida Activity." *International Journal of Environmental Research and Public Health* 16 (11). MDPI: 1913.

Ferrer, C. M., E. Olivete, S. L. Orias, M. R. Rocas, S. Juan, J. Z. Dungca, T. Mahboob, S. Barusrux, and V. Nissapatorn. 2018. "A Review on Streptomyces Spp. as Plant-Growth Promoting Bacteria (PGPB)." *Asian Journal of Pharmacognosy* 2 (3): 32–40.

Gajalakshmi, R. Madhumitha Dr P, M. Aesha Bi, and E. Udhayasudhan. 2022. "Exploratory Of Actinomycetes And Its Biotechnology Applications-A Review." *NeuroQuantology* 20 (9). NeuroQuantology: 5943.

Goodwin, Sara, John D. McPherson, and W. Richard McCombie. 2016. "Coming of Age: Ten Years of next-Generation Sequencing Technologies." *Nature Reviews Genetics* 17 (6). Nature Publishing Group: 333–51.

Goredema, N., T. Ndowora, R. Shoko, and E. Ngadze. 2020. "Morphological and Molecular Characterisation of Streptomyces Spp. Which Suppress Pathogenic Fungi." *African Crop Science Journal* 28 (4): 555–66.

Goris, Johan, Konstantinos T. Konstantinidis, Joel A. Klappenbach, Tom Coenye, Peter Vandamme, and James M. Tiedje. 2007. "DNA–DNA Hybridization Values and Their Relationship to Whole-Genome Sequence Similarities." *International Journal of Systematic and Evolutionary Microbiology* 57 (1). Microbiology Society: 81–91.

Gosse, Jessica Thandara, Soumya Ghosh, Amanda Sproule, David Overy, Naowarat Cheeptham, and Christopher N Boddy. 2019. "Whole Genome Sequencing and

Metabolomic Study of Cave Streptomyces Isolates ICC1 and ICC4." *Frontiers in Microbiology* 10. Frontiers: 411994.

Hamid, Mohamed E., Thomas Reitz, Martin R. P. Joseph, Kerstin Hommel, Adil Mahgoub, Mogahid M. Elhassan, François Buscot, and Mika Tarkka. 2020. "Diversity and Geographic Distribution of Soil Streptomycetes with Antagonistic Potential against Actinomycetoma-Causing Streptomyces Sudanensis in Sudan and South Sudan." *BMC Microbiology* 20. Springer: 1–13.

Hannigan, Geoffrey D., David Prihoda, Andrej Palicka, Jindrich Soukup, Ondrej Klempir, Lena Rampula, Jindrich Durcak, Michael Wurst, Jakub Kotowski, and Dan Chang. 2019. "A Deep Learning Genome-Mining Strategy for Biosynthetic Gene Cluster Prediction." *Nucleic Acids Research* 47 (18). Oxford University Press: e110–e110.

Hassler, Gabriel W., Andrew F. Magee, Zhenyu Zhang, Guy Baele, Philippe Lemey, Xiang Ji, Mathieu Fourment, and Marc A Suchard. 2023. "Data Integration in Bayesian Phylogenetics." *Annual Review of Statistics and Its Application* 10. Annual Reviews: 353–77.

Henao, Laura, Ramin Shirali Hossein Zade, Silvia Restrepo, Johana Husserl, and Thomas Abeel. 2023. "Genomes of Four Streptomyces Strains Reveal Insights into Putative New Species and Pathogenicity of Scab-Causing Organisms." *BMC Genomics* 24 (1). Springer: 143.

Herron, Matthew D., Peter L. Conlin, and William C. Ratcliff. 2022. *The Evolution of Multicellularity*. CRC Press.

Huang, Xiaolong, Shuangqing Zhou, Dongyi Huang, Jing Chen, and Weiming Zhu. 2016. "Streptomyces Spongiicola Sp. Nov., an Actinomycete Derived from Marine Sponge." *International Journal of Systematic and Evolutionary Microbiology* 66 (2). Microbiology Society: 738–43.

Ikeda, Haruo, Jun Ishikawa, Akiharu Hanamoto, Mayumi Shinose, Hisashi Kikuchi, Tadayoshi Shiba, Yoshiyuki Sakaki, Masahira Hattori, and Satoshi Omura. 2003. "Complete Genome Sequence and Comparative Analysis of the Industrial Microorganism Streptomyces Avermitilis." *Nature Biotechnology* 21 (5). Nature Publishing Group: 526–31.

Kapli, Paschalia, Ziheng Yang, and Maximilian J. Telford. 2020. "Phylogenetic Tree Building in the Genomic Age." *Nature Reviews Genetics* 21 (7). Nature Publishing Group UK London: 428–44.

Karagoz, Kenan, Fatih Dadasoglu, Burak Alaylar, and Recep Kotan. 2024. "Evaluation of Molecular Typing Methods for Some Scab-Causing Streptomyces Strains from Turkey." *World Journal of Microbiology and Biotechnology* 40 (4). Springer: 122.

Katz, E., C. J. Thompson, D. A. Hopwood, Stefan Schwarz, I. Perez, Shivali Patel, Charles V. Preuss, et al. 2017. "Metabolic Network Model Guided Engineering Ethylmalonyl-CoA Pathway to Improve Ascomycin Production in Streptomyces Hygroscopicus Var. Ascomyceticus." *The Journal of Antibiotics* 16 (1). Elsevier: 1–13.

Kaweewan, Issara, Hikaru Hemmi, Hisayuki Komaki, and Shinya Kodani. 2020. "Isolation and Structure Determination of a New Antibacterial Peptide Pentaminomycin C from Streptomyces Cacaoi Subsp. Cacaoi." *The Journal of Antibiotics* 73 (4). Nature Publishing Group UK London: 224–29.

Khadayat, Karan, Dawa Dindu Sherpa, Krishna Prakash Malla, Sunil Shrestha, Nabin Rana, Bishnu P. Marasini, Santosh Khanal, Binod Rayamajhee, Bibek

Raj Bhattarai, and Niranjan Parajuli. 2020. "Molecular Identification and Antimicrobial Potential of Streptomyces Species from Nepalese Soil." *International Journal of Microbiology* 2020. Hindawi.

Khushboo, and Kashyap Kumar Dubey. 2023. "Enhanced Production of Lipstatin through NTG Treatment of Streptomyces Toxytricini KD18 at 5 L Bioreactor Level." *Applied Biochemistry and Biotechnology* 195 (11). Springer: 6881–92.

Komaki, Hisayuki. 2023. "Recent Progress of Reclassification of the Genus Streptomyces." *Microorganisms* 11 (4). MDPI: 831.

Komaki, Hisayuki, and Tomohiko Tamura. 2021. "Differences at Species Level and in Repertoires of Secondary Metabolite Biosynthetic Gene Clusters among Streptomyces Coelicolor A3 (2) and Type Strains of S. Coelicolor and Its Taxonomic Neighbors." *Applied Microbiology* 1 (3). MDPI: 573–85.

Kremer, Laura S., Saskia B. Wortmann, and Holger Prokisch. 2018. "'Transcriptomics': Molecular Diagnosis of Inborn Errors of Metabolism via RNA-Sequencing." *Journal of Inherited Metabolic Disease* 41. Springer: 525–32.

Kumar, Nishitha R., Tejashree A. Balraj, Swetha N. Kempegowda, and Akila Prashant. 2024. "Multidrug-Resistant Sepsis: A Critical Healthcare Challenge." *Antibiotics* 13 (1). MDPI: 46.

Kusuma, Ali B., Imen Nouioui, and Michael Goodfellow. 2021. "Genome-Based Classification of the Streptomyces Violaceusniger Clade and Description of Streptomyces Sabulosicollis Sp. Nov. from an Indonesian Sand Dune." *Antonie Van Leeuwenhoek* 114. Springer: 859–73.

Lautru, Sylvie, Robert J. Deeth, Lianne M. Bailey, and Gregory L. Challis. 2005. "Discovery of a New Peptide Natural Product by Streptomyces Coelicolor Genome Mining." *Nature Chemical Biology* 1 (5). Nature Publishing Group US New York: 265–69.

Law, Jodi Woan-Fei, Hooi-Leng Ser, Acharaporn Duangjai, Surasak Saokaew, Sarah I. Bukhari, Tahir M. Khan, Nurul-Syakima Ab Mutalib, Kok-Gan Chan, Bey-Hing Goh, and Learn-Han Lee. 2017. "Streptomyces Colonosanans Sp. Nov., a Novel Actinobacterium Isolated from Malaysia Mangrove Soil Exhibiting Antioxidative Activity and Cytotoxic Potential against Human Colon Cancer Cell Lines." *Frontiers in Microbiology* 8. Frontiers: 260348.

Law, Jodi Woan-Fei, Kei-Xian Tan, Sunny Hei Wong, Nurul-Syakima Ab Mutalib, and Learn-Han Lee. 2018. "Taxonomic and Characterization Methods of Streptomyces: A Review." *Progress In Microbes & Molecular Biology* 1 (1).

Lee, Namil, Soonkyu Hwang, Jihun Kim, Suhyung Cho, Bernhard Palsson, and Byung-Kwan Cho. 2020a. "Mini Review: Genome Mining Approaches for the Identification of Secondary Metabolite Biosynthetic Gene Clusters in Streptomyces." *Computational and Structural Biotechnology Journal* 18. Elsevier: 1548–56.

Lee, Namil, Woori Kim, Soonkyu Hwang, Yongjae Lee, Suhyung Cho, Bernhard Palsson, and Byung-Kwan Cho. 2020b. "Thirty Complete Streptomyces Genome Sequences for Mining Novel Secondary Metabolite Biosynthetic Gene Clusters." *Scientific Data* 7 (1). Nature Publishing Group UK London: 55.

Li, Ruiqiang, Hongmei Zhu, Jue Ruan, Wubin Qian, Xiaodong Fang, Zhongbin Shi, Yingrui Li, Shengting Li, Gao Shan, and Karsten Kristiansen. 2010. "De Novo Assembly of Human Genomes with Massively Parallel Short Read Sequencing." *Genome Research* 20 (2). Cold Spring Harbor Lab: 265–72.

Li, Yisong, and Ying Huang. 2022. "Distribution and Evolutionary History of Sialic Acid Catabolism in the Phylum Actinobacteria." *Microbiology Spectrum* 10 (1). Am Soc Microbiol: e02380–21.

Lim, Ju-Hyeon, Chang-Ro Lee, Vijayalakshmi Dhakshnamoorthy, Jae Seon Park, and Soon-Kwang Hong. 2016. "Molecular Characterization of Streptomyces Coelicolor A (3) SCO6548 as a Cellulose 1,4-β-Cellobiosidase." *FEMS Microbiology Letters* 363 (3). Oxford University Press: fnv245.

Madhaiyan, Munusamy, Venkatakrishnan Sivaraj Saravanan, Wah-Seng See-Too, Camila Gazolla Volpiano, Fernando Hayashi Sant'Anna, Fábio Faria da Mota, Iain Sutcliffe, Vartul Sangal, Luciane Maria Pereira Passaglia, and Alexandre Soares Rosado. 2022. "Genomic and Phylogenomic Insights into the Family Streptomycetaceae Lead to the Proposal of Six Novel Genera." *International Journal of Systematic and Evolutionary Microbiology* 72 (10). Microbiology Society: 5570.

Majer, Haley M., Rachel L. Ehrlich, Azad Ahmed, Joshua P. Earl, Garth D. Ehrlich, and Joris Beld. 2021. "Whole Genome Sequencing of Streptomyces Actuosus ISP-5337, Streptomyces Sioyaensis B-5408, and Actinospica Acidiphila B-2296 Reveals Secondary Metabolomes with Antibiotic Potential." *Biotechnology Reports* 29. Elsevier: e00596.

Maldonado, Luis A., William Fenical, Paul R. Jensen, Christopher A. Kauffman, Tracy J. Mincer, Alan C. Ward, Alan T. Bull, and Michael Goodfellow. 2005. "Salinispora Arenicola Gen. Nov., Sp. Nov. and Salinispora Tropica Sp. Nov., Obligate Marine Actinomycetes Belonging to the Family Micromonosporaceae." *International Journal of Systematic and Evolutionary Microbiology* 55 (5). Microbiology Society: 1759–66.

McCullough, Erin L., Lauren Verdeflor, Alaina Weinsztok, Jason R. Wiles, and Steve Dorus. 2020. "Exploratory Activities for Understanding Evolutionary Relationships Depicted by Phylogenetic Trees: United but Diverse." *The American Biology Teacher* 82 (5). University of California Press: 333–37.

Melo, Ricardo Rodrigues, De, Geizecler Tomazetto, Gabriela Felix Persinoti, Hélia Harumi Sato, Roberto Ruller, and Fabio Marcio Squina. 2018. "Unraveling the Cellulolytic and Hemicellulolytic Potential of Two Novel Streptomyces Strains." *Annals of Microbiology* 68. Springer: 677–88.

Mitra, Debasis, Rittick Mondal, Bahman Khoshru, Ansuman Senapati, T. K. Radha, Bhaswatimayee Mahakur, Navendra Uniyal, et al. 2022. "Actinobacteria-Enhanced Plant Growth, Nutrient Acquisition, and Crop Protection: Advances in Soil, Plant, and Microbial Multifactorial Interactions." *Pedosphere* 32 (1): 149–70. DOI:10.1016/S1002-0160(21)60042-5

Moran, Mary Ann, Laura T. Rutherford, and Robert E. Hodson. 1995. "Evidence for Indigenous Streptomyces Populations in a Marine Environment Determined with a 16S RRNA Probe." *Applied and Environmental Microbiology* 61 (10). Am Soc Microbiol: 3695–3700.

Niu, Mingshan, Yan Sun, Xuejiao Liu, Li Tang, and Rongguo Qiu. 2013. "Tautomycetin Induces Apoptosis by Inactivating Akt through a PP1-Independent Signaling Pathway in Human Breast Cancer Cells." *Journal of Pharmacological Sciences* 121 (1). Elsevier: 17–24.

Osama, Nada, Walid Bakeer, Mai Raslan, Hanan A. Soliman, Usama Ramadan Abdelmohsen, and Mohamed Sebak. 2022. "Anti-Cancer and Antimicrobial

Potential of Five Soil Streptomycetes: A Metabolomics-Based Study." *Royal Society Open Science* 9 (2). The Royal Society: 211509.

Park, Ok-Kyung, Ha-Young Choi, Geon-Woo Kim, and Won-Gon Kim. 2016. "Generation of New Complestatin Analogues by Heterologous Expression of the Complestatin Biosynthetic Gene Cluster from Streptomyces Chartreusis AN1542." *ChemBioChem* 17 (18). Wiley Online Library: 1725–31.

Paulus, Constanze, Yuriy Rebets, Bogdan Tokovenko, Suvd Nadmid, Larisa P. Terekhova, Maksym Myronovskyi, Sergey B. Zotchev, Christian Rückert, Simone Braig, and Stefan Zahler. 2017. "New Natural Products Identified by Combined Genomics-Metabolomics Profiling of Marine Streptomyces Sp. MP131-18." *Scientific Reports* 7 (1). Nature Publishing Group UK London: 42382.

Pérez-Victoria, Ignacio, Daniel Oves-Costales, Rodney Lacret, Jesús Martín, Marina Sánchez-Hidalgo, Caridad Díaz, Bastien Cautain, Francisca Vicente, Olga Genilloud, and Fernando Reyes. 2019. "Structure Elucidation and Biosynthetic Gene Cluster Analysis of Caniferolides A–D, New Bioactive 36-Membered Macrolides from the Marine-Derived Streptomyces Caniferus CA-271066." *Organic & Biomolecular Chemistry* 17 (11). Royal Society of Chemistry: 2954–71.

Qadir, Syamand Ahmed Qadir. 2017. "Isolation, Identification and Molecular Characterization of Streptomyces from Soil in Halabja, Iraq." *Fen Bilimleri Enstitüsü.*

Rajivgandhi, Govindan Nadar, Govindan Ramachandran, Jia-Ling Li, Lingzi Yin, Natesan Manoharan, Moorthy Rajesh Kannan, Arockiam Antony Joseph Velanganni, Naiyf S Alharbi, Shine Kadaikunnan, and Jamal M Khaled. 2020. "Molecular Identification and Structural Detection of Anti-Cancer Compound from Marine Streptomyces Akiyoshiensis GRG 6 (KY457710) against MCF-7 Breast Cancer Cells." *Journal of King Saud University-Science* 32 (8). Elsevier: 3463–69.

Seipke, Ryan F., Martin Kaltenpoth, and Matthew I. Hutchings. 2012. "Streptomyces as Symbionts: An Emerging and Widespread Theme?" *FEMS Microbiology Reviews* 36 (4). Blackwell Publishing Ltd Oxford, UK: 862–76.

Shao, Lei, Jiachen Zi, Jia Zeng, and Jixun Zhan. 2012. "Identification of the Herboxidiene Biosynthetic Gene Cluster in Streptomyces Chromofuscus ATCC 49982." *Applied and Environmental Microbiology* 78 (6). Am Soc Microbiol: 2034–38.

Shepherdson, Evan M. F., Christine R. Baglio, and Marie A. Elliot. 2023. "Streptomyces Behavior and Competition in the Natural Environment." *Current Opinion in Microbiology* 71. Elsevier: 102257.

Shivlata, L., and Satyanarayana Tulasi. 2015. "Thermophilic and Alkaliphilic Actinobacteria: Biology and Potential Applications." *Frontiers in Microbiology* 6. Frontiers: 156396.

Simar, Shelby R., Blake M. Hanson, and Cesar A. Arias. 2021. "Techniques in Bacterial Strain Typing: Past, Present, and Future." *Current Opinion in Infectious Diseases* 34 (4). LWW: 339–45.

Singh, Varsha. 2021. "Whittaker's Five Kingdoms of Organisms: Minor Revisions Suggested by Considerations of the Origin of Mitosis." *Resonance*, 579.

Singhvi, Nirjara, Vipin Gupta, Namrata Dhaka, and Kashyap Kumar Dubey. 2023. "Draft Genome Sequence of Streptomyces Sp. KD18, Isolated from Industrial Soil." *3 Biotech* 13 (1). Springer: 34.

Stadler, Marc, and Veronika Hellwig. 2004. "PCR-Based Data and Secondary Metabolites as Chemotaxonomic Markers in High-Throughput Screening for Bioactive Compounds from Fungi." In *Handbook of Industrial Mycology*, 288–327. CRC Press.

Subramaniam, Gopalakrishnan, Vivek Thakur, Rachit K. Saxena, Srinivas Vadlamudi, Shilp Purohit, Vinay Kumar, Abhishek Rathore, Annapurna Chitikineni, and Rajeev K Varshney. 2020. "Complete Genome Sequence of Sixteen Plant Growth Promoting Streptomyces Strains." *Scientific Reports* 10 (1). Nature Publishing Group UK London: 10294.

Tanaka, Yoshitake, and Satoshi Omura. 1990. "Metabolism and Products of Actinomycetes-an Introduction." *Actinomycetologica* 4 (1). The Society for Actinomycetes Japan: 13–14.

Thakur, Mony, Punit Kumar, Deepanshi Rajput, Vinod Yadav, Namrata Dhaka, Rishikesh Shukla, and Kashyap Kumar Dubey. 2023. "Genome-Guided Approaches and Evaluation of the Strategies to Influence Bioprocessing Assisted Morphological Engineering of Streptomyces Cell Factories." *Bioresource Technology* 376. Elsevier: 128836.

Thakur, Mony, Ankita Yadav, Kashyap Kumar Dubey, Tikam Chand Dakal, and Vinod Yadav. 2024. "Antimicrobial Activity against Antibiotic-Resistant Pathogens and Antioxidant Activity and LCMS/MS Phytochemical Content Analysis of Selected Medicinal Plants." *Journal of Pure & Applied Microbiology* 18 (1).

Timmis, Kenneth, and Juan Luis Ramos. 2021. "The Soil Crisis: The Need to Treat as a Global Health Problem and the Pivotal Role of Microbes in Prophylaxis and Therapy." *Microbial Biotechnology 14*(3), 769–797. Wiley Online Library.

Tiwari, Kavita, and Rajinder K. Gupta. 2013. "Diversity and Isolation of Rare Actinomycetes: An Overview." *Critical Reviews in Microbiology* 39 (3). Taylor & Francis: 256–94.

Tizabi, Daniela, Tsvetan Bachvaroff, and Russell T. Hill. 2022. "Comparative Analysis of Assembly Algorithms to Optimize Biosynthetic Gene Cluster Identification in Novel Marine Actinomycete Genomes." *Frontiers in Marine Science* 9. Frontiers Media SA: 914197.

Vestergaard, Lau K., Douglas N. P. Oliveira, Claus K. Høgdall, and Estrid V. Høgdall. 2021. "Next Generation Sequencing Technology in the Clinic and Its Challenges." *Cancers* 13 (8). MDPI: 1751.

Waksman, Selman A. 1953. "Streptomycin: Background, Isolation, Properties, and Utilization." *Science* 118 (3062). American Association for the Advancement of Science: 259–66.

Xu, Xiao-Na, Liang-Yu Chen, Chao Chen, Chun Su, and Xin-Qing Zhao. 2018. "Genome Mining of the Marine Actinomycete Streptomyces Sp. DUT11 and Discovery of Tunicamycins as Anti-Complement Agents." *Frontiers in Microbiology* 9. Frontiers: 380459.

Yadav, Ajar Nath, Priyanka Verma, Sunil Kumar, Vinod Kumar, Manish Kumar, Thankappan Chellammal Kumari Sugitha, Bhim P Singh, Anil Kumar Saxena, and Harcharan Singh Dhaliwal. 2018. "Actinobacteria from Rhizosphere: Molecular Diversity, Distributions, and Potential Biotechnological Applications." In *New and Future Developments in Microbial Biotechnology and Bioengineering*, 13–41. Elsevier.

Yuan, Fang, Shouliang Yin, Yang Xu, Zilong Li, Keqiang Fan, and Guohui Pan. 2021. "The Richness and Diversity of Catalases in Bacteria." *Frontiers in Microbiology* 12. Frontiers: 645477.

Zhang, Chunyan, Wenjuan Ding, Xiangjing Qin, and Jianhua Ju. 2019. "Genome Sequencing of Streptomyces Olivaceus SCSIO T05 and Activated Production of Lobophorin CR4 via Metabolic Engineering and Genome Mining." *Marine Drugs* 17 (10). MDPI: 593.

Zhang, Haihan, Ben Ma, Tinglin Huang, and Yinjie Shi. 2021. "Nitrate Reduction by the Aerobic Denitrifying Actinomycete Streptomyces Sp. XD-11-6-2: Performance, Metabolic Activity, and Micro-Polluted Water Treatment." *Bioresource Technology* 326. Elsevier: 124779.

Zhang, Hongyu, Hongbo Wang, Yipeng Wang, Hongli Cui, Zeping Xie, Yang Pu, Shiqian Pei, Fuchao Li, and Song Qin. 2012. "Genomic Sequence-Based Discovery of Novel Angucyclinone Antibiotics from Marine Streptomyces Sp. W007." *FEMS Microbiology Letters* 332 (2). Blackwell Publishing Ltd Oxford, UK: 105–12.

Zhou, Ling-Feng, Jun Wu, Shuai Li, Qi Li, Li-Ping Jin, Cai-Ping Yin, and Ying-Lao Zhang. 2021. "Antibacterial Potential of Termite-Associated Streptomyces Spp." *ACS Omega* 6 (6). ACS Publications: 4329–34.

Zhou, Zhan, Jianying Gu, Yong-Quan Li, and Yufeng Wang. 2012. "Genome Plasticity and Systems Evolution in Streptomyces." In BMC Bioinformatics, 13:1–17. BioMed Central.

Chapter 2

Introduction of *Streptomyces* as microbial drug factory of actinomycetes and rich source of novel secondary metabolites

Aarzoo
University College Dublin, Dublin, Ireland

Naveen Kumar
Kurukshetra University, Haryana, India

2.1 INTRODUCTION

The family *Actinomycetaceae* recognizes *Streptomyces*, a genus of Gram-positive, filamentous bacteria, as a microbial drug factory and a rich source of novel secondary metabolites (SMs). These remarkable microorganisms have played a key role in pharmaceuticals and biotechnology due to their ability to produce a diverse array of bioactive compounds with therapeutic potential. These bacteria may live in a range of settings, including mangroves, symbionts, severe and understudied ecosystems, terrestrial and marine locations, and more. *Streptomyces* has been researched in over 850 species. 10^9 CFU of bacteria, of which 10^7 CFU are Actinobacteria, are thought to be present in one gram of soil.

Streptomyces species demonstrate an intricate process of multicellular development that starts with the growth of hyphae from spores and concludes with the creation of septa containing a chain of uninucleated spores. Additionally, these species have the ability to generate large branching substrates and aerial mycelia. Among other unique characteristics, streptomyces have linear, relatively big genomes (8–10 Mb) and an extraordinarily high G + C% (>70%). Multiple BGCs are found on each genome of *Streptomyces*, and these clusters are the source of many bioactive substances used in agriculture and medicine (Alam et al., 2022).

2.2 MICROBIAL SOURCES: WHY THE *STREPTOMYCES*?

It is well known that microorganisms are crucial for the discovery of new drugs from natural products. The vast variety of structurally different natural goods may have their origins in species abundance. It will be more affordable to harvest these chemicals from microbial sources than to use

 DOI: 10.1201/9781032694634-2

the chemical synthesis method. Consequently, a large number of microbes have been examined for their ability to produce anticancer chemicals or leads. The anticancer effect of these naturally occurring microbial chemicals is thought to control the immune response, stop cell division, and trigger apoptosis. The phylum Actinobacteria, one of the most extensive taxonomic groupings in the Bacteria domain, is among the most well-known major lineages. It has a wide range of species that have been reported, some of which are well-studied model organisms. There are a lot of different types of Gram-positive bacteria in this phylum. They are split into six main groups, which are *Acidimicrobiia, Actinobacteria, Coriobacteria, Nitriliruptoria, Rubrobacteria*, and *Thermoleophilia* (Law et al., 2020). Actinobacteria have been recognized as the primary natural product manufacturers of bioactive substances since the 1950s. Approximately 50% of the SMs identified to date, such as enzymes, antibiotics, immunosuppressive medications, and anti-tumor medicines, are produced by actinomycetes. *Streptomyces* is a well-known representative genus of the class Actinobacteria, and it contains more than 70% of the commercially useful antibiotics. Notably, 80% of the actinobacterial natural compounds that have been documented to date originate from the genus *Streptomyces*.

Streptomycetes outnumber several other microbial groupings in terms of diversity and bioactive potential. For example, streptomycetes are widely dispersed in both terrestrial and aquatic environments; over 900 species of this bacterial group have been identified with a valid publication name as of yet. Furthermore, a growing body of research suggests that streptomycetes can be found in harsh environments. Because streptomycetes produce a large number of easily distributed spores, their distribution is thought to be cosmopolitan. Because some of them have genomes with more than 20 gene clusters encoding known or expected biosynthetic pathways and their genomes are very large (>7 Mbp), streptomycetes still have the best ability to make bio substances. This improved streptomycetes' remarkable ability to generate a range of novel and/or advantageous SMs with antibacterial, antioxidant, and other qualities. As a result, this became the main attraction for scientists looking into the genus' bioactive potential. Anticancer medications such as bleomycin (glycopeptides group), dactinomycin (non-ribosomal peptides group), mitomycin C (quinones group), and doxorubicin (anthracyclines group) are known to be produced by streptomycetes. Only a small number of substances, nevertheless, have the potential to be employed as drugs while present in their natural state. Furthermore, naturally existing microbial compounds can be enhanced or modified to produce important therapeutic leads (Law et al., 2020).

2.3 SECONDARY METABOLITES FROM ACTINOMYCETES

Streptomyces is the producer of 70–80% of naturally occurring bioactive chemicals with pharmacological or agrochemical use. From *Streptomyces* strains, novel compounds with various biological properties are consistently identified. Antibiotics are the primary and most significant product of *Streptomyces*. Since 1955, the primary source of novel antibiotics has been the genus *Streptomyces*. They are also the source of medicines such as immunomodulators (which can either weaken or strengthen the immune system), vasoactive substances, neurological agents, and substances that kill bacteria, fungi, tumors, parasites, viruses, insects, pests, and herbs. After antibiotics, *Streptomyces* is primarily known for its enzymes, which include lipases, proteases, and cellulases. It is well known that marine Actinobacteria are capable of producing a large range of SMs. It is probable that every Actinobacteria strain possesses the genetic capacity to produce between 10 and 20 SMs. Approximately 23,000 antibiotics derived from microbes have been identified. Antibiotics are among the many SMs that Actinobacteria, primarily those of the genus *Streptomyces*, can create as bioactive substances. Among other microbial groups, the group's huge metabolic potential is unopposed. The primary factors that draw researchers to the discovery of new metabolites are their vast diversity and underutilization. Figure 2.1 shows a schematic of the procedures taken in order to develop a pharmacological agent.

2.3.1 Antibacterial activity

An antibacterial chemical is a substance that either prevents or eliminates germs. Infectious disorders remain one of the main causes of death due to bacteria that are resistant to antibiotics. The prevalence of resistance in microbial diseases is escalating rapidly on a global scale. Reduced efficacy and antibiotic resistance in infections necessitate the development of novel alternatives. It is vital to discover better, more effective medications with no side effects in order to solve this issue. The antibacterial activity of marine Actinobacteria has generally been the subject of much research. Abyssomicin C, a novel polycyclic polyketide antibiotic, is produced by a marine strain of *Verrucosispora*. It inhibits folic acid formation earlier than the well-known synthetic sulfa medicines because it is a potent inhibitor of para-aminobenzoic acid production. Abyssomicin C exhibits strong activity against Gram-positive bacteria, encompassing vancomycin- and multiple-resistant *Staphylococcus aureus* clinical isolates. It is possible to create Abyssomicin C or its analog as an antibacterial agent to combat diseases that are resistant to drugs (Rath et al., 2005; Bister et al., 2004). A novel antibiotic named Essramycin, belonging to the triazolopyrimidine class, was found in *Streptomyces* sp. The minimum inhibitory concentration (MIC) of 2–4 g/mL suggests that the chemical possesses antibacterial properties against both Gram-positive and Gram-negative bacteria, as reported by El-Gendy et al. in (2008). Lynamicins, also known as chlorinated bisindole pyrroles, are derived from the *Marinispora* sp. The antimicrobial spectrum of these compounds was evaluated against a panel of 11 pathogens, revealing their broad-spectrum activity against both

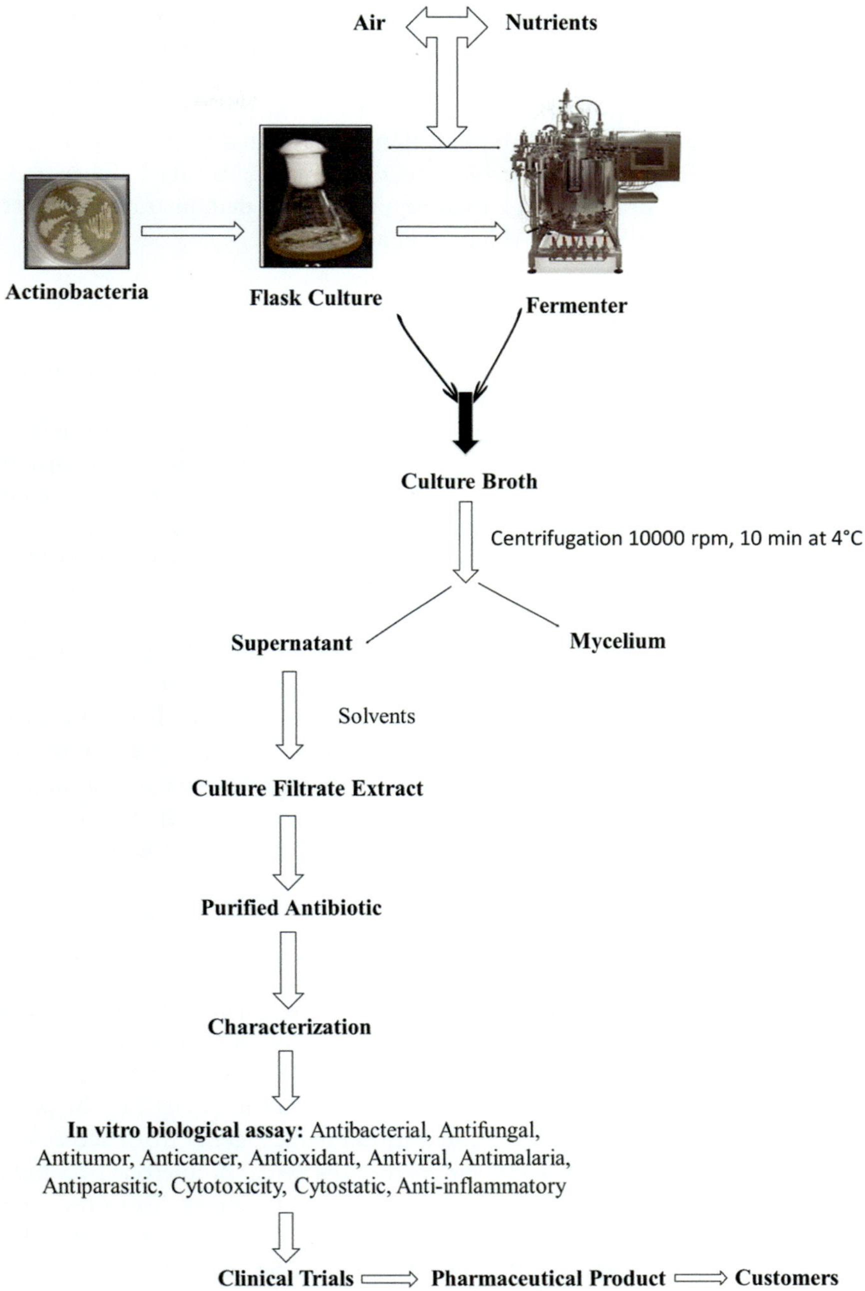

Figure 2.1 Flow chart illustrating the main steps involved in isolating and growing Actinobacteria, extracting compounds, and characterizing them chemically and biologically in order to produce pharmaceuticals.

Gram-positive and Gram-negative organisms. Surprisingly, the compounds worked against bacteria that are not easily killed with medicine. For example, they killed *Enterococcus faecium* that was not susceptible to vancomycin and *S. aureus* that was not susceptible to methicillin (McArthur et al., 2008). Marinapyrroles are compounds that were isolated from *Streptomyces* sp. They have a unique bispyrrole structure and are highly halogenated. Strong antibiotic properties of marinopyrroles are demonstrated against methicillin-resistant *S. aureus* (Hughes et al., 2008).

2.3.2 Antifungal activity

Several antibiotics have been derived from various microorganisms; however, ongoing research is being conducted to discover novel antibiotics that effectively combat pathogenic fungi. Marine Actinobacteria are useful biological organisms that create chemicals that inhibit fungus. *Streptomyces* species are commonly saprophytic and are typically present in soils, where they have a significant impact on the decomposition of complex biopolymers and the production of antibiotics. Han et al. (2009) reported that marine *Streptomyces* sp. DA11 was obtained from South China and associated with the sponge *Craniella australiensis*. This strain produced the enzyme chitinase and had antifungal activity against *Candida albicans* and *Aspergillus niger*. In addition to cellulose and starch, chitin is one of the three most abundant polysaccharides found in nature. The substance is a homopolymer of N-acetylglucosamine, with a linear β-1,4 linkage. Chitinase and its derivatives are particularly helpful for biomedical applications like wound healing, cartilage tissue synthesis, medicine distribution, and neuron formation because of their great biocompatibility and antifungal effects.

2.3.3 Antitumor activity

Thousands of antibiotics have been successfully used in clinical settings over the past 50 years to prevent and treat illnesses caused by germs. Therefore, it should come as no surprise that an increasing number of bacteria and tumor cells are becoming resistant to antibiotics or even several drugs. In the past, creating innovative antibacterial and antitubercular agents from accessible resources was the main strategy for solving the issue. Numerous accounts exist of physiologically active and architecturally fascinating SMs produced by bacteria, many of which are being exploited as potent antibiotics by the pharmaceutical industry. Marine Actinobacteria have been shown to include additional type-I polyketide-derived chemicals that are anti-cancer. Arenicolides, which are 26-membered polyunsaturated macrolactones, are one type of such material. They are produced by *Salinispora arenicola* strain CNR-005, the required marine Actinobacteria, which was isolated from marine sediments 20 meters below the Guam shoreline. In particular, arenicolide A was found to be moderately cytotoxic to the human colon cancer cell line HCT-116, with an IC50 of 30 g/mL (Williams et al. 2007).

2.3.4 Anticancer activity

Millions of people die from the terrible disease cancer every year. *Streptomyces* also produces certain well-known anticancer drugs, such as bleomycin and doxorubicin. It has been demonstrated that selina-4(14),7(11)-diene-8,9-dio from *Streptomyces* spp. QD518 possesses anticancer properties. *Streptomyces* sp. M491 produces two terpene antibiotics, T-muurolol and Amorphane. T-Muurolol exhibits a moderate level of cytotoxicity. A marine invertebrate was used to isolate *Streptomyces* spp. BL-49-58-005, which secretes indoles A and B with cytotoxic action (Alam et al., 2022).

2.3.5 Anti-inflammatory compounds

It has been found that the marine actinomycetes that produce lipomycin and saphenic acid have anti-inflammatory qualities. Additionally, it was discovered that *Micromonospora* sp. generated bioactive compounds with anti-inflammatory and antibacterial characteristics. Furthermore, the anti-inflammatory metabolites cyclomarin A and C are produced by *Streptomyces aparinicola*; cyclomarin A has been demonstrated to possess anti-malaria and anti-tuberculosis qualities. It has been observed that administering cyclomarin A topically or intraperitoneally reduces edema (Alam et al., 2022).

2.3.6 Production of enzymes

Numerous enzymes from different actinomycetes genera have been found to be useful in biomicrobial fields and biotechnological endeavors (Table 2.1). Utilizing information from genome and protein sequencing, scientists are always searching for novel actinomycetes to produce enzymes such as cellulases, xylanases, proteases, chitinases, and amylases. Two examples

Table 2.1 Enzymes produced by various Actinobacteria

Streptomyces sp.	*Enzyme*
S. erumpons	Amylase
S. cyaneofuscatus	Tyrosinase
S. luteogriseus R10	Phytase
S. rimosus	Peptide hydrolase
S. lydicus	Pectinase
S. globisporus	*N*-Acetylmuramidase
S. brahimensis	Laccase

of industrial applications that are utilized in the commercial production of biotechnology products, such as hydrolyzed proteins from diverse protein sources, are protease from *S. griseus* and kerase from *S. fradiae*. One benefit of *Streptomyces* proteases is that they may easily remove the mycelium using filtration or straightforward centrifugation. It has been shown that Actinobacteria make it easier for different types of bacteria to make L-asparginase. These bacteria are mostly found in soil and include *S. griseus, S. karnatakensis, S. albidoflavus*, and *Nocardia* spp (Narayana et al., 2008).

2.3.7 Antibiotics

Since the first antibiotics were found, actinomycetes have been one of the most prolific sources for the development of novel antibiotics. In 1940, actinomycetes containing a culture of *S. antibioticus* delivered the first antibiotic discovery: streptomycin. This was followed in 1942 by streptothricin from *S. lavendulae* and in 1944 by streptomycin from *S. griseus*. *Streptomyces* species have been the dominant source of clinical antibiotics; approximately 80% of all antibiotics derived from actinomycetes have originated from this one genus. Actinomycetes are the only microbes that produce ten of the antibiotic classes that are obtained from microbes. These include aminoglycosides, actinomycins, streptothricins, anthracyclines, daunomycin-type anthracyclines, nigericin-type polyether antibiotics, polyene macrolides, oligomycin-type large-membered macrolides, and quinoxaline peptides (Subramani and Sipkema, 2019). Polyketides have numerous pharmaceutical uses, making them highly significant natural compounds. Examples of these polyketides are the antibacterial erythromycin, the antifungal nystatin, and the antiparasitic avermectin. *Streptomyces* sp. is regarded as the primary generator of antibiotics, having created all preceding antibiotics (Table 2.2) (Hasani et al., 2014).

Table 2.2 List of various antibiotics that *Streptomyces* sp. produces

Streptomyces sp.	*Antibiotic*
S. nodosus	Amphotricin B
S. noursei	Nistatin
S. griseus	Streptomycin
S. viridifaciens	Tetracycline
S. erythraeus	Erythromycin
S. vensuella	Chloramphenicol
S. vinaceus	Viomycin
S. roseosporus	Daptomycin
S. clavuligerus	Cephalosporin

2.3.8 Vitamins

Since vitamin B12 is found in nature, microorganisms or Actinobacteria may also produce it. Significant interest in the prospect of generating vitamins through microbial fermentations was aroused by the discovery of vitamin B12 from Actinobacteria fermentations. Cobalt salts appear to be a prerequisite for all Actinobacteria to generate vitamins when added to the medium. Cobalt is a fairly potent bactericidal agent; therefore use caution when adding this precursor. The fermentation mechanisms that provide the antibiotics streptomycin, aureomycin, grisein, and neomycin also create some vitamin B12 (Das et al., 2022).

2.3.9 Biosurfactants

Microbially produced compounds with hydrophilic and hydrophobic moieties are surface-active biosurfactants that don't require mineral oil as a feedstock, in contrast to surfactants made chemically. Biosurfactants are widely used in scientific research disciplines, including mining, oil recovery, nutrition, cosmetics, textiles, varnishes, and medications. The lipopeptide antibiotic daptomycin, which Cubist Pharmaceuticals markets under the brand name Cubicin, has garnered a lot of interest for the treatment of Gram-positive bacterial infections. Actinobacteria are a type of microorganism that includes a wide range of bioemulsifiers and biosurfactants. Glycolipids based on glucose are one class of biosurfactants that have been extensively researched. The majority of them are composed of glucose units connected to glycosides to form a trehalose moiety, and they feature a hydrophilic backbone (García-Bernal et al., 2018).

2.3.10 Single-cell protein feed

Fishmeal can be made from marine actinomycetes. Certain SMs produced by actinomycetes may aid in the development of juvenile shrimp and fish. Aplasmomycin, boromycin, and typical amino acids such as azaleucine, 4-oxalysine, amino dichlobutyric acid, and alanosine are a few examples of these metabolites. The development rate, food conversion efficiency, and protein content of juvenile prawns and shrimp fed on actinomycetes contained in the feed were shown to be greater (Nakamura et al., 1977). As a result, single-cell protein of microbial origin appears to be a promising alternative among the rare protein sources, replacing up to 25–50% of fishmeal in aquaculture operations.

2.4 FERMENTATION STRATEGIES FOR THE BIOTECHNOLOGICAL PRODUCTION OF SECONDARY METABOLITES

In streptomycetes cultures, oxygen is just as important as food selection; both must be carefully regulated to direct biosynthesis toward secondary metabolism rather than biomass creation. Increasing agitation speed or aeration fluxes has historically been used to ensure an adequate oxygen transfer rate. Research that integrates medium formulation with agitation speed modulation is often published. For example, Wentzel et al. (2012) discovered that the agitation speed utilized, which varies from 900 to 1300 revolutions per minute (rpm), affects the generation of red-pigmented undecylprodigiosin and blue-pigmented actinorhodin from *S. coelicolor* A3(2) M145. Non-stressful agitation conditions, like 900–1100 rpm, yielded the highest production values, and dissolved oxygen (DO) content of 50.0% was necessary to provide a highly reproducible process. During the fermentation process, *streptomycetes* mycelia, pellet differentiation, and SM synthesis are required at high DO percentages in the vessel. However, increasing agitation speed to attain a high oxygen rate may have the opposite effect of what is intended since it destroys and strains the mycelium, which lowers the output of SMs. As a result, research into new impeller designs for the vessel has been done over the years, along with experiments using immobilized systems in place of liquid cultures and oxygen vectors added to the fermentation medium. It has been demonstrated that the synthesis of SMs depends on morphology. According to Chakravarty and Kundu's (2016) findings, the morphology of *Streptomyces roseosporus* is determined by several factors, including the source of carbon and nitrogen, the inoculum's size, the air flow rate, and the subsequent addition of precursors. Due to its filamentous nature, *S. roseosporus* free cell or pellet fermentations may produce less daptomycin; additionally, during fermentation, a dense biomass network may form, increasing the broth's viscosity and limiting the volumetric oxygen transfer coefficients (Chakravarty and Kundu, 2016). These reasons have led to the increasing use of immobilizing or encapsulating filamentous strains in submerged fermentation procedures as an alternative to classical fermentations for boosting antibiotic production. Compared to free-cell liquid cultures, these procedures produce larger mycelium densities, aggregates, or pellets with a higher percentage of dead hyphae in the core and higher proportions of viable hyphae (Scaffaro et al., 2017).

In a recent study, using *S. roseosporus* NBRC 12910 cells immobilized on unusual supports like ultra-porous refractory brick flakes or silk sachets, very high daptomycin concentrations, between 4.89 and 3.62 g/L, were reached in a 2.0-L airlift bioreactor in 6–8 repeated batches of 132 h each (Chakravarty and Kundu, 2016). *S. coelicolor* M145 produced actinorhodin, a blue-pigmented, polyketide, redox-active antibiotic, up to 3.17 g/L (5 mM) in 0.5-L shake flasks and in 192 hours by encasing cells in calcium alginate (López-García et al., 2014). In a different study, the strain was used to produce both red-pigmented tripyrrole undecylprodigiosin, an antibiotic with antimalarial, antiulcer, anticancer, and apoptotic properties, and blue-pigmented actinorhodin in submerged cultures of immobilized cells on electrospun PLA membranes that were either treated with O2-plasma or not (Scaffaro et al., 2017). In 168 hours, the *S. coelicolor* M145 mycelium was able to adhere to membranes and produced increased biomass (up to two times) and actinorhodin and undecylprodigiosin concentrations (up to four and ten times higher, respectively) in contrast to the control cultures (Scaffaro et al., 2017).

2.5 CONCLUSION

Despite their immense potential, the exploration of *Streptomyces* as microbial drug factories faces challenges such as low yields, complex fermentation processes, and competition from alternative sources. Overcoming these challenges requires interdisciplinary approaches that combine microbiology, genetics, bioinformatics, and process engineering. The future holds promise for the discovery of new bioactive compounds and the optimization of *Streptomyces*-based bioproduction processes.

Understanding the intricacies of *Streptomyces* secondary metabolite production has not only contributed to the development of numerous clinically important drugs but also paved the way for the exploration of new bioactive compounds with potential applications in medicine, agriculture, and industry.

In conclusion, *Streptomyces* stands as a cornerstone in the realm of microbial drug discovery and biotechnology. *Streptomyces* are very useful for the pharmaceutical industry because they can be genetically modified, have unique morphological traits, and make a lot of secondary metabolites. They also have a large repertoire of bioactive compounds that can be used to make new medicines.

ABBREVIATIONS

SMs	secondary metabolites
BGCs	biosynthetic gene clusters
DO	dissolved oxygen
PLA	polylactic acid
CFU	colony-forming units

REFERENCES

Alam, K., Mazumder, A., Sikdar, S., Zhao, Y. M., Hao, J., Song, C., ... & Li, A. 2022. Streptomyces: The biofactory of secondary metabolites. *Frontiers in Microbiology* 13:968053.

Bister, B., Bischoff, D., Strobele, M., Riedlinger, J., Reicke, A., Wolter, F., ... & Sussmuth, R. D. 2004. Abyssomicin CA polycyclic antibiotic from a marine Verrucosispora strain as an inhibitor of the p-aminobenzoic acid/tetrahydrofolate biosynthesis pathway. *Angewandte Chemie-International Edition 43*(19), 2574–2582.

Chakravarty, I., & Kundu, S. 2016. Improved production of Daptomycin in an airlift bioreactor by morphologically modified and immobilized cells of Streptomyces roseosporus. *AMB Express*, *6*(1), 101.

Das, K. K., Pattnaik, S., & Behera, S. K. 2022. Application of actinomycetes in secondary metabolites production: an overview. *Journal of Emerging Technologies and Innovative Research*, *9*(9), 150–169.

El-Gendy, M., Shaaban, M., Shaaban, K. A., El-Bondkly, A. M., & Laatsch, H. 2008. Essramycin: a first triazolopyrimidine antibiotic isolated from nature. *The Journal of Antibiotics*, *61*(3), 149–157.

García-Bernal, M., Medina-Marrero, R., Rodríguez-Jaramillo, C., Marrero-Chang, O., Campa-Córdova, Á. I., Medina-García, R., & Mazón-Suástegui, J. M. 2018. Probiotic effect of Streptomyces spp. on shrimp (Litopenaeus vannamei) postlarvae challenged with Vibrio parahaemolyticus. *Aquaculture Nutrition*, *24*(2), 865–871.

Han, Y., Yang, B., Zhang, F., Miao, X., & Li, Z. 2009. Characterization of antifungal chitinase from marine Streptomyces sp. DA11 associated with South China Sea sponge Craniella australiensis. *Marine Biotechnology*, *11*, 132–140.

Hasani, A., Kariminik, A., & Issazadeh, K. 2014. Streptomycetes: characteristics and their antimicrobial activities. *International Journal of Advanced Biological and Biomedical Research*, *2*(1), 63–75.

Hughes, C. C., Prieto-Davo, A., Jensen, P. R., & Fenical, W. 2008. The marinopyrroles, antibiotics of an unprecedented structure class from a marine Streptomyces sp. *Organic Letters*, *10*(4), 629–631.

Law, J. W. F., Law, L. N. S., Letchumanan, V., Tan, L. T. H., Wong, S. H., Chan, K. G., ... & Lee, L. H. 2020. Anticancer drug discovery from microbial sources: The unique mangrove streptomycetes. *Molecules* 25:5365.

López-García, M. T., Rioseras, B., Yagüe, P., Álvarez, J. R., & Manteca, Á. 2014. Cell immobilization of Streptomyces coelicolor: effect on differentiation and actinorhodin production. *International Microbiology: The Official Journal of the Spanish Society for Microbiology*, *17*(2), 75.

McArthur, K. A., Mitchell, S. S., Tsueng, G., Rheingold, A., White, D. J., Grodberg, J., ... & Potts, B. C. 2008. Lynamicins a– e, chlorinated bisindole pyrrole antibiotics from a novel marine actinomycete. *Journal of Natural Products*, *71*(10), 1732–1737.

Nakamura, H., Iitaka, Y., Kitahara, T., Okazaki, T., & Okami, Y. 1977. Structure of aplasmomycin. *The Journal of Antibiotics*, *30*(9), 714–719.

Narayana, K. J. P., Kumar, K. G., & Vijayalakshmi, M. 2008. L-asparaginase production by Streptomyces albidoflavus. *Indian Journal of Microbiology* 48:331–336.

Rath, J. P., Kinast, S., & Maier, M. E. 2005. Synthesis of the fully functionalized core structure of the antibiotic abyssomicin C. *Organic Letters*, *7*(14), 3089–3092.

Scaffaro, R., Lopresti, F., Sutera, A., Botta, L., Fontana, R. M., & Gallo, G. 2017. Plasma modified PLA electrospun membranes for actinorhodin production intensification in Streptomyces coelicolor immobilized-cell cultivations. *Colloids and Surfaces B: Biointerfaces*, *157*, 233–241.

Subramani, R., & Sipkema, D. 2019. Marine rare actinomycetes: a promising source of structurally diverse and unique novel natural products. *Marine Drugs*, *17*(5), 249.

Wentzel, A., Bruheim, P., Øverby, A., Jakobsen, Ø. M., Sletta, H., Omara, W. A., ... & Ellingsen, T. E. 2012. Optimized submerged batch fermentation strategy for systems scale studies of metabolic switching in Streptomyces coelicolor A3 (2). *BMC Systems Biology*, 6, 1–16.

Williams, P. G., Miller, E. D., Asolkar, R. N., Jensen, P. R., & Fenical, W. 2007. Arenicolides A– C, 26-membered ring macrolides from the marine actinomycete Salinispora arenicola. *The Journal of Organic Chemistry*, 72(14), 5025–5034.

Chapter 3

Bioprocess optimization and scale-up to enhance the production of secondary metabolites from *Streptomyces*

Mohit Kumar
Sharda University, Greater Noida, U.P., India

Rabab Anjum, Ekta, Kashish Zoya and Tanisha Sharma
IMS Engineering College, Ghaziabad, U.P., India

3.1 INTRODUCTION

The soil-dwelling bacterium *Streptomyces* has long been a focus of biotechnological and pharmaceutical research because of its abundant and wide production of a variety of secondary metabolites. These secondary metabolites, which include immunosuppressants, anticancer agents, and antibiotics, have enormous potential for a wide range of industrial uses. However, improving the bioprocess and increasing production are essential steps to fully utilize *Streptomyces* in the integration of secondary metabolites. The methods, difficulties, developments in bioprocess optimization, and scale-up for augmenting *Streptomyces* secondary metabolite production are explained in this book chapter.

Streptomyces is a common and significant microorganism that makes up a significant fraction of the microbial population in soil ecosystems (Antoraz et al., 2015). The genus is distinguished by its high genomic diversity, with each strain capable of yielding a unique set of secondary metabolites (Zotchev, 2012). Because of its extraordinary capacity to generate a broad range of secondary metabolites, the filamentous gram-positive bacterium *Streptomyces* has attracted a lot of interest in the biotechnological and pharmaceutical industries (Watve et al., 2001). These organic substances, known as secondary metabolites or bioactive compounds, do not directly participate in the growth and reproduction of the organism but frequently carry out vital ecological tasks like defense against rival microbes (Demain & Sanchez, 2009). The bacterium *Streptomyces*, which is frequently found in soil environments, has been thoroughly investigated and used for the synthesis of bioactive substances, especially antibiotics (Bérdy, 2005).

Streptomyces is renowned for its production of antibiotics, a class of secondary metabolites that inhibit the growth of bacteria or kill bacteria. The production of secondary metabolites from *Streptomyces* species that are beneficial in

 DOI: 10.1201/9781032694634-3

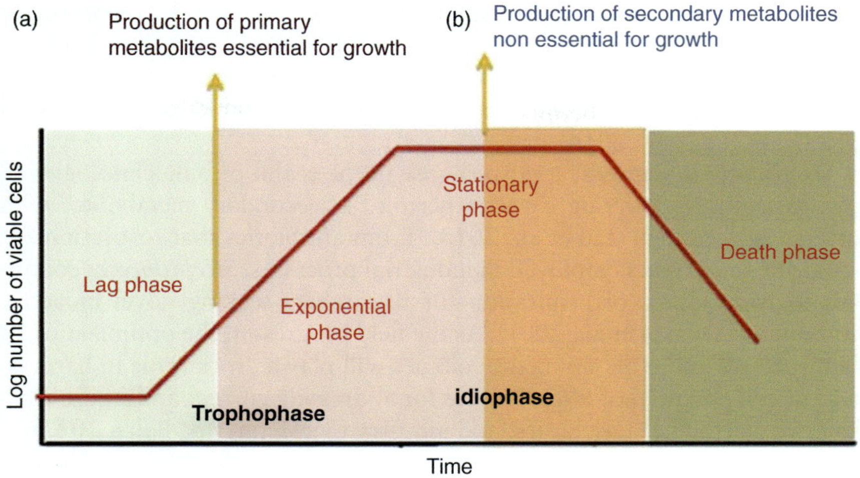

Figure 3.1 Production of metabolites.

Harir, M., Bendif, H., Bellahcene, M., Fortas, Z., & Pogni, R. (2018). *Streptomyces* secondary metabolites. *Basic biology and applications of actinobacteria*, 6, 99–122.

many ways are specifically produced in the stationary phase as shown in Figure 3.1. The first antibiotic, *streptomycin*, which was obtained from *Streptomyces* in 1943, established the commencement of the antibiotic era and revolutionized medicine (Schatz et al., 1944). While antibiotics are the most well-known secondary metabolites, *Streptomyces* produces a broad collection of bioactive compounds with diverse functions. Secondary metabolites from *Streptomyces* have demonstrated antifungal (Liu et al., 2013), anticancer (Watve et al., 2001), immunosuppressive (Chen & Du, 2016), and antiviral properties (Jaki et al., 2000). Beyond pharmaceuticals, *Streptomyces* has found applications in various biotechnological processes. Enzymes produced by *Streptomyces* strains are employed in industrial products such as the synthesis of detergents and biofuels (Anné et al., 2017). *Streptomyces* remains a valuable source of naturally derived products for drug discovery (Watve et al., 2001). Ongoing research focuses on exploring novel secondary metabolites with potential therapeutic applications (Rokem et al., 2007), addressing the growing challenge of antibiotic resistance.

Streptomyces possesses a complex genetic and metabolic machinery responsible for the biosynthesis of bioactive compounds (Liu et al., 2013). The study of *Streptomyces* genetics and genomics has enabled researchers to manipulate strains for enhanced metabolite production (McHenney & Baltz, 2022). Despite its significance, unlocking the full potential of *Streptomyces* for biotechnological and pharmaceutical applications poses challenges. Issues such as low yields, long fermentation times, and complex downstream processing have prompted ongoing efforts in bioprocess optimization and scale-up (Escalante-Aburto et al., 2022). *Streptomyces* contribute to soil health and play a crucial role in nutrient cycling in terrestrial ecosystems (Antoraz et al., 2015). Understanding the ecological interactions involving *Streptomyces* is essential for sustainable biotechnological practices (Tyc et al., 2014).

Streptomyces stands as a powerhouse in the realm of biotechnology and pharmaceuticals, offering a vast repertoire of secondary metabolites with diverse applications (Liu et al., 2013). From antibiotics that revolutionized medicine to enzymes employed in industrial processes, *Streptomyces* continues to be a source of inspiration for researchers seeking novel bioactive compounds (Watve et al., 2001). As the field progresses, the optimization of bioprocesses and scale-up methodologies will play a crucial role in harnessing the full potential of *Streptomyces* for addressing current and future challenges in medicine, agriculture, and industry (McHenney & Baltz, 2022).

3.2 *STREPTOMYCES* STRAIN SELECTION AND GENETIC MANIPULATION

The foundation of a successful bioprocess lies in the careful selection of *Streptomyces* strains with high secondary metabolite–producing potential. Genetic manipulation techniques play a pivotal role in enhancing strains for optimal performance. Overexpression of key biosynthetic genes and targeted genetic modifications contribute to increased yields and improved production efficiency. One very important strain of *Streptomyces* named *Streptomyces avermitilis* was unearthed by Satoshi Omura, a Japanese biochemist, in Shizuoka Prefecture, Japan (Omura et al., 2001). This soil-living, gram-positive bacteria with characterized branching filamentous morphology not only carries out a complex morphological differentiation but is also renowned for its ability to synthesize a diverse array of bioactive compounds (Ikeda et al., 2003). Avermectin, which is used as an anthelmintic and insecticidal agent, is one of the significant secondary metabolites formed by this bacterium and is commercially vital in human and veterinary medicine (Burg et al., 1979). *Streptomyces* produce about 75% of natural bioactive which are crucial for agrochemical and industrial applications (Watve et al., 2001). Approximately two-thirds of the prominent antibiotics in existence originate from the *Streptomyces* genus. Streptomycin, kanamycin, gentamicin, doxorubicin, and so on are some of the important antibiotics that originated from different species of *Streptomyces* (Challis & Hopwood, 2003). In addition, various antifungal treatments like *S. padanus* strain PMS-702 can be used to treat the deteriorating condition of the leaf of cabbage caused by the pathogenic fungus *Rhizoctonia solani* AG-4 (Jogaiah et al., 2018). From about 23000 recognized secondary metabolites, roughly 80% are attributed to various *Streptomyces* species, making them prolific producers of bioactive compounds (Bérdy, 2005).

3.3 GENERATION OF DIVERSE SECONDARY METABOLITES FROM *STREPTOMYCES* SPECIES AND THEIR USES IN DIFFERENT FIELDS

Due to their complex and powerful secondary metabolites production, *Streptomyces* species have been contemplated as a warehouse of a wide range of natural products. They are the largest current producer of new and innovative secondary metabolites (Donald et al., 2022). The pathway affiliated to the secondary metabolites (see Figure 3.2) that are synthesized from *Streptomyces* include antibiotics, antifungals, antibacterials, antitumors, and enzyme inhibitors (Manivasagan et al., 2014). The aerial species of *Streptomyces* are a more essential source of unique antibiotics in comparison with the vegetative species. This is because of the adaptations to extreme conditions and unique compounds found in marine *Streptomyces*. The presence of extreme conditions in marine environments like high salinity and pressure leads to the development of unique metabolic pathways and bioactive compounds, which make them valuable for bioprospecting.

Secondary metabolites are known to be produced during the lag phase and to reach a stationary phase; however, in the case of *Streptomyces*, bioactive compounds can also be produced during the exponential, stationary,

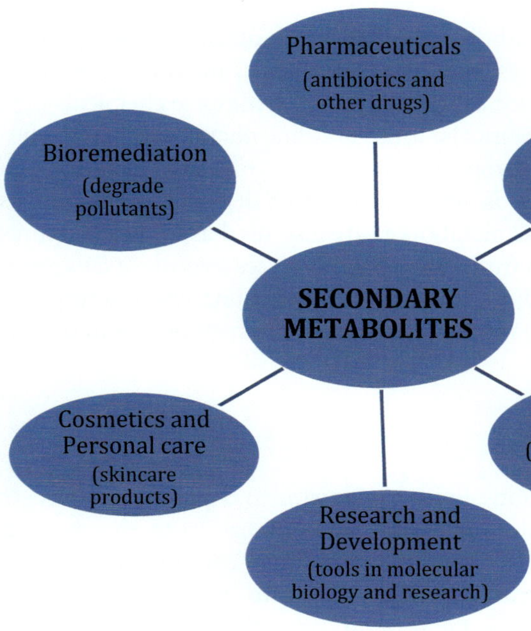

Figure 3.2 Diverse application of secondary metabolites produced from *Streptomyces*.

and death phases (Harir et al., 2018). Low-molecular-weight organic compounds, known as secondary metabolites, are synthesized by different bacteria and viruses. These substances do not play a direct role in the primary metabolic functions related to the growth and development of the organism. However, they have demonstrated significant importance for human health, the environment, and the economic well-being of society. These metabolites play an essential role in various aspects of science and industry. Their uses in various fields are given in the following sections.

3.3.1 Pharmaceuticals

Bacterial secondary metabolites have yielded numerous antibiotics and other drugs that are crucial for treating infectious diseases. The excessive use of antibiotics and other drugs like antifungals and antibacterials in human therapy and for farm animals have created a pathogenic bacterial resistance to multiple drugs. Therefore, secondary metabolites are continuously required to develop new antibiotics and other drugs to fight antibiotic-resistant bacteria and fungal and viral diseases. For example, penicillin (from *Penicillium* bacteria) and vancomycin (from *Streptomyces*) are some useful antibiotics that are used widely.

3.3.2 Biotechnology

Bacterial secondary metabolites are used in biotechnological processes for the production of enzymes, bioactive compounds, and other bio-based products. The enzymes are used as biocatalysts to synthesize chemicals in various industries like pharmaceuticals, food, and fine chemicals (Khushboo et al., 2022). Some *Streptomyces* species produce enzymes used in various biotechnological processes, such as amylase for starch degradation or proteases for protein hydrolysis. These metabolites are also used as immunosuppressants, as compounds like rapamycin from *Streptomyces hygroscopicus* are used in organ transplant procedures to suppress the immune system's response. Like all these, there are many more uses of secondary metabolites in the vast field of biotechnology.

3.3.3 Bioremediation

In bioremediation, various living organisms and plants are used to neutralize or detoxify pollutants and contaminants from the environment. Microorganisms generate metabolites, which can metabolize or break down contaminants, transforming them into less harmful substances. In comparison with other microorganisms, *Streptomyces* stands out as a prime choice for bioremediation due to its exceptional resilience and versatility in coping with diverse environmental stress factors (Romano-Armada et al., 2020).

3.3.4 Agriculture

Biopesticides, biofertilizers, and plant growth promoters are products of bacterial metabolites. They lessen the need for artificial fertilizers and pesticides and aid in the control of plant diseases. Pesticide-resistant bacteria have emerged as a result of the widespread use of synthetic pesticides. Because they are environmentally friendly, using biopesticides and biofertilizers derived from secondary metabolites of *Streptomyces* species improves agricultural production's sustainability. With its extensive biotechnological potential, the genus *Streptomyces* is a strong contender for the biological control of phytopathogenic microbes (Pacios-Michelena et al., 2021). Along with functioning as nutrient enhancers and a plentiful supply of bioactive secondary metabolites utilized in agriculture, *Streptomyces* species interact with a multitude of parameters that regulate soil fertility (Vurukonda et al., 2018). These interactions are all influenced by chemical insecticides or pesticides with intricate chemical structures. *S. roseolus* is used to prevent aflatoxin B1 contamination. According to Shakeel et al. (2018), *S. yanglinensis* may also prevent *Aspergillus flavus mycelium* from growing, safeguarding food and feed from aflatoxin contamination. For the biocontrol of soilborne diseases, specifically *Fusarium oxysporum* sp. *lycopersici* (FOL) and tomato verticillium, as well as black scorch on date palms caused by *Ceratocystis radicicola*, *S. griseoviridis K61* (Mycostop) was employed. In all, nine (ix) distinct formulations of *Streptomyces—S. violaceus niger YCED9*, *S. saraceticus KH400*, *S. lydicus WYEC108*, and *S. griseoviridis K61*—were employed for the biocontrol of bacterial and fungal soil diseases (Alam et al., 2022).

3.3.5 Cosmetics and personal care

Some secondary metabolites have cosmetic and skincare applications, including antimicrobial properties and ingredients for antiaging products. So, the cosmetic industry and researchers have extensively explored the creation of bioactive cosmetic formulations to reduce the signs of aging and enhance skin beauty and health in response to the strong consumer demand for skincare products that have proven effectiveness (Saraiva et al., 2023). For example, algae-derived secondary metabolites have been found to offer skin benefits, including shielding against UV radiation and decreasing skin laxity, uneven texture, and wrinkles. Similarly, different species of *Streptomyces* also produce secondary metabolites that have some antimicrobial and antioxidant properties and are used in the cosmetics industry. One of the examples of this is *Streptomyces chattanoogensis* THA-metabolites, which have impressive antiaging activity on the skin (Zheng et al., 2022).

3.4 BIOACTIVITY OF *STREPTOMYCES*

Streptomyces are responsible for generating approximately 70–80% of natural bioactive compounds renowned for their applications in pharmaceuticals and agrochemicals. Continuously, novel metabolites with diverse biological properties are extracted from diverse strains of *Streptomyces*. Since 1955, the *Streptomyces* genus has been the primary source of fresh antibiotics. *Streptomyces* serve as the origin of antibacterial, antifungal, antitumor, antiviral, antiparasitic, insecticidal, pesticidal, and herbicidal agents, in addition to pharmacological compounds like immunomodulators, vasoactive substances, and neuroactive agents. The second most significant products of *Streptomyces* after antibiotics are enzymes, for example cellulases, proteases, pectinases, lipases, amylases, and many others.

3.4.1 Production of antibiotics from *Streptomyces*

It has been well researched that *Streptomyces* is accountable for the manufacture of the majority of antibiotics and secondary compounds with biological activity. Approximately half of the identified *Streptomyces* species are acknowledged as antibiotic producers. *Actinomycetes* are responsible for the production of around two-thirds of microbial antibiotics, and roughly 80% of these antibiotics are sourced from the *Streptomyces* genus. When considering other secondary metabolites, *actinomycetes* still stand out as the primary contributors, accounting for approximately 60% of them. Some of the examples are aminoglycosides (streptomycin, neomycin, and gentamicin), ansamycins (rifamycin), and macrolides (azithromycin, erythromycin, and clarithromycin) (Vaishnav and Demain, 2011).

Streptomyces and its diverse strains have played a critical role in the production of a majority of antibiotics (see Figure 3.3), as these microorganisms generate antibiotics to eliminate potential competitors. *Streptomycin* was among the initial antibiotics discovered, and it is synthesized by *S. griseus*. Presently, a range of *Streptomyces* species contribute to roughly 75% of both medical and commercial antibiotics, effectively serving these fields. In the quest for new antibiotics, research has shifted toward the isolation of *Streptomyces* and the meticulous screening of their utilization in various environments. Table 3.1 shows the examples of various antibiotics generated by varied species of *Streptomyces* and their implementations. Additionally, studies have revealed that variables like nutrient availability, cultivation methods, and other factors can influence the development of antibiotics in *Streptomyces* (Arasu et al., 2014). Considering this, the composition of the growth medium and the metabolism of any organism can influence the biosynthesis of antibiotics. Further research on actinomycetes has demonstrated that they have the capability to produce more than one antibiotic and the same antibiotics can be fabricated by different species of Actinobacteria (Yun et al., 2018).

3.4.2 Formulation of enzymes from *Streptomyces*

A wide array of enzymes with applications in microbial and biotechnological fields can be derived from various actinomycete genera. After thoroughly

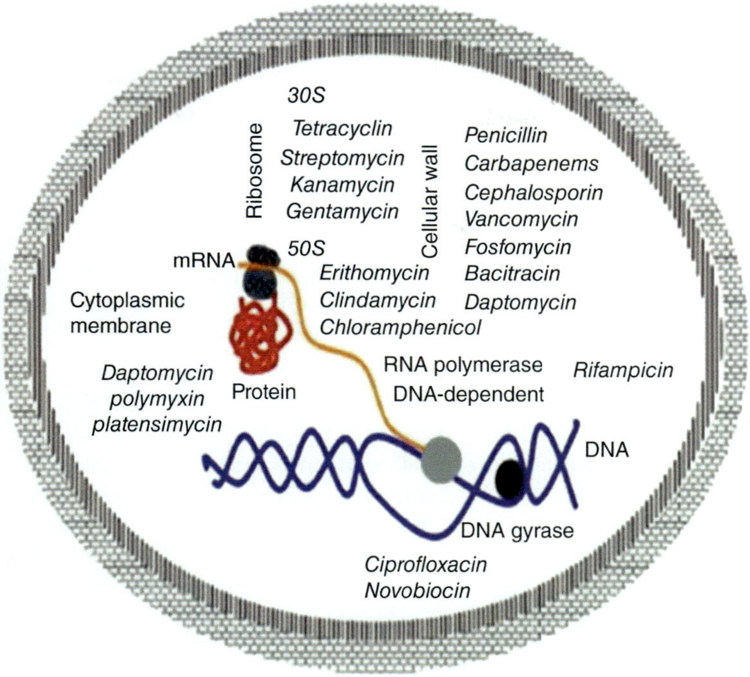

Figure 3.3 Production of antibiotics from streptomyces.

Procópio, R. E. D. L., Silva, I. R. D., Martins, M. K., Azevedo, J. L. D., & Araújo, J. M. D. (2012). Antibiotics produced by *Streptomyces*. *Brazilian Journal of Infectious Diseases, 16*, 466–471.

Table 3.1 Examples of antibiotics generated by various species of *Streptomyces* and their implementations (Yun et al. 2018)

Antibiotic compound	*Streptomyces species*	*Application*
Avermectin	*S. avermitilis*	Antiparasitic
Chinikomycin	*Streptomyces species*	Anticancer
Daryamides	*Streptomyces species*	Antifungal; Anticancer
Piericidins	*Streptomyces species*	Antitumor
Pristinamycine	*S. pristinaespiralis*	Antibacterial
Streptomycin	*S. griseus*	Antimicrobial

analyzing the genomic and protein sequencing data, actinomycetes are perpetually examined and harnessed for the production of amylases, proteases, xylanases, cellulases, and various other enzymatic products. Some species of *Streptomyces* produce extracellular enzymes (Table 3.2) that have the potential to put back the toxic chemicals that are being used in various industries (Kumar et al., 2022).

Table 3.2 Some examples of enzymes produced using actinobacteria and their industrial utilization (Kumar et al., 2022)

Enzyme	*Industry*	*Use*	*Species used*
Amylase	Textile	Starch removal from woven fabrics	*S. erumpons*
	Baking	Bread softening; volume	
	Starch	Production of glucose; syrups	
Lipase	Detergent	Stain removal	*Streptomyces griseus*
	Baking	Stability of dough	
	Dairy	Flavoring of cheese	
	Textile	Cleaning, Deinking	
Pectinase	Beverage	Clarification, mashing	*S. lydicus*
	Textile	Scouring	
Chitinase	Bioremediation	Chitin waste utilization	*Streptomyces griseus*
Penicillin amidase	Commercial significance	Industrial production of 6-aminopenicillanic acid	*Streptomyces clavuligerus*
Xylanase	Baking	Dough-conditioning	*Streptomyces spp.*
	Animal feeds	Digestibility	
	Pulp and paper	Boosting of bleach	

3.4.3 Production of probiotics from *Streptomyces*

"Probiotics are live microbial feed supplements which beneficially affect the host animal by improving microbial balance". Since *Streptomyces* can synthesize antibiotic compounds to combat infectious illnesses and release exoenzymes that improve feed utilization and digestion, they are used as probiotics in the aquaculture sector (Cuozzo et al., 2023). Extensive literature is available on the application of *Streptomyces* species to enhance the growth of tiger shrimp. Additionally, it turned out that an antibiotic that was introduced to the feed and produced from marine actinobacteria was successful in detecting the feed's effectiveness against the white spot syndrome virus in black tiger shrimp as well as showing its effects on the feed in vivo (Cuozzo et al., 2023).

3.4.4 *Streptomyces* used in control of plant diseases

Actinobacteria can play a crucial role in controlling plant diseases. They offer significant potential in the agricultural industry by serving as a lavish store of agro-active compounds for the growth of plants and their biocontrol as listed in Table 3.3. Over the past five years, around 60% of recently explored insecticides and herbicides derived from *Streptomyces* have highlighted the importance of their growth. Kasugamycin, identified in *Streptomyces kasugaensis*, serves as a potent bactericidal and fungicidal metabolite,

Table 3.3 Several examples of antibiotics generated by actinobacteria that reduce a range of plant diseases

Plant Disease	*Strains of Streptomyces*	*Antibiotic generated*
Damping-off of cabbage	*Streptomyces padanus*	Fungichromin
Rice blast	*Streptomyces kasugaensis*	Kasugamycin
Powdery mildew	*Sterptoverticillium rimofaciens*	Mildiomycin
Asparagus root	*Streptomyces griseus*	Faeriefungin
Potato scab	*S. melanosporofaciens*	Geldanamycin
Root rot of pea geldanus	*Streptomyces hygroscopicus*	Geldanamycin
Blotch of wheat	*S. malaysiensis*	Malayamycin

inhibiting microbial protein biosynthesis without affecting mammals, making it remarkably safe (Khan et al., 2023). Evaluation of the *Streptomyces* sp. culture supernatant demonstrated its efficacy in inhibiting plant pathogenic *Rhizoctonia solani* cultured in in vitro conditions, exhibiting the ability of actinobacteria to mitigate damping off stringency in tomato plants (Pacios-Michelena et al., 2021). To effectively control certain plant diseases, definite *Streptomyces*-derived products have been developed and evaluated. For example, to manage *Ceratocystis radicicola*, which causes black burns on date palms, *Streptomyces griseoviridis* strain K61 (MycostopR) has been employed (Pacios-Michelena et al., 2021). *Streptomyces* strains exhibit antiviral and antibiofilm properties, and they produce antagonistic compounds like siderophores that inhibit bacterial infections. Furthermore, they generate a diverse array of biologically active secondary metabolites, including antioxidants, anticancer agents, and enzymes with industrial applications (Butt et al., 2024). Microbial inoculants, such as these bacteria, offer significant potential for enhancing agricultural sustainability. As Qiu et al. (2019) suggested, these microorganisms can promote plant health and improve nutrient availability, making them valuable tools for sustainable agriculture.

According to research conducted by Mitrovic et al. (2021), the soil segregate potency of *Streptomyces hygroscopicus* was explored for the biological control of storage apple pathogens.

3.5 OPTIMIZATION STRATEGIES FOR SECONDARY METABOLITE PRODUCTION FROM *STREPTOMYCES*

3.5.1 Optimization of media composition

It is widely recognized that creating a suitable fermentation medium is essential to the synthesis of secondary metabolites. Prior experience creating an appropriate base medium might be very important for further media optimization. A variety of environmental elements, including carbon sources, phosphorus, and nitrogen, have an impact on the fermentation process that produces secondary metabolites (Mondal et al., 2022). Additionally, the strains of *Streptomyces* used in fermentation determine both the qualitative and quantitative differences in the synthesis of useful metabolites, underscoring their crucial function as suppliers of precursors and energy for the generation of secondary metabolites. Thus, evaluating the impact of medium constituents and ambient conditions is a basic and crucial first step in augmenting metabolite production in the *Streptomyces* genus (Al-Ghazali and Omran, 2017).

In accordance with a research paper (Al-Ghazali and Omran, 2017), the optimization of culture conditions was done to enhance the generation of antimicrobial metabolites from *Streptomyces* sp. LH9 through batch culture fermentation. To increase the synthesis of antimicrobial metabolites from *Streptomyces* sp. LH9 by batch culture fermentation, the culture conditions were optimized. With the objective of optimizing the sources of carbon, the isolate was grown in triplicate in a basal medium containing a range of carbon sources (sucrose, glycerol, dextrose, maltose, glucose, sorbitol, and galactose) at varying concentrations of sugar from 1% to 3.5%. The results aimed to identify the ultimate carbon source and its optimal concentration for maximizing antibiotic production. This study also optimized the nitrogen source for antibiotic synthesis from *Streptomyces* sp. LH9 by adding various nitrogen sources (0.5%) individually to the primary fermentation media. These sources incorporated soybean, peptone, casein, and beef extract. According to research findings, peptone was the most efficient source of nitrogen for stimulating the production of antibiotic chemicals in *Streptomyces* sp. LH9. Antibiotic levels have been shown to be influenced by the kind and availability of nitrogen molecules in the fermentation media; high yields are favored by a close temporal linkage between biomass increase and the production of antibacterial chemicals. Additionally, the same isolate of *Streptomyces* sp. LH9 was grown under optimal circumstances, with the basal medium having the right amounts of nitrogen and carbon sources in addition to the other previously listed nutritional elements. Phosphate, however, came from several substances, including Na_2HPO_4, $(NH_4)H_2PO_4$, NaH_2PO_4, KH_2PO_4, and $(NH_4)_2HPO_4$, and a concentration of 0.2% Na_2HPO_4 was recognized as the optimal phosphate source for antibacterial metabolite production (Al-Ghazali & Omran, 2017).

Optimizing culture media helps the researchers increase the productivity of their cultures; the efficiency of the biotechnological processes can be maximized, and the production of desired products such as pharmaceuticals, enzymes, or bio-fuels can be enhanced. It is a fascinating area where science and creativity come together to develop the best possible environment that can be achieved for the organisms being cultured. This optimization is necessary because the composition of the culture media copiously influences the growth rate, yield, and quality of the cultured organisms or cells (Arul Jose et al., 2013).

3.5.2 Optimization of culture conditions

Culture optimization is a procedure of refining and enriching the composition of nutrient-rich media that are utilized to grow and sustain diverse types of cells, microorganisms, or tissues in a laboratory setup. However, the environment has an enormous effect on the manner in which microbes produce metabolites. Therefore, further aim was to develop an environment that supplies optimal conditions for the productivity, growth, and viability of the organisms or cells being cultured. Various factors (pH, temperature, nutrient concentrations, carbon sources, amino acids, vitamins, and growth factors) are carefully altered and tuned up in this process to promote the desired cellular or microbial growth (Khattab et al., 2016).

3.5.2.1 Optimization of incubation period

Another research conducted in the year 2015 by Bundale et al. involved introducing the *Streptomyces* isolates into a basal medium and letting them develop for 20 days. The cultivation procedure transpired within a rotating shaker operating at a rate of 130 revolutions per minute (rpm) while preserving a temperature of 28 °C. The zone of inhibition was used to measure the bioactive metabolites' antibacterial activity in order to gauge their synthesis (Bundale et al., 2015). It was found that within 48 hours, all four isolates started to produce bioactive metabolites. On the fourth day, isolate R1 produced the most (19 mm), remained stable until the tenth day, then gradually reduced until producing more on the eighteenth day. Similar patterns were seen in isolate R5, with the highest output occurring on days 6 and 14. In contrast, isolate R3 peaked in production on day eight (19 mm), but from day ten onward, it steadily declined. Production of isolate Y8 began on the second day and rose gradually until the tenth day, when it peaked at 28 mm. It was rather stable until the sixteenth day, at which point on the eighteenth day, there was a dramatic drop (16 mm) (Bundale et al., 2015).

3.5.2.2 Optimization of pH

In a different study (Rakesh et al., 2014), which was done for the investigation of the effect of pH on the production of antibacterial metabolites, the *Streptomyces* sp. SRDPTK-07 strain was introduced into flasks containing sterile SCN broth. The pH of the broth was adjusted to different levels, specifically 3, 5, 7, 9, and 11, using 0.1N HCl and 0.1N NaOH. These flasks were then placed in an incubator and maintained at a temperature of 30±2 °C for a duration of 7 days.

After the incubation period, the ingredients of the flasks underwent filtration using sterile Whatman No. 1 filter paper. The resulting filtrate of the culture was subjected to extraction using ethyl acetate (EtOAc) in a separation funnel. The EtOAc extracts were subsequently concentrated, and their antibacterial activity was assessed using the agar well diffusion method.

Results depicted that no inhibitory activity was observed at pH 3, while the extract from the medium adjusted to pH 9 exhibited maximum inhibitory activity. However, beyond pH 9, a decrease in antibacterial activity was noted. In summary, the study found that pH had a significant impact on the antibacterial activity of the extract. pH 9 was identified as the optimal condition for the highest inhibitory activity, but activity declined when the pH exceeded 9 (Rakesh et al., 2014).

3.5.2.3 Optimization of temperature

In an experiment (Rakesh et al. 2014) that was performed to investigate the impact of temperature on the production of antibacterial metabolites, the *Streptomyces* sp. SRDP-TK-07 strain was introduced into flasks containing sterile SCN broth. These flasks were then incubated at different temperatures: 4, 20, 30, and 45 °C.

Interestingly, no production of inhibitory agents was observed when the broth was incubated at 4 °C. However, as the temperature of incubation increased, a higher level of antibacterial activity was observed for the isolate. The extract obtained from the culture incubated at 45 °C displayed particularly high inhibitory activity.

The study explored the influence of temperature on the production of antibacterial metabolites. The SRDP-TK-07 strain was incubated at various temperatures, and the resulting culture filtrates were extracted and tested for antibacterial activity. Notably, no inhibitory agents were produced at 4 °C, but increasing the incubation temperature led to higher antibacterial activity. The extract obtained from the culture incubated at 45 °C exhibited notable inhibitory activity (Rakesh et al., 2014).

3.5.3 Methods used for optimization of culture media and cultivation conditions for increased secondary metabolites production from *Streptomyces*

For secondary metabolites that are produced by microorganisms, the media components and the optimum levels of culture conditions are very much essential. In a microbial system, the production of microbial secondary metabolites can be enhanced using fine-tuning of physical criteria and nutritional composition of the production media (Harir et al., 2018).

The exploration for optimization of culture conditions is usually performed using non-statistics, using only single-variable approach and statistical experimental framework approaches. As the former is exceedingly strenuous and resource draining than the statistical methods, techniques of data-driven experimental design that particularly includes Plackett–Burman design and response surface methodology (RSM) are extensively utilized to select the pertinent variable and acquire the pinnacle levels, sequentially. As compared to the conventional practice of optimization of single factor, the implementation of these data-driven modeling techniques for experimental optimization of media can lead to enhanced production of the products, diminished process fluctuation, reduced time, and all-encompassing costs. To select the most influential feature among the constituents of an intricate medium, many researchers have applied the Plackett–Burman design (Korayem et al., 2015).

3.5.3.1 Response surface methodology and Plackett–Burman design

Response surface methodology (RSM) is a versatile statistical tool used to optimize processes where the outcome is influenced by several variables. It helps in understanding the relationships between these variables and the desired response, making it easier to identify the best conditions for achieving optimal results. RSM typically uses designs like central composite or Box-Behnken to develop a quadratic model. This model reveals how variables interact and helps pinpoint the optimal settings for maximum efficiency. In the context of bioprocess optimization, RSM is particularly valuable for fine-tuning conditions to boost the production of important secondary metabolites from microorganisms.

Plackett–Burman design is a straightforward and efficient method for identifying the most influential factors out of many possible variables. It's especially useful at the beginning of an experimental study when you need to determine which variables significantly affect the outcome. By varying each variable between two levels, the Plackett–Burman design allows researchers to quickly screen and highlight the key factors with a minimal number of experiments. This design is ideal for initial screenings, making it easier to focus on the most critical variables before applying more detailed optimization methods like RSM. In bioprocess optimization, it helps to identify essential medium components and environmental conditions that significantly impact microbial growth and metabolite production.

In a study performed in 2017 (Ahsan et al., 2017), *Streptomyces diastatochromogenes* KX852460 was selected as a therapeutic strain for the fact that it can synthesize secondary metabolites, which are designed to avoid the development of spot disease on tobacco leaves triggered by *Rhizoctonia solani* AG-3. Optimal antimycotic effect against *Rhizobacter solani* AG-3 was achieved by the incorporation of medium components including peanut meal, soluble starch, sodium chloride, autolyzed yeast, and ammonium sulfate into the process.

To enhance secondary metabolite production, a central mixed plan within the response surface methodology was implemented in subaqueous fermentation. The optimized concentrations of medium components including peanut meal, soluble starch, sodium chloride, yeast extract, and ammonium sulfate were identified using RSM (Chen et al., 2022). The results are also demonstrated in Figure 3.4.

Metabolic amendments during subaqueous fermentation were examined. The levels of reducing sugar and total sugar consumed increased, indicative of bacterial growth. A decrease in amino nitrogen and total sugar suggested a shift in metabolism. Furthermore, the production of secondary and other metabolites played a role in altering the pH of the fermentation batch.

In another research (Yun et al., 2018) the antibacterial potential of *Streptomyces* spp., isolated from cassava rhizosphere soil, was assessed against *Fusarium oxysporum* f.sp. *cubense race 4* (FOC4). It was examined that out of sixty-three different strains, thirteen displayed robust antibacterial activity and underwent evaluation against several fungal pathogens. Through 16S rDNA sequencing, the strain identified as *Streptomyces* spp., exhibiting zenith inhibition antagonistic to all test pathogens, was selected for in-depth studies.

To optimize the growth of *Streptomyces* sp. *1–14* and enhance metabolite production, response surface methodology was employed. using different optimization techniques, including Plackett–Burman design, path of steepest ascent determinations, and Box-Behnken design, optimized fermentation conditions were identified including glucose at 38.877 g/L, $CaCl2\bullet2H2O$ at 0.161 g/L, temperature at 29.97 °C, and inoculation amount at 8.93%, which gave antibacterial activity of 56.13% against FOC4, representing a 12.33% improvement compared to pre-optimization levels (43.80%).

The application of response surface methodology in medium optimization significantly impacted the production of important secondary metabolites by *Streptomyces* sp. *1–14*. Additionally, it was emphasized that parameters such as pH, light exposure, and storage temperature must be vigilantly monitored during fermentation and storage to minimize the formation of products with diminished antibacterial activity.

This methodology not only proves valuable for further investigations into anti-FOC4 substance production but also holds potential for developing bio-control agents aimed at suppressing or controlling banana fusarium wilt.

The application of response surface methodology is widespread in the optimization of cryoprotectants for biocontrol strains during preparation processes. To enhance the viability of biocontrol agents, post freeze-drying, a variety of cryoprotectants and centrifugal conditions were systematically analyzed. In the context of optimizing the freeze-drying process for *Bacillus amyloliquefaciens* B1408, experiments generated by Box-Behnken design were performed to determine the ideal concentrations of three cryoprotectants: glucose, trehalose, and xylitol.

The results of these experiments pointed to the effectiveness of centrifugation conditions at 5000 r/min for 10 minutes. The optimal concentrations of cryoprotectants were identified as 1.00% glucose, 4.74% trehalose, and

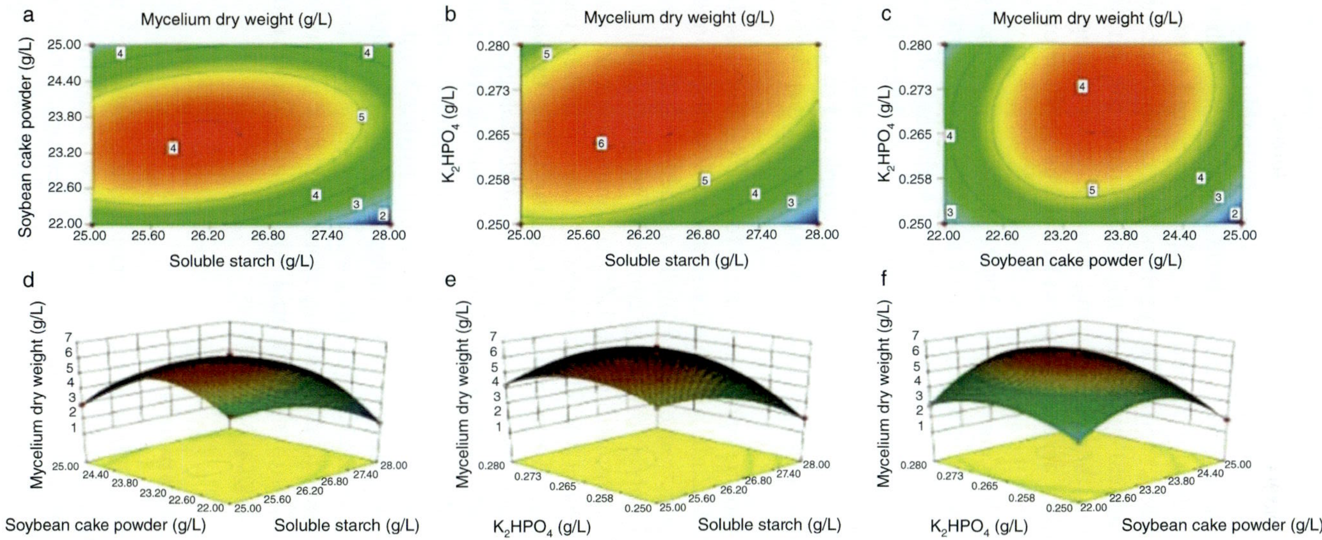

Figure 3.4 Response surface methodology in *Streptomyces alfalfae* XN-04.

***Microorganisms*, *10*(9), 1854. Chen, J., Lan, X., Jia, R., Hu, L., & Wang, Y. (2022). Response surface methodology (RSM) mediated optimization of medium components for mycelial growth and metabolites production of *Streptomyces*.**

1.45% xylitol. Remarkably, the established survival rate of cells after freeze-drying reached 91.24%. This compellingly demonstrates the feasibility of utilizing freeze-drying as a preparation method for the biocontrol strain B1408.

The outcomes of this investigation not only offer valuable discernment into the preparation of biocontrol strain B1408 but also establish a theoretical foundation for potential commercial applications and formulation development. The identified optimal conditions and cryoprotectant concentrations provide practical guidelines for maximizing the survival of biocontrol agents during the freeze-drying process, contributing to the broader field of biocontrol agent utilization in various applications (Han et al., 2018).

The study performed by Gao et al. in 2009+ utilized response surface methodology to optimize the composition of the medium for producing avermectin B1a by *Streptomyces avermitilis* 14-12A through shaker flask cultivation. Through the Plackett–Burman design, it was determined that corn starch and yeast extract significantly influenced the production of avermectin B1a. The steepest ascent method was then employed to identify the optimal region for medium composition, and subsequently, response surface analysis was applied.

The analysis pinpointed the optimized values for the variables as 149.57 g/L for corn starch and 8.92 g/L for yeast extract. In the verification experiment, the observed production of 5128 mg/L aligned with the predicted values. This represented a notable increase compared to the original production level of 3528 mg/L, demonstrating a 1.45-fold improvement.

These findings underscore the effectiveness of the optimized medium composition in enhancing avermectin B1a production. The utilization of statistical approaches such as response surface methodology, Plackett–Burman design, and the steepest ascent method not only identified influential factors but also facilitated the attainment of an optimal production environment. The successful verification experiment further validated the accuracy of the predicted optimal values, emphasizing the reliability of the applied optimization strategies. Overall, this research showcases a systematic and effective approach to maximizing avermectin B1a production in shaker flask cultivation, with the potential for broader implications in industrial and biotechnological applications.

3.5.3.2 Artificial neural networks

Artificial neural networks (ANNs) are computational algorithms crafted to repeat data patterns, inspired by the complex framework of the biological nervous systems. Consisting of interconnected neurons, ANNs establish a web of processing elements that cooperate in a synchronized manner to tackle tangled problems. This neural arrangement empowers ANNs to shine in situations where selective trends or identifying patterns presents a challenge. Despite being in existence for almost five decades, ANNs have recently encountered a surge in popularity. The foundational principles of ANNs have endured, but it is the extensive embrace of robust computational tools in contemporary society that has triggered a rebirth, offering advantages to experts, engineers, and consumers alike. ANN technologies have been extensively used for the optimization of bioprocess parameters which help in enhanced production of acetic acid as mentioned in Figure 3.5 (Upadhyay et al., 2023).

The revival of ANNs is characterized by their growing significance and relevance across diverse fields. These networks now play a vital role in unwinding complex problems that were previously deemed insuperable. The widespread and universal integration of ANNs into various applications underscores their flexibility and efficiency. As we navigate the intricacies of our modern world, the reinvigoration of ANNs emphasizes their transformative influence on technology and problem-solving methods. This renewed enthusiasm aligns with the evolving panorama of computational capabilities, charting the course for ongoing progress in artificial intelligence and machine learning (Abdolrsol et al., 2021).

In these times, the widespread utilization of artificial neural networks has diversely impacted the landscape of media production across various bioprocesses and enzyme synthesis. Diverse studies have deliberately presented a thoroughly comparative analysis, contrasting the efficiency of response surface methodology with that of artificial neural networks. Across various

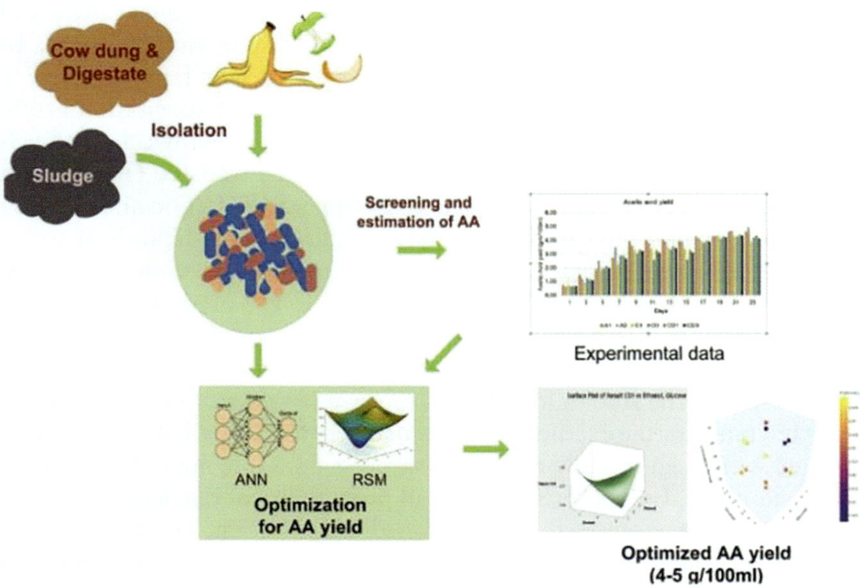

Figure 3.5 Enhance production of acetic acid using ANN technology (Upadhyay et al., 2023).

literature, several reports have designed the superior outcomes in terms of enzyme and biomolecule production achieved through the application of ANNs when compared with RSM.

Beyond the realm of microbiology, ANNs have established themselves as vital tools in diverse scientific optimization processes, demonstrating significant success. This review serves as an introductory exploration into the multiplex applications of ANNs, providing a perceptive primer on their fundamental principles. As the narrative unfolds, specific studies take center stage, highlighting the prowess of ANNs in the microbial production of biomolecules. The comparative assessment with RSM within these studies not only underscores the advantages of ANNs but also contributes to the clarifying of methods employed in bioproduction.

The integration of ANNs in these studies represents a paradigm shift, leading to advanced computational techniques to increase precision and efficiency. As we navigate the intricacies of these applications, it becomes apparent that ANNs have immense potential as a transformative force in optimizing and revolutionizing microbial bioproduction methods, creating a course toward a future where the accord between artificial intelligence and bioprocessing reaches unrivaled levels (Upadhyay et al., 2023).

3.5.3.3 Genetic algorithms

Streptomyces, being a thread-like bacterium, has an indispensable dependence on oxygen as a substrate for the production of its metabolites. However, oxygen having low solubility causes toil in its transfer. This toil is further compounded by a distinctive morphology of the mycelium. To overcome this problem, elevating the stirrer speed is a familiar strategy employed to enhance the supply of oxygen. In case of fermentation broth, which exhibits high viscosity due to presence of dispersed mycelia and pellets, robust agitation becomes necessary for oxygen transfer. However, an exquisite balance must be maintained, as intemperate agitation and aeration can negatively influence productivity by subjecting cells to high shear stress, resulting in cellular damage. During the exploration of various solutions, a higher aeration rate was found to be critical to the process and resulted in the highest level of biocontrol agents that were effective against storage apple diseases belonging to the *Alternaria* and *Fusarium* genera.

Interestingly, the studies found that concurrently augmenting the aeration rate facilitated for the exertion of a lower stirrer speed. This tempering, incorporated with a significantly reduced bioprocess duration, not only optimized production but also bolstered the overall profitability of the defined production process. In essence, this multifaceted optimization, attained through genetic algorithms, not only intensified biocontrol agent production but also negotiated the challenge of shear stress, offering a more sustainable and economically viable approach to *Streptomyces* sp. metabolite production in the context of combating *Alternaria* and *Fusarium* pathogens in stored apples.

3.6 LARGE SCALE PRODUCTION OF BIOPHARMACEUTICAL FROM *STREPTOMYCES*

Streptomyces stands out as a resourceful genus with remarkable applications in biotechnology, particularly as a pivotal platform for drug production. Despite the complicated life cycle and metabolism of *Streptomyces*, existing review papers often concentrate on isolated aspects, such as isolating new strains, identification of metabolites, morphological studies, and genetic or metabolic investigations.

The study comprehensively outlines cutting-edge advancements in biotechnological manufacturing and bioconversion operations, specifically focusing on anticancer drugs, antibiotics, steroidal drugs, and immunosuppressant and anthelmintic agents produced by *Streptomyces*. Emphasis is placed on process development aspects, illuminating diverse approaches and technologies employed to enriched production yields. The investigation explores the impact of nutrients and oxygen on *Streptomyces* metabolism, introduces novel fed-batch fermentation tactics, innovative precursor argumentation methods, and distinct bioreactor designs. It also includes biotechnological strategies coupled with metabolic engineering and genetic tools targeted at enhancing strain performance.

The investigation extends to the utilization of whole, free, and restrained cells on unconventional supports for bioconversion processes of drugs. The significant topics include the influence of carbon, nitrogen, oxygen, and nutrients on *Streptomyces* metabolism, advancements in fed-batch fermentation, pioneering techniques to precursor supplementation, and the design of specialized bioreactors.

Microbial secondary metabolites are widely used in a variety of industries, including food additives, insecticides, and pharmaceuticals. To formulate these compounds, diverse fermenting techniques are employed across numerous scales, spanning from laboratory settings to manufacturing at larger scales, each manifesting distinct advantages and shortcomings. There are four cultivation techniques.

- Traditional immobilization
- Solid-state fermentation
- Microparticle-enhanced cultivation
- Membrane gradostat reactor.

The focus is on the significance of physical support, both in terms of its presence and its impact on microbial processes. The genus *Streptomyces* was selected for its critical role in the generation of bioactive compounds and its sensitivity to cultivation conditions. The methods discussed are explained and compared based on research findings linked to this genus. Instead of identifying a technique that works for everyone, the goal is to provide a critical evaluation of their current state and upcoming obstacles.

3.6.1 Case study on scale-up processes

L-asparaginase stands out as a promising enzyme with a broad array of noteworthy applications, spanning cancer therapy and the realms of starchy food industries (Patel et al., 2022). The optimized value of *L-asparaginase* produced from *Streptomyces brollosae* NEAE-115 was undertaken by utilization of statistical methodologies, specifically the Plackett–Burman design and face-centered central composite design methodology. Consequently, the optimal medium for generating *L-asparaginase* in the filtration extract of *Streptomyces brollosae* NEAE-115 was identified, comprising the following components: 2.0g maltose, 20.0g maize starch, 10.0g L-glutamine, 1.0g potassium nitrate, 1.0g dipotassium hydrogen phosphate, 0.50g magnesium sulfate heptahydrate, 0.10g sodium chloride, pH set at 7, fermentation duration of 7 days, temperature set at 30 °C, inoculant quantity at 4% (v/v), rotation speed at rpm of 150, and inoculum age of 48 hours.

The process of cell growth, the utility of carbohydrates, and the synthesis of *L-asparaginase* were further examined under various growing settings in a Stirred Tank Reactor (Volume = 7L). Under uncontrolled pH circumstances, a significant increment in cellular proliferation and carbohydrate consumption was seen with a stirring velocity increase of 200 to 600 rpm. After 96 hours at 400 rpm, the peak *L-asparaginase* activity of 108.46 U/mL was attained. In contrast, under uncontrolled pH circumstances, the particular enzyme synthesis peaked at approximately 20.3 U/mg cells. The subsequent refining of enzyme synthesis was accomplished by keeping the pH at 7 and using 400 rpm as the specified stirring speed. In cultures with regulated pH, the amount of enzyme produced reached 162.11 U/mL, which is almost 50% more than the value obtained from uncontrolled pH (108.46 U/mL) (El-Naggar et al., 2019).

L-glutaminase (L-glutamic acid hydrolase, EC3.5.1.1) emerges as an enzyme of signifying therapeutic importance, playing a crucial role in certain cancer treatment applications, particularly acute lymphoblastic leukemia (ALL). This tetrameric enzyme facilitates the breakdown of L-glutamine into L-glutamic acid and ammonia. The efficacy of *L-glutaminase* in combating neoplastic cells hinges on their heightened demand for L-glutamine as a vital amino acid to sustain rapid malignant proliferation. Neoplastic cells, lacking L-glutamine synthetase, rely on external sources derived from dietary intake, bodily absorption, and serum availability to secure L-glutamine for survival and propagation. In contrast, normal healthy cells can internally synthesize L-glutamine, rendering them resilient to L-glutamine deprivation. Intravenous administration of L-glutaminase significantly reduces free L-glutamine levels in the bloodstream, selectively inducing the demise of neoplastic cells due to L-glutamine scarcity. Beyond its therapeutic applications, *L-glutaminase* finds utility in various sectors, such as food processing, where it converts L-glutamine to L-glutamic acid. Acrylamide, a known carcinogen, is successfully prevented from forming by pretreating starchy foods with enzymes before frying or baking.

Microbial *L-glutaminases*, extensively investigated for their chemotherapeutic potential in treating human cancer, are derived from diverse sources like fungi, yeast, bacteria, and *actinomycetes* through submerged fermentation. Presently, clinical use involves *L-glutaminase* from two bacterial strains, *Erwinia chrysanthemi* and *E. coli*, for ALL treatment. While some *actinomycetes*, including *Streptomyces karnatakensis* and *Streptomyces albidoflavus*, produce cell-bound L-glutaminase, others release it extracellularly. Large-scale batch fermentations employing stirred tank bioreactors have proven effective for enzyme production.

The effect of carbon sources on the optimal production of *L-asparaginase* was also examined by employing the face-centered central composite design to identify the major variables influencing *L-asparaginase* synthesis through Plackett–Burman design. A statistical analysis of the data revealed elevated *L-glutaminase* synthesis. Therefore, the purpose of this study was to refine many procedural parameters that affect *Streptomyces brollosae* NEAE-115's production of *L-asparaginase* through submerged fermentation in agitation flasks and a 7-L stirred tank bioreactor (El-Naggar et al., 2019).

3.7 FUTURE PROSPECTS

As the global shift from fossil-based to bio-based economies persists, there"s a growing utilization of industrial bioprocesses for the production of biofuels and materials and healthcare products. A bioprocess involves using living cells or their constituents to create enhanced products. Artificial Intelligence (AI) is a field of computer science that seeks to replicate the thinking of humans for problem-solving. AI systems can autonomously make a verdict based on predetermined protocols or identifying patterns in data. In today's times, an increased number of AI advancements has been implemented to optimize as well as manage the bioprocesses, thereby enhancing their performance (Singh & Singhal, 2023). Figure 3.6 shows several applications of AI that could be used for optimizing bioprocess parameters.

AI proves highly valuable in enhancing the optimization of bioprocess parameters by adeptly managing intricate data, discerning patterns, and iteratively enhancing solutions. The subsequent delineation outlines precise methods through which AI actively contributes to optimizing bioprocess parameters.

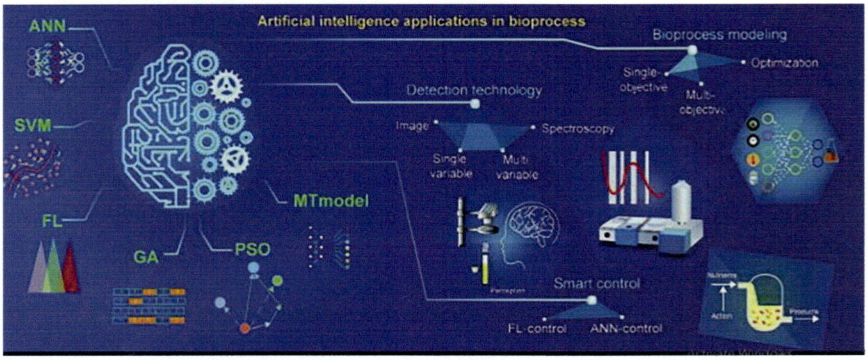

Figure 3.6 Artificial intelligence applications in bioprocess (Cheng et al., 2023).

3.8 DATA ANALYSIS AND PATTERN RECOGNITION

Handling Large Datasets: Bioprocesses generate vast amounts of data. AI algorithms, particularly machine learning techniques, can efficiently process and analyze large datasets to identify patterns and correlations.

Hidden Relationships: AI can uncover hidden relationships between various bioprocess parameters, helping researchers understand complex interactions that may not be immediately apparent through traditional methods.

3.9 PROCESS MODELING

Dynamic Simulation: AI facilitates the development of dynamic process models that simulate the behavior of the bioprocess under different conditions. These models enable researchers to predict outcomes, test hypotheses, and explore the effects of parameter changes before implementing them in the actual process.

Continuous Learning: Machine learning models integrated into process models can continuously learn from new data, improving the accuracy and adaptability of the model over time.

3.10 OPTIMIZATION ALGORITHMS

Exploring Parameter Space: Optimization algorithms, driven by AI, can efficiently explore the vast parameter space of bioprocesses. Genetic algorithms, for instance, mimic natural selection to iteratively refine parameters, finding optimal conditions that maximize desired outcomes.

Iterative Improvement: AI algorithms allow for iterative improvements, adapting strategies based on previous results and refining the search for optimal parameter sets.

3.11 REAL-TIME MONITORING AND CONTROL

Dynamic Adjustment: AI–based systems can monitor bioprocess parameters in real-time using sensors and automated control systems. This enables dynamic adjustments to conditions, ensuring that the process operates at optimal levels.

Adaptive Responses: AI allows for adaptive responses to fluctuations in raw material quality, environmental conditions, or other variables, maintaining optimal process performance.

3.12 DECISION SUPPORT SYSTEMS

Informed Decision-Making: AI–driven decision support systems analyze historical data, current conditions, and optimization algorithms to provide informed recommendations for parameter adjustments or process modifications.

Reducing Human Bias: AI minimizes the impact of human bias in decision-making, providing objective and data-driven insights.

3.13 BIOPROCESS TROUBLESHOOTING

Anomaly Detection: AI can identify anomalies and deviations from expected patterns, facilitating quick issue identification and troubleshooting.

Proactive Problem Resolution: By detecting issues early, AI contributes to proactive problem resolution, reducing downtime and ensuring the stability of the bioprocess.

3.14 ADAPTIVE CONTROL STRATEGIES

Responding to Changes: AI enables the development of adaptive control strategies that respond to changes in real-time. This ensures optimal performance even in dynamic and evolving conditions.

Enhancing Robustness: Adaptive control strategies contribute to the robustness of bioprocesses, making them more resilient to variations in input parameters.

3.15 MULTI-OBJECTIVE OPTIMIZATION

Balancing Conflicting Goals: AI is well suited for handling multi-objective optimization problems where conflicting goals, such as maximizing yield while minimizing costs, need to be balanced.

Pareto Front Optimization: AI algorithms can identify Pareto-optimal solutions, representing the trade-offs between multiple objectives.

The elucidation of complicated, intricate relations within living organisms is beyond the capabilities of many sophisticated mathematical models, including design of experiments, partial least square regression, principal component analysis, multivariate data analysis, and so on. These models are widely used for optimizing engineering methodologies. Hence, a broad spectrum of machine learning models, including principal coordinate analysis, GP, Rain Forest, Support Vector Machine, and ANN, are now discovering extensive use in bioengineering and scaling up of economically significant bio products. Hence, the current chapter covers the implementation of ML in bioengineering via a variety of case studies and also posts current obstacles and future aspects in order to interpret an encompassing view on this.

In summary, AI's capabilities in data analysis, modeling, optimization, real-time monitoring, and decision support make it a powerful tool for optimizing bioprocess parameters. Its ability to handle complexity and adapt to changing conditions contributes significantly to the efficiency, productivity, and robustness of bioprocessing industries.

3.16 CONCLUSION

From the preceding discussion it could be concluded that diverse species of *Streptomyces* are responsible for generating large varieties of secondary metabolites that can be used as antimicrobials, enzymes, and in agriculture. Several cosmetic and personal uses have also been established by several scientists. To produce these metabolites in large amounts from different *Streptomyces* species, it is essential to optimize the media and cultivation conditions wherein these microorganisms can be grown. Optimization of fermentative parameters can be done by various techniques, some of which are mentioned in this book chapter, including response surface methodology linked with Plackett–Burman design, central composite design, artificial neural networks linked with genetic algorithms, and many more. Nowadays, artificial intelligence has evolved to be used for the optimization of appropriate media and cultivation conditions for increased production of desired metabolites. Optimization of bioprocess parameters would further open the possibilities of large-scale production of desired metabolites by cultivating *Streptomyces* strains, including their production on an industrial scale. With the advancement of the processes involved during optimization studies, bioprocesses, and scale-up methodologies, it will play a pivotal role in harnessing the full potential of *Streptomyces* for addressing current and future challenges in medicine, agriculture, and industry.

REFERENCES

Abdolrasol, M. G., Hussain, S. S., Ustun, T. S., Sarker, M. R., Hannan, M. A., Mohamed, R., ... & Milad, A. (2021). Artificial neural networks based optimization techniques: A review. *Electronics*, *10*(21), 2689.

Ahsan, T., Chen, J., Wu, Y., & Irfan, M. (2017). Application of response surface methodology for optimization of medium components for the production of secondary metabolites by Streptomyces diastatochromogenes KX852460. *AMB Express*, 7, 1–10.

Alam, K., Mazumder, A., Sikdar, S., Zhao, Y. M., Hao, J., Song, C., ... & Li, A. (2022). Streptomyces: The biofactory of secondary metabolites. *Frontiers in Microbiology*, *13*, 968053.

Al-Ghazali, L. H., & Omran, R. (2017). Optimization of medium composition for antibacterial metabolite production from Streptomyces sp. *Optimization*, *10*(9).

Anné, J., Maldonado, B., Van Impe, J., Van Mellaert, L., & Bernaerts, K. (2017). Recombinant protein production and streptomycetes. *Applied Microbiology and Biotechnology*, 101(3), 915–929.

Antoraz, S., Santamaría, R. I., Díaz, M., Sanz, D., & Rodríguez, H. (2015). Towards a new focus in antibiotic and drug discovery from the Streptomyces arsenal. *Frontiers in Microbiology*, 6.

Arasu, M. V., Rejiniemon, T. S., Al-Dhabi, N. A., Duraipandiyan, V., Ignacimuthu, S., Agastian, P. K. S., ... & Choi, K. C. (2014). Nutritional requirements for the production of antimicrobial metabolites from Streptomyces. *African Journal of Microbiology Research*, *8*(8), 750–758.

Arul Jose, P., Sivakala, K. K., & Jebakumar, S. R. D. (2013). Formulation and statistical optimization of culture medium for improved production of antimicrobial compound by Streptomyces sp. JAJ06. *International Journal of Microbiology*, *2013*.

Bérdy, J. (2005). Bioactive microbial metabolites. *The Journal of Antibiotics*, *58*(1), 1–26.

Bundale, S., Begde, D., Nashikkar, N., Kadam, T., & Upadhyay, A. (2015). Optimization of culture conditions for production of bioactive metabolites by Streptomyces spp. isolated from soil. *Advances in Microbiology*, *5*(06), 441.

Burg, R. W., Miller, B. M., Baker, E. E., Birnbaum, J., Currie, S. A., Hartman, R., Kong, Y. L., Monaghan, R. L., Olson, G., Putter, I., Tunac, J. B., Wallick, H., Stapley, E. O., Oiwa, R., & Omura, S. (1979). Avermectins, new family of potent anthelmintic agents: producing organism and fermentation. *Antimicrobial Agents and Chemotherapy*, *15*(3), 361–367.

Butt, U. D., Khan, S., Liu, X., Sharma, A., Zhang, X., & Wu, B. (2024). Present status, limitations, and prospects of using Streptomyces bacteria as a potential probiotic agent in aquaculture. *Probiotics and Antimicrobial Proteins*, *16*(2), 426–442.

Challis, G. L., & Hopwood, D. A. (2003). Synergy and contingency as driving forces for the evolution of multiple secondary metabolite production by Streptomyces species. *Proceedings of the National Academy of Sciences*, *100*(2), 14555–14561.

Chen, J., Lan, X., Jia, R., Hu, L., & Wang, Y. (2022). Response surface methodology (RSM) mediated optimization of medium components for mycelial growth and metabolites production of Streptomyces alfalfae XN-04. *Microorganisms*, *10*(9), 1854.

Chen, Y., & Du, L. (2016). Iterative polyketide biosynthesis by modular polyketide synthases in bacteria. *Applied Microbiology and Biotechnology*, 100(2), 541–557.

Cheng, Y., Bi, X., Xu, Y., Liu, Y., Li, J., Du, G., ... & Liu, L. (2023). Artificial intelligence technologies in bioprocess: opportunities and challenges. *Bioresource Technology*, *369*, 128451.

Cuozzo, S., de LeBlanc, A. D. M., LeBlanc, J. G., Hoffmann, N., & Tortella, G. R. (2023). Streptomyces genus as a source of probiotics and its potential for its use in health. *Microbiological Research*, *266*, 127248.

Demain, A. L., & Sanchez, S. (2009). Microbial drug discovery: 80 years of progress. *The Journal of Antibiotics*, *62*(1), 5–16.

Donald, L., Pipite, A., Subramani, R., Owen, J., Keyzers, R. A., & Taufa, T. (2022). Streptomyces: Still the biggest producer of new natural secondary metabolites, a current perspective. *Microbiology Research*, *13*(3), 418–465.

El-Naggar, N. E. A., Moawad, H., El-Shweihy, N. M., El-Ewasy, S. M., Elsehemy, I. A., & Abdelwahed, N. A. (2019). Process development for scale-up production of a therapeutic L-asparaginase by Streptomyces brollosae NEAE-115 from shake flasks to bioreactor. *Scientific Reports*, *9*(1), 13571.

Escalante-Aburto, A., Ramírez-Malule, H., Cruz-Borbolla, J., Lizardi-Mendoza, J., & Rios-Estepa, R. (2022). Strategies for improving *Streptomyces* secondary metabolites production. *Frontiers in Bioengineering and Biotechnology*, *10*, Article 914345. https://doi.org/10.3389/fbioe.2022.914345

Gao, H., Liu, M., Liu, J., Dai, H., Zhou, X., Liu, X., ... & Zhang, L. (2009). Medium optimization for the production of avermectin B1a by Streptomyces avermitilis 14-12A using response surface methodology. *Bioresource Technology*, *100*(17), 4012–4016.

Han, L., Pu, T., Wang, X., Liu, B., Wang, Y., Feng, J., & Zhang, X. (2018). Optimization of a protective medium for enhancing the viability of freeze-dried Bacillus amyloliquefaciens B1408 based on response surface methodology. *Cryobiology*, 81, 101–106.

Harir, M., Bendif, H., Bellahcene, M., Fortas, Z., & Pogni, R. (2018). Streptomyces secondary metabolites. *Basic Biology and Applications of Actinobacteria*, 6, 99–122.

Ikeda, H., Ishikawa, J., Hanamoto, A., Shinose, M., Kikuchi, H., Shiba, T., Sakaki, Y., Hattori, M., & Ōmura, S. (2003). Complete genome sequence and comparative analysis of the industrial microorganism Streptomyces avermitilis. *Nature Biotechnology*, 21(5), 526–531.

Jaki, B., Heilmann, J., & Sticher, O. (2000). New antibacterial metabolites from Streptomyces interferens. *Journal of Natural Products*, *63*(8), 1093–1095.

Jogaiah, S., Abdelrahman, M., Tran, L. S. P., & Ito, S. I. (2018). Different mechanisms of Trichoderma virens-mediated resistance in tomato against Fusarium wilt involve the jasmonic and salicylic acid pathways. *Molecular Plant Pathology*, 19(4), 868–883.

Khan, S., Srivastava, S., Karnwal, A., & Malik, T. (2023). Streptomyces as a promising biological control agents for plant pathogens. *Frontiers in Microbiology*, *14*, 1285543.

Khattab, A. I., Babiker, E. H., & Saeed, H. A. (2016). Streptomyces: isolation, optimization of culture conditions and extraction of secondary metabolites. *International Current Pharmaceutical Journal*, *5*(3), 27–32.

Khushboo, Kumar P., Dubey, K. K., Usmani, Z., Sharma, M., & Gupta, V. K. (2022). Biotechnological and industrial applications of Streptomyces metabolites. *Biofuels, Bioproducts and Biorefining*, *16*(1), 244–264.

Korayem, A. S., Abdelhafez, A. A., Zaki, M. M., & Saleh, E. A. (2015). Optimization of biosurfactant production by Streptomyces isolated from Egyptian arid soil using Plackett–Burman design. *Annals of Agricultural Sciences*, *60*(2), 209–217.

Kumar, M., Kumar, P., Das, P., Solanki, R., & Kapur, M. K. (2022). Proactive role of *Streptomyces* spp. in plant growth stimulation and management of chemical pesticides and fertilizers. *International Journal of Environmental Science and Technology*, *19*(10), 10457–10476.

Liu, G., Chater, K. F., Chandra, G., Niu, G., & Tan, H. (2013). Molecular regulation of antibiotic biosynthesis in streptomyces. *Microbiology and Molecular Biology Reviews*, *77*(1), 112–143.

Manivasagan, P., Venkatesan, J., Sivakumar, K., & Kim, S. K. (2014). Pharmaceutically active secondary metabolites of marine actinobacteria. *Microbiological Research*, *169*(4), 262–278.

McHenney, M. A., & Baltz, R. H. (2022). Streptomyces genetics and secondary metabolite discovery. *Chemical Reviews*, 122(17), 9759–9834.

Mitrović, I., Lukić, N., Grahovac, M., Jokić, A., Dodić, J., & Grahovac, J. (2021). Optimization of Streptomyces hygroscopicus Cultivation Parameters in a Lab-scale Bioreactor. *Chemical Engineering & Technology*, *44*(2), 349–358.

Mondal, P. P., Galodha, A., Verma, V. K., Singh, V., Show, P. L., Awasthi, M. K., ... & Jain, R. (2022). Review on machine learning-based bioprocess optimization, monitoring, and control systems. *Bioresource Technology*, 128523.

Ōmura, S., Ikeda, H., Ishikawa, J., Hanamoto, A., Takahashi, C., Shinose, M., ... & Hattori, M. (2001). Genome sequence of an industrial microorganism Streptomyces avermitilis: deducing the ability of producing secondary metabolites. *Proceedings of the National Academy of Sciences*, *98*(21), 12215–12220.

Pacios-Michelena, S., Aguilar Gonzalez, C. N., Alvarez-Perez, O. B., Rodriguez-Herrera, R., Chávez-González, M., Arredondo Valdes, R., ... & Ilyina, A. (2021). Application of Streptomyces antimicrobial compounds for the control of phytopathogens. *Frontiers in Sustainable Food Systems*, *5*, 696518.

Patel, P. G., Panseriya, H. Z., Vala, A. K., Dave, B. P., & Gosai, H. B. (2022). Exploring current scenario and developments in the field of microbial L-asparaginase production and applications: A review. *Process Biochemistry*, *121*, 529–541.

Qiu, Z., Egidi, E., Liu, H., Kaur, S., & Singh, B. K. (2019). New frontiers in agriculture productivity: optimised microbial inoculants and in situ microbiome engineering. *Biotechnology Advances*, *37*(6), 107371.

Rakesh, K. N., Dileep, N., Junaid, S., & Prashith, K. T. R. (2014). Optimization of culture conditions for production of antibacterial metabolite by bioactive Streptomyces species srdp-tk-07. *International Journal of Advanced Pharmaceutical Sciences*, *5*, 1809–1816.

Rokem, J. S., Lantz, A. E., & Nielsen, J. (2007). Systems biology of antibiotic production by microorganisms. *Natural Product Reports*, 24(6), 1262–1287.

Romano-Armada, N., Yañez-Yazlle, M. F., Irazusta, V. P., Rajal, V. B., & Moraga, N. B. (2020). Potential of bioremediation and PGP traits in Streptomyces as strategies for bio-reclamation of salt-affected soils for agriculture. *Pathogens*, *9*(2), 117.

Saraiva, S. M., Miguel, S. P., Araujo, A. R., Rodrigues, M., Ribeiro, M. P., & Coutinho, P. (2023). Cosmetic Industry: Natural Secondary Metabolites for Beauty and Aging. In *Natural Secondary Metabolites: From Nature, Through Science, to Industry* (pp. 853–891). Cham: Springer International Publishing.

Schatz, A., Bugle, E., & Waksman, S. A. (1944). Streptomycin, a substance exhibiting antibiotic activity against gram-positive and gram-negative bacteria. *Proceedings of the Society for Experimental Biology and Medicine*, *55*(1), 66–69.

Shakeel, Q., Lyu, A., Zhang, J., Wu, M., Li, G., Hsiang, T., & Yang, L. (2018). Biocontrol of Aspergillus flavus on peanut kernels using *Streptomyces yanglinensis* 3–10. *Frontiers in Microbiology*, *9*, 1049.

Singh, A., & Singhal, B. (2023). Role of machine learning in bioprocess engineering: Current perspectives and future directions. In R. Malik, R. Kumar, & R. P. Singh (Eds.), *Design and applications of nature-inspired optimization: Contribution of women leaders in the field* (pp. 39–54). Springer.

Tyc, O., van den Berg, M., Gerards, S., van Veen, J. A., Raaijmakers, J. M., de Boer, W., & Garbeva, P. (2014). Impact of interspecific interactions on antimicrobial activity among soil bacteria. *Frontiers in Microbiology*, *5*, Article 567. https://doi.org/10.3389/fmicb.2014.00567

Upadhyay, A., Kovalev, A. A., Zhuravleva, E. A., Pareek, N., & Vivekanand, V. (2023). Enhanced production of acetic acid through bioprocess optimization employing response surface methodology and artificial neural network. *Bioresource Technology*, *376*, 128930.

Vaishnav, P., & Demain, A. L. (2011). Unexpected applications of secondary metabolites. *Biotechnology Advances*, *29*(2), 223–229.

Vurukonda, S. S. K. P., Giovanardi, D., & Stefani, E. (2018). Plant growth promoting and biocontrol activity of Streptomyces spp. as endophytes. *International Journal of Molecular Sciences*, 19(4), 952.

Watve, M. G., Tickoo, R., Jog, M. M., & Bhole, B. D. (2001). How many antibiotics are produced by the genus Streptomyces? *Archives of Microbiology*, *176*(5), 386–390.

Yun, T. Y., Feng, R. J., Zhou, D. B., Pan, Y. Y., Chen, Y. F., Wang, F., ... & Xie, J. H. (2018). Optimization of fermentation conditions through response surface methodology for enhanced antibacterial metabolite production by Streptomyces sp. 1-14 from cassava rhizosphere. *PloS One*, 13(11), e0206497.

Zheng, S., Oh, S., Fang, M., Bellere, A. D., Jung, J., Nguyen, T. T. M., ... & Yi, T. H. (2022). *Streptomyces* spp. isolated from Rosa davurica rhizome for potential cosmetic application. *Cosmetics*, *9*(6), 126.

Zotchev, S. B. (2012). Marine actinomycetes as an emerging resource for the drug development pipelines. *Journal of Biotechnology*, *158*(4), 168–175.

Chapter 4

Potential biotechnological and industrial applications of *Streptomyces* spp. metabolites

Anushree Pant, Saheli Sabnam and A. Ranu
National Institute of Technology, Jalandhar, India

Harmanpreet Meehnian
Chandigarh University, Mohali, India

Asim Kumar Jana
National Institute of Technology, Jalandhar, India

4.1 INTRODUCTION

The current situation is characterized by widespread contamination produced by pollutants released into water bodies from textiles, dyes, pesticides, tanneries, pharmaceuticals, distillery units and paper and pulp industries. Not only mankind but aquatic and terrestrial animals are struggling with survival challenges due to toxic environmental conditions. Different physical and chemical methods have been employed but they failed in bioremediation as they are neither economical nor environment friendly. So, to overcome these problems conventional methods are switching over to biological methods. Enzymes as a biocatalyst offer such benefits as they are a sustainable and eco-friendly alternative to toxic chemicals used in various industries. They also have the advantages of efficiency, selectivity and low environmental effect. Therefore, the use of microorganisms for enzymes production is the best alternative for different industries.

Streptomyces spp. are gram-positive bacteria that exist in a variety of environments and could colonize in aquatic and terrestrial niches. They are saprophytic in nature and pathogenic to plants and animals. *Streptomyces* are comparable to filamentous fungi in terms of morphology and microbiology as they can adapt to comparable ecological niches (de Lima Procópio et al., 2012). The genus is responsible for producing more than 80% of secondary metabolites and enzymes that can be used in different industries including, detergent, pharmaceuticals, textile, leather and biodiesel and ester production (Alokika & Singh, 2019; Lim et al., 2015; Sharma et al., 2016; Thomas et al., 2016; Wink et al., 2017). Utilization of enzymes instead of chemical synthesis brings economic and environmental benefits in multiple

DOI: 10.1201/9781032694634-4

industries (Choi et al., 2015; Sheldon & Pereira, 2017). Cellulase, chitinase and xylanase are the main enzymes produced from *Streptomyces* spp. that are essential in the textile and paper pulp industries. They are also used for lignocellulosic biomass conversion into biofuels. Biofuel production from lignocellulosic biomass has given a new hope to overcome the limitations of fossil fuels as well as waste management (M. Kumar et al., 2020). Apart from this, biocatalysts like lipases are reported to catalyze esterification, transesterification and hydrolysis reactions. There has been a lot of interest in the production of flavor ester and aroma compounds employed in the dairy, food and cosmetics industries (Ahorsu et al., 2018). Enzymes from *Streptomyces* were also used for polythene and nonbiodegradable waste degradation and showed dependency on extracellular enzymes rather than toxic chemicals (Shao et al., 2019).

Large-scale enzyme synthesis is now more efficient due to the development of molecular approaches (Adrio & Demain, 2014). Genetic engineering techniques including combinatorial biosynthesis and metabolic engineering have contributed to creating efficient and stable enzymes that enhance metabolic processes (Fisher et al., 2014).

Streptomyces spp. have a special ability to create a wide variety of enzymes for a wide range of uses that have been designed to get beyond the constraints of chemicals. The production and utilization of enzymes facilitates pollutants degradation efficaciously. Furthermore, this chapter highlights enzymes produced from actinomycetes, *Streptomyces* as an alternative biological agent apart from fungi and yeast as a prospective sink for different applications. It also describes the concept and techniques of immobilization and large-scale production of biocatalysts. Molecular strategies employed for metabolic processes improvement to increase the yield of enzymes have been discussed.

4.2 LIFE CYCLE OF *STREPTOMYCES*

The life cycle of *Streptomyces* can be divided into vegetative and reproductive, which can develop in solid (with substrate, solid state) and submerged growth. Spores and spore chains typically form in liquid cultures whereas solid cultures are responsible for vegetative mycelium and aerial hyphae growth (Yagüe et al., 2013). Different biotechnological applications are dependent on different life cycle phases and metabolism stages. In solid cultures, the life cycle starts with spore germination followed by the formation of vegetative mycelium or substrate mycelium (a complex branched web of compartmentalized hyphae with widely spaced septa) which remains present throughout the entire life cycle stages (Manteca et al., 2005). After programmed cell death the branching hyphae form "aerial mycelium" that grows into the air and has a hydrophobic cover made of two proteins—chaplin and rodlin—that facilitates hyphae piercing through the soil surface (van Dissel et al., 2014). At the end, after another programmed cell death hyphae compartmentalization and spore formation occurs (Manteca & Yagüe, 2018).

Liquid media (submerged cultures) have similar germination and substrate mycelium development processes as in solid culture. However, subsequent growth has different morphologies; freely dispersed mycelia (single cells), open mycelial networks (mats or clumps), and pellets (dense structures with a nutrient-deficient center) (van Dissel et al., 2014). In pellets, programmed cell death occurs in the center, while secondary mycelium differentiates at the periphery. Aerial mycelium and spores are not formed in most species when grown in liquid media, but they still produce secondary metabolites (Manteca & Yagüe, 2018). During the initial vegetative growth phase, from a single spore to aerial hyphae germination, the whole cell is responsible for primary metabolism to produce vitamins and enzymes such as amylase, lipase, cellulase and xylanase (Claessen et al., 2006; Granozzi et al., 1990). These enzymes have characteristics to contribute in regio-selective and stereo-specific bioconversion processes. The immobilized cell applications require biomass in full vegetative phase (Figure 4.1). Through secondary metabolism beneficial bioactive components of high commercial importance, like antibiotics, anti-microbial agents, secondary metabolites and pigments are produced.

Secondary metabolites are natural compounds that function as antitumor, antihypertensive, immunosuppressive, antimicrobial (Singh and Khajuria, 2018; Nguyen et al., 2020), antioxidative, antiviral, cytotoxic, herbicidal, insecticide and plant growth-promoting agents (Salwan and Sharma, 2020; Pham et al., 2021; Sharma et al., 2021). Four classes of secondary metabolites are, compounds that are agrobiologicals (insecticides, pesticides and herbicides); regulatory agents (siderophores, morphogenic agents and growth factors); antagonistic agents (anti-protozoans, antibacterials, antifungals and antivirals); and pharmacological drugs (neurological agents, immunomodulators, enzyme inhibitors and antitumor agents). Secondary metabolites are produced during the stationary phase (idiophase) of *Streptomyces* species development, which are not essential for their growth but provide a competitive edge. Some metabolites assist bacterial cells with siderophores (sequestering metals like iron), which protect against UV light through pigmentation; reducing competition via antibiotics; and facilitates communication with other species. They also play crucial roles in microbe-microbe and host-microbe interactions in animal guts, regulating biosynthetic pathways (Bolourian and Mojtahedi, 2018; Wang et al., 2019). Additionally, these metabolites contribute to humus formation, soil biodegradation, and produce volatile substances like geosmin, which gives soil a specific "wet earth odor" characteristic. Figure 4.2 shows different bacterial growth phases and production of secondary metabolites.

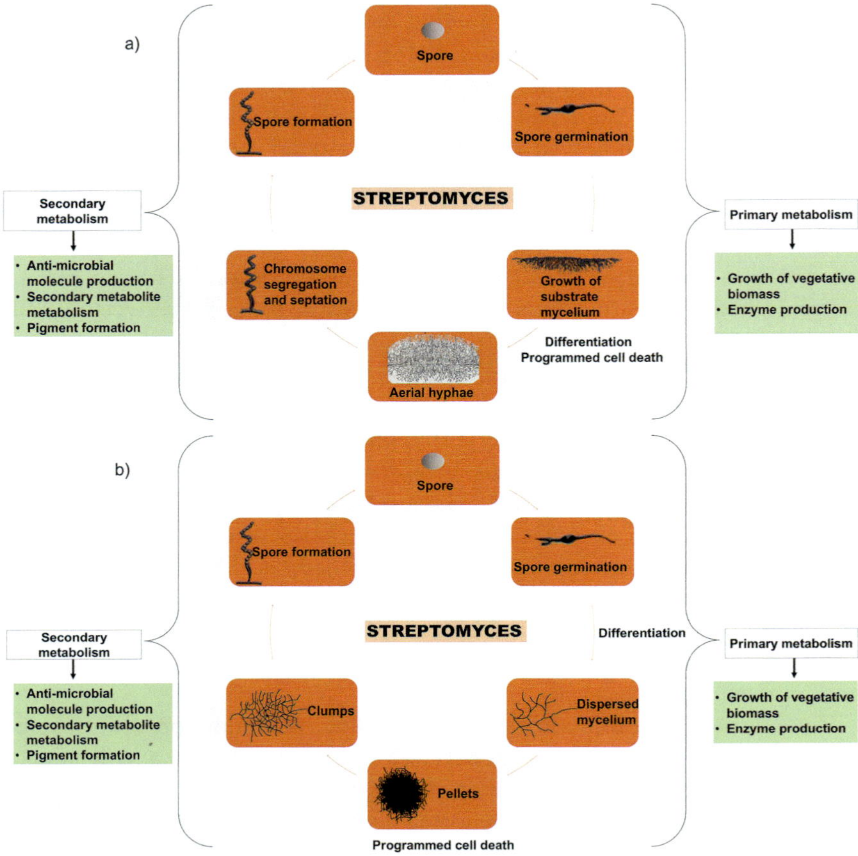

Figure 4.1 Life cycle of *Streptomyces* (a) solid (b) liquid cultures and production of different metabolites-based vegetative and reproductive phase.

4.3 EXTRACELLULAR ENZYMES FROM *STREPTOMYCES* AND THEIR UTILIZATION IN DIFFERENT INDUSTRIES

The most ubiquitous bacterial genus found in the environment are *Streptomyces* spp. with the production of diverse natural products through their complex primary and secondary metabolism. Macrolides, terpenes, tetracyclines, glycopeptides and aminoglycosides are some of the intracellular and extracellular secondary metabolites with high structural diversity and a wide range of bioactivities (Alam et al., 2022). Apart from metabolites, *Streptomyces* spp. like *Streptomyces coelicolor*, *Streptomyces lividans* and *Streptomyces griseus* produces a wide range of extracellular enzymes like cellulase, xylanase, laccase, chitinase, lipase, alpha-amylase, phospholipase and others that has a major role in various industry sectors for ecological

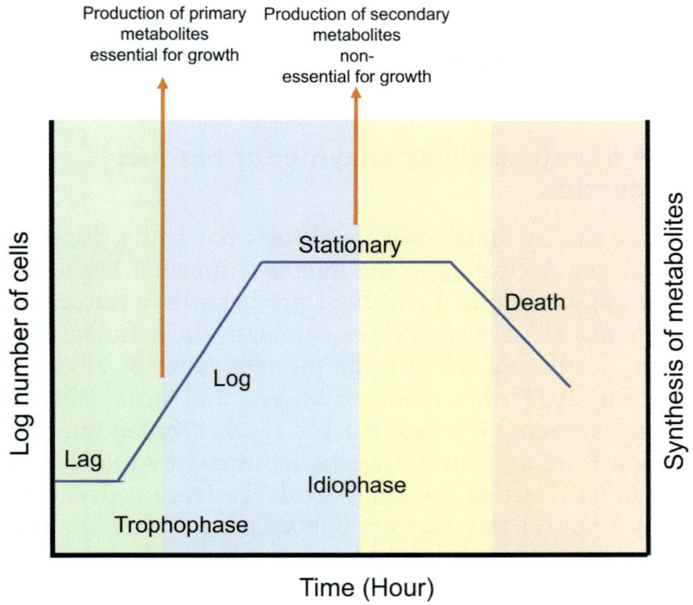

Figure 4.2 Different bacterial growth phases and metabolites production. Primary metabolites production typically happens during the late lag phase and middle of the exponential phase. In contrast, secondary metabolites production takes place at the end of the stationary phase and throughout the persistent phase.

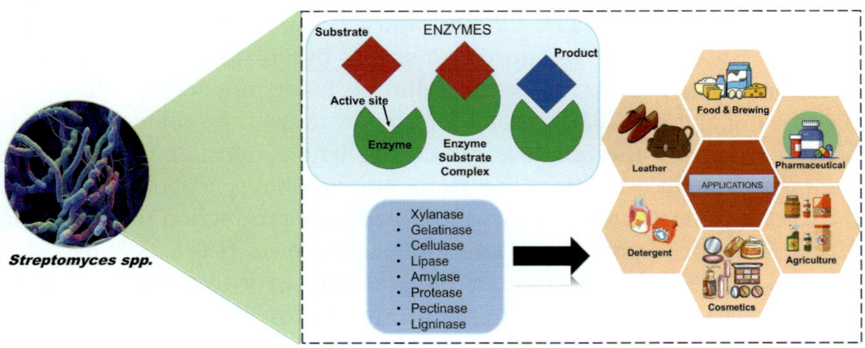

Figure 4.3 Applications of various extracellular enzymes produced from *Streptomyces* sp.

benefits (Figure 4.3) (Talamantes et al., 2016). Additionally, groups of these actinomycetes that produce enzymes demonstrated efficacy in waste management and biodegradation (Sarkar et al., 2011). Enzymes produced by *Streptomyces* spp. of actinomycetes have been described to have the ability to degrade nonbiodegradable pollutants such as polythene, in addition to managing biodegradable wastes (Sangale, 2012). Extracellular enzymes are being used in a variety of industries in place of harmful chemicals (Singh et al., 2016). Table 4.1 shows the role of different types of enzymes produced by *Streptomyces* spp. in diverse applications.

4.3.1 Role of extracellular enzyme for bio-fuel production

Over decades, the combustion of fossil fuels has had a disastrous effect on the ecosystem. According to the Intergovernmental Panel on Climate Change, the combustion of fossil fuels produces the greatest amount of GHG emissions. These changes are cumulatively harming human and other animal health and are also the primary cause of biodiversity loss (Dawson et al., 2011). Bioethanol is an essential sustainable substitute for petroleum in motor vehicles that has sparked broad interest. Due to its production from agricultural waste, industrial by-products and non-food raw materials, second-generation biofuel offers an advantage (Robak et al., 2018). Extracellular enzymes like xylanase and cellulase produced by the mangrove-associated *Streptomyces olivaceus* in a fermentation process are used to produce bioethanol from agro-residues by saccharification (Muthusamy et al., 2019). Actinomycetes produce extracellular enzymes that are active throughout a wide range of pH and temperature, making them a superior option for the synthesis of bioethanol. Enzymatic hydrolysis is a crucial process for the generation of biofuel (Rulli et al., 2016). Among different enzymes, cellulases hydrolyze cellulose to produce reducing sugars, for example glucose by breaking β-1,4 glucan bonds (Mihajlovski et al., 2021). On the other hand, xylanase enhances the process of hydrolyzing cellulose into sugars by breaking down the β-1,4 backbone of xylan that prevents cellulase from functioning effectively (Brito-Cunha et al., 2013). Pectin and lignin restrict the cellulase enzyme from hydrolyzing the cellulose present in lignocellulosic biomass, hence pectinase and laccase breaks the pectin and lignin respectively to enhance the release of fermentable sugars for bio-ethanol production (Zheng et al., 2017). Thus, it is possible to lessen dependence on fossil fuels by using enzymes like cellulase, laccase and xylanase to hydrolyze the lignocellulosic biomass and ultimately by fermenting hexose and pentose sugar to produce bioethanol.

4.3.2 Role of extracellular enzymes in food industry for the replacement of hazardous chemicals

Food industries use chemicals such as potassium bromate, nitrogen dioxide, azodicarbonamide and others as preservatives or oxidizing agents (Yamaguchi et al., 2008; Dubey et al., 2017). If the quantity of these chemicals deviates from standardizing guidelines, the chemicals can have harmful effects on human health, including neurotoxicity, nephrotoxicity,

Table 4.1 Production of different enzymes from *Streptomyces* spp. and their role in diverse applications

Industries	Streptomyces *spp.*	*Enzymes*	*Role/purpose*	*Reference*
Biofuel production industry	*Streptomyces olivaceus/ Streptomyces enissocaesilis* NRRLB-16365/ *Streptomyces coelicolor* A3(2)/ *Streptomyces fulvissimus* CKS7.	Xylanase	Breaks xylan (hemicelluloses) to reduce sugar.	(Muthusamy et al., 2019) (Celaya-Herrera et al., 2021) (Mihajlovski et al., 2021)
		Cellulase	Breaks down the cellulose into oligosaccharides.	
		Laccase	Removes lignin to expose cellulose for hydrolysis	
Food industry	*Streptomyces* GHBA1/ *Streptomyces albidoflavus* NA297	Pectinase	Clarification of fruit juice.	(Kohli & Gupta, 2015) (Murayama et al., 2013)
	Streptomyces lividans TK24	Phospholipase	Flavoring cheese.	
Papermaking industry	*Streptomyces psammoticus* MTCC 7334	Manganese peroxide	Degrades the phenolic subunits of lignin.	(Niladevi & Prema, 2008) (Luo et al., 2016)
	Streptomyces althioticus LMZM	Xylanase	Used in biobleaching process by degrading xylan to improve the brightness of pulp.	
Textile industry	*Streptomyces albaduncus*	Cellulase	Biopolishing of ramie fibers.	(Kalia & Sheoran, 2011) (Kalantzi et al., 2019)
	Streptomyces griseus	Pectate lyase	Bioscouring of cotton fabrics.	

cancer and allergies (Nakamura, 2006). The lives of consumers may be at risk if they ingest any of the aforementioned chemicals . Alternatively, extracellular enzymes have found widespread application in many food processing phases, such as brewing, cheese-making, wine-making, meat-tenderizing, and so on (Ravindran & Jaiswal, 2016; Raveendran et al., 2018). Extracellular enzymes produced by *Streptomyces* spp. are directly used in the food industry for a variety of applications and are environmentally benign. In baking industries, xylanase plays a vital role in improving dough quality that target arbinoxylan fibers. Xylanase breaks down these fibers and improves dough's structure, air holding capacity and expansion rate (Moers et al., 2005). Hydrolyzed hemicellulose forms a tight binding with water in the dough, reducing its stiffness and increasing its volume while producing more consistent and finer crumbs (de Queiroz Brito Cunha et al., 2018). Pectinase enzymes are extensively employed in the brewing and fruit juice clarifying industries. It affects pectin and disrupts the glycosidic linkages that bind the galacturonic acid monomers together, reducing pectin's ability to hold water; thus, a high juice production is achieved (Singh Jadaun, 2018). Various enzymes are utilized in the food industries to replace toxic chemicals while still being environmentally benign, such as catalase used in cheese production, α-amylase in the brewing industry to break down starch and laccase used for the removal of polyphenol from wine (M. Kumar et al., 2020).

4.3.3 Extraction of lignin by enzymes in paper industries

Wood, which is mostly composed of cellulose, hemicellulose, lignin and other materials such as oils, fats, resins, waxes, tannins, phenolics and alkaloid compounds, is used in paper manufacture. To separate the fibers from the lignin, the primary step is to pulp the woody material; this process is known as delignification (Costa et al., 2017). Many nations use chemical treatment as a standard method of delignification, such as peracetic acid and acidified sodium chlorite (R. Kumar et al., 2013). In addition to the traditional approach, hypochlorites (HClO, $CaOCl_2$, NaOCl) are employed as an oxidizer to remove the lignin that was earlier utilized for sulphate pulps (Daraei, 2016). These chemicals are used in pulping and bleaching, which produces a lot of hazardous effluent that is harmful to aquatic life. The pulp and paper sector has been looking for a different approach to replace the chemicals used in the paper-making process (Walia et al., 2017). Extracellular xylanase and laccase produced by *Streptomyces* spp. and fungi have eliminated the incorporation of hazardous chemicals in bleaching and pulping (Kapoor et al., 2007). Laccase targets lignin directly for breakdown, cleaving the non-phenolic and phenolic subunits by disrupting the aryl ether bond (β-O-4', β-5' and β-β) (A. Kumar & Chandra, 2020). Xylanase is utilized to break the bond between hemicellulose and lignin

(Christopher et al., 2014), and the combination of laccase and xylanase improved delignification (Woolridge, 2014). Government and environmental protection organizations should encourage citizens and businesspeople to increase the usage of this type of enzymes in the paper industry (Zainith et al., 2019; Chowdhury et al., 2019).

4.3.4 Role of enzymes in processing textile industries

The textile business contributes to meeting one of humanity's basic requirements. A series of procedures known as "wet processing" is used in conventional textile manufacture (Cavaco-Paulo et al., 2003; D. Kumar et al., 2021a). Many hazardous chemicals, including carbon tetrachloride, as well as solubilizing, cleaning, lubricating, dyeing, defoaming and surfactant agents are utilized excessively in the various stages of textile production (Gulzar et al., 2019). It results in the production of massive amounts of harmful chemical-filled effluents, posing a serious threat to the environment and having harmful effects on human health, that is,carcinogenesis and mutagenesis (Drumond Chequer et al., 2013; Kabir et al., 2019). To address this type of environmental issue, the textile sector is turning to bio-based solutions. The textile industry uses oxidoreductases and hydrolases as key enzymes. The hydrolase class enzymes, which are utilized in the desizing, bioscouring, biopolishing and bleaching of textiles, include alpha amylase, catalase, cellulase, lipase and pectinase (Kumar et al., 2021b). Laccases used in dye decolorization (Mate & Alcalde, 2017) and transferases like transglutaminases, used for the surface modification of synthetic fibers (Shen, 2019), are examples of oxidoreductase enzymes. Extracellular enzymes produced by *Streptomyces* spp. are commonly employed in the textile industry due to their wide pH and temperature range comparing to other sources (M. Kumar et al., 2020).

4.3.5 Role of extracellular enzymes in the replacement of chemicals used for insecticides and fungicides

The following groups of chemicals are utilized as fungicides and insecticides: carbamates, pyrethroids, neonicotinoids, organochlorines and organophosphates (Nicolopoulou-Stamati et al., 2016). Because of their toxicity, they poison the ecosystem and enter the food chain that causes harm to human health (Özkara et al., 2016). In addition to their detrimental effects on the environment and human health, chemical pesticides also cause insect resistance (Zhu et al., 2016). Fungi and insects contain chitin layers in their bodies, and researchers are optimistic about stable chitinase enzyme from fungi and bacteria. Stable chitinase enzyme and bioactive chemicals produced by *Streptomyces* spp. are utilized as efficient pesticides in agriculture (Vurukonda et al., 2018). Because they are naturally occurring soil organisms and do not harm the environment or public health, *Streptomyces* spp. are touted as a "green" alternative to pesticides (Edelvio et al., 2018). The extremely active and stable chitinase that *Streptomyces* secretes are stable over a wide pH and temperature range.

4.3.6 Role of extracellular enzyme in waste management and biodegradation

Two conventional approaches are used for solid waste management in many developing countries such as burning of waste and landfills which lead to greenhouse gas emissions in the atmosphere and cause climate changes. A large portion of municipal solid waste is kitchen waste. Instead of dumping in landfills and open burning, researchers have developed a systematic approach in the degradation of kitchen waste (Sarkar et al., 2011). The major components in kitchen waste are cellulose, starch, fat, lipids, protein and other components which can be degraded by the extracellular enzymes like cellulase, lipase, laccase, amylase, xylanase and others produced from various bacteria and fungi. *Streptomyces* spp. like *Streptomyces coelicolor* and *Streptomyces olivaceus* are preferably used in waste management due to the production of a wide range of extracellular enzymes and their complex secondary metabolism (A. Kumar & Chandra, 2020). Hydrolysis of cellulose creates hemicellulose-rich solid waste by an extracellular enzyme that releases glucose, xylose and xylan; it can then be fermented for ethanol production and manage waste without hazard to the environment (Althuri & Venkata Mohan, 2019).

4.4 APPLICATION OF IMMOBILIZED *STREPTOMYCES* ENZYMES IN BIOCATALYST SYSTEMS

The growing use of *Streptomyces* enzymes in biocatalysis is based on their capability to carry out several reactions. It has better stability and reusability and it is more efficient when immobilized (Hassan et al., 2019). However, we must first consider the immobilization techniques and then the applications of immobilized enzyme. The incorporation of new strategies such as covalent binding, enzyme adsorption and entrapment of biocatalyst enzymes greatly broadened the field of application. A recent example has been provided by an enzymatic analysis carried out by Zhang et al. (2021) that illustrates the development of this method toward higher operational efficiency and stability of the enzyme over a range of environments. For immobilizing enzymes, a number of methods have been explored so far including encapsulation, cross-linking, entrapment, covalent binding and adsorption. The well-known method of immobilizing enzymes through covalent binding involves attaching the enzymes to support materials (such as polyacrylamide, porous silica, porous glass, agarose etc.) by strong, highly stable bonds (Wong et al., 2008). Covalent binding has several benefits, including the production of robust enzymes and an adequate enzyme recovery for subsequent usage. Covalent binding also improves the stability of the enzymes by enhancing their stereospecificity (Fernández-Lorente et al., 2001). Some disadvantages of covalent binding include lengthy incubation times and somewhat complex steps of immobilization (Marrazza, 2014). Using a technique known as cross-linking immobilization, enzymes are bound by covalent bonds in the absence of carriers. In order to complete the intermolecular cross-linking, linker agents are present and function as bridges between two nearby enzyme molecules (Gao et al., 2010). A robust bond is formed between the enzymes during cross-linking immobilization, causing better stabilities (Wong et al., 2008). But this method is very costly as it requires a crystallized form of the enzyme (Cao et al., 2003). Reversible immobilization is achieved using the straightforward carrier-bound method of adsorption that principally depends on physical adsorption. Alumina, activated carbon, ion-exchange resins and many other materials are used to facilitate adsorption (Jesionowski et al., 2014). This technique is simple to use and somewhat inexpensive, but it constitutes a weak bond between the enzyme and the carrier, such as ionic, hydrophobic, salt linkage, or hydrogen bonds (Brady & Jordaan, 2009). Entrapment is a process of immobilization that cannot be reversed. Enzymes are only confined within a web of fibers by covalent or noncovalent linkages (Won et al., 2005). Several advantages of the entrapment process include cheap production cost, reduced mass transfer, high loading capacity and mechanical stability of the enzymes (Costa et al., 2004). Using this immobilization strategy has a number of drawbacks, including a poor enzyme loading capacity and the possibility of support disintegration at the time of polymerization (Nguyen & Kim, 2017). Numerous polymeric matrices are used to entrap different biomolecules in the encapsulation immobilization process (Beatriz et al., 2013). Entrapment and encapsulation are similar in that the cells and the enzymes remain free in solutions but in a confined area. But encapsulation is a more sensitive process (Kurzbaum et al., 2020). Increased morphological stability of enzymes, planned physicochemical permeability and reduced enzyme leakage are some benefits of this technique. A few drawbacks of this method include severe diffusion issues and the rupture of membranes in the event of rapid reaction product accumulation (Rother & Nidetzky, 2014). In the current chapter, a few examples of the utilization of immobilized *Streptomyces* enzymes as a biocatalyst are demonstrated.

4.4.1 Industrial biocatalysis

The efficiency and cost effectiveness of the industrial processes will be improved if a biocatalyst approach using immobilized *Streptomyces* enzymes is introduced. Lipases produced by *Streptomyces* have higher effectiveness on various matrices as compared to many commercially available enzymes for biodiesel production (Bhan & Singh, 2020).

Wang et al. (2017a) have reported the applications of immobilized *Streptomyces* lipases for biodiesel production. This involved the immobilization of lipases onto Amberlite XAD1180 resin using covalent linkage. In this study, using immobilized lipase systems provided a more promising strategy for transesterification reactions than batch reactors, thus providing a feasible technique for biodiesel production.

4.4.2 Environmental remediation

Utilizing *Streptomyces* enzyme immobilization as an alternative for environmental remediation toward the reduction of pollutants is very promising. In fact, immobilization system is one of the specific methods used in increasing the effectiveness of enzymes that will aid in breaking down the recalcitrant pollutants. Compared to others, *Streptomyces* enzymes break down pesticides and dyes more effectively (Sharma et al., 2018). Salwan et al. (2023) showed the efficiency of immobilized enzymes produced by *Streptomyces* for the discoloration of different dyes like Bromophenol blue, B-chrome azurol, methylene blue, Coomassie brilliant blue G250, rose Bengal and methylene blue. The laccase enzymes that were immobilized could be used up to 2–3 times without remarkable drop-in activity. This showed the significance of immobilized laccase in dye discoloration, which could be useful with many different types of wastewater treatment processes.

4.4.3 Food and beverage industry

The use of immobilized *Streptomyces* enzymes could revolutionize the operations of the food and beverage industries. Incorporation of immobilized *Streptomyces* esterase in flavor ester synthesis can produce food product having exclusive flavors (Sharma et al., 2022). Immobilized *Streptomyces* are highly valuable for the food and beverage industry for enhancement in quality and productivity. According to Jørgensen et al. (1988), use of immobilized glucose isomerase from a *Streptomyces* strain improved high fructose corn syrup (HFCS) production by tenfold. Anisha & Prema (2008) reported the enhanced production of the enzyme alpha galactosidase with the use of alginate entrapped *Streptomyces griseoloalbus*. Alpha galactosidases are used to reduce or eliminate antinutritive galacto-oligosaccharides like raffinose family sugars that induce flatulence, which significantly enhances the nutritional content of legume-based food. Another study (Afzali et al., 2021) used immobilized *Streptomyces griseobrunneus* for producing cyclodextrin glucotransferase enzyme which hydrolyses starch as a substrate to cyclodextrin. It demonstrated the superior qualities of polymeric immobilization, which include protecting the bacterial cells from toxins and inhibitors that improved the production of enzymes.

4.4.4 Pharmaceutical synthesis

Including immobilized *Streptomyces* enzymes in pharmaceutical synthesis is an excellent strategy for efficient and sustainable drug synthesis. Immobilized *Streptomyces* antibiotics simplify the complex pharmaceutical intermediate synthesis (Benítez-Mateos et al., 2022). Such approach helps in the generation of major drugs with a notable improvement in yields along with the pursuit of green chemistry undertakings by pharmaceutical sectors. Wang et al. (2017b) investigated the immobilization of a lipase MAS1 from a *Streptomyces* strain for thermal stability. Resin xad1180 was employed as the support material in the immobilization process. This caused increased thermal stability of the immobilized enzyme. Lipase was seven times heat-stable and reusable up to five times.

With the advent of technology, processes for the *Streptomyces* enzyme immobilization may become viable and efficient for use in different industries ranging from environment remediation to pharmaceuticals.

4.5 GROWTH OF *STREPTOMYCES* AT FERMENTER LEVEL

Streptomyces are common bacteria that play an important role in biotechnology, agriculture and pharmaceuticals, being rich in several bioactive elements (Salwan & Sharma, 2020). Producing these important compounds with efficiency requires highly optimized *Streptomyces* growth in the fermenter. We will therefore discuss some essential factors such as media composition, oxygen input, pH control, and scale-up procedures that are required for better understanding the growth of *Streptomyces* in fermenters systems (Singh et al., 2017).

4.5.1 Media composition and nutrient optimization

It is essential to note that constituents of growth media impact *Streptomyces* growth in fermenter systems. Therefore, for optimal development and synthesis of metabolites, *Streptomyces* requires specific nutrients due to its extensive secondary metabolism (Hwang et al., 2014). Customized media provide essential carbon and nitrogen compounds and different vitamins and minerals, favoring good microbial growth (Yeboah et al., 2023). Nutrient optimization adds to the refinement of the requirements for bioactive compound production. Improving the formulation or the composition of the media may contribute not only to increased biomass yield but also to higher production of bioactive compounds which can be applicable on the industrial scale (Kashiwagi et al., 2017). Such an approach creates a foundation for the progression of drug discovery, biomanufacturing, and other biotechnological approaches. Khattab et al. (2016) reported the optimum growth conditions for *Streptomyces* strain isolated from soil sediments as: 2.5 g/L of soybean meal concentration, 1 g/L of $MgSO_4.7H_2O$ conc., 2.5g/L of K_2HPO_4 conc., incubation time 7 days, pH 7.5 and temperature 30 °C. In another study (Ju et al., 2018) the optimal growth conditions for *Streptomyces rimosus* AG-P1441 were reported to be: 3% glucose, 2.5% soybean meal, 3.5% corn starch, 5.9 mM glutamate and 1.2 mM $MgCl_2$. An optimum growth condition of *Streptomyces* sp. CMSTAAHL-3 was reported to be: temperature (30 °C), incubation period (7 days), pH (7), mineral source ($MgSO_4$), carbon source (starch), nitrogen source (urea) and NaCl (3%) (Selvaraj et al., 2023).

4.5.2 Oxygen supply and aeration strategies

To achieve better *Streptomyces* growth in fermenter systems, there is a need for enhancing oxygen supply as well as aeration methods used during production. The metabolism and biosynthesis activities depend on its efficiency in transporting oxygen because it is aerobic (Zhou et al., 2018). The insufficiency in oxygen supply leads to lower production levels of desired metabolites, which also hinder normal growth (Barbuto Ferraiuolo et al., 2021). Optimized respiratory conditions by adjusting aeration rates yield increased biomass and high production of bioactive substances (Dobson et al., 2008). These innovations not only make fermentation a more efficient process, they are also important to the production of enzymes, antibiotics and other secondary metabolic products through large-scale cultivations of *Streptomyces* (Kumar et al., 2021a).

Filamentous microorganisms such as *Streptomyces* display morphologies between disseminated mycelia and pellets during submerged fermentation, based on the type of microbial strain and culture conditions. Pellet formation happens through interwoven hyphae, where individual hyphae are connected to form a tight network. The pellet's surface contains loose mycelia that serve as growth sites. During agitation, these mycelia break, creating a new pellet. The morphologies of the pellets impact downstream processing, heterogeneous mass transfer and medium rheology (P. Kumar et al., 2023). The rheological properties of the culture medium, such as surface tension and viscosity, affect the volumetric oxygen transfer coefficient (kLa). As a result, maintaining sufficient amounts of dissolved oxygen concentration without any limitation is required to maintain appropriate oxygen transfer rates for sustaining the microorganism's respiration rate during the bioprocess (Ribeiro et al., 2021).

4.5.3 pH control

Fermentation system development for enhanced growth of *Streptomyces* necessitates optimal pH control (Jiang et al., 2018). *Streptomyces* are very susceptible to changes in pH values that affect enzyme activity and uptake of nutrients as well as overall metabolic functions (Hodgson, 2000). It may lead to poor growth and reduced production of bioactive compounds (Hwang et al., 2014). For medical, agricultural and other industrial processes, accurate control of pH is important because it gives the most favorable environmental conditions, allowing robust growth of *Streptomyces* with consequent production of the maximum number of bioactive compounds. As shown by Dzhavakhiya et al. (2016), the optimum pH was 6.8-7.0 for the strain *Streptomyces virginiae* VKM Ac-2738D in case of virginiamycin production. Baskaran & Muthukumarasamy (2017) studied the various pH conditions for the growth medium of *Streptomyces* for higher amylase activity and found that maximum growth of *Streptomyces* sp. can be achieved by maintaining it in an optimum pH of 7. Shivalee et al. (2018) found the maximum chitinase production from *Streptomyces pratensis* strain KLSL55 can be achieved by using a growth medium of pH 8. Shrestha et al. (2023) reported maximum pH activity from *Streptomyces* sp. to be pH 7 after studying enzyme production in a wide pH range (3–9).

4.5.4 Scale-up strategies

Scale-up plays a crucial role in the development of *Streptomyces* that involves the shift of processes from the laboratory to the commercial level. Scale-up becomes essential in meeting the requirements of large-scale bioprocessing of drugs, agrochemicals and other biotechnological materials (Meyer et al., 2017). In a large set-up, the factors that can influence *Streptomyces* growth and synthesis of bioactive compounds are pH, incubation temperature, media composition, aeration, oxygen level and agitation. El-Moslamy et al. (2023) employed batch and fed-batch fermentation techniques for scaling up *Streptomyces albus* E56 biomass yield for the production of ZnO nanoparticles. Effective scale-up practices allow for reproducible and efficient *Streptomyces* fermentations which also influences sustainability and cost-effectiveness (Burke, 2008). Significant design modifications and condition optimization are required to scale up *Streptomyces* biomass production. A larger reactor is mandatory for proper mixing and aeration to prevent oxygen and nutrient depletion. Upgraded spargers and impeller systems are needed to boost mass transfer. To sustain constant microbial activity, different process variables need to be maintained precisely. Prevention of foam formation also becomes essential to ensure high yield (de Mello et al., 2023). During scale-up, increase in shear forces from rapid mixing can harm *Streptomyces* cells, lowering the number of enzymes produced. Therefore, shear-protective measures become imperative (Zhou et al., 2018).

There has been a significant contribution to the production of bioactive compounds that has been made through the understanding of fermenter-level *Streptomyces* growth. The discussion of the key aspects that influence *Streptomyces* growth induces more efficient and sustainable production procedures. These revelations in fact add to our biotechnological understanding of *Streptomyces* physiology and, furthermore, industries who rely on those bioactive components produced by these versatile bacteria. Fermenter systems have great promise for optimal cultivation of *Streptomyces* that may generate novel drugs, chemicals or bio-related products.

4.6 GENETIC ENGINEERING TECHNIQUES TO DEVELOP TAILORED *STREPTOMYCES* STRAIN

Genetic engineering is important in changing the genes of *Streptomyces* spp. to produce multi-enzymes and medicines by UV irradiation to boost yield and stability. In a case study, the cholesterol-lowering medication pravastatin was created by modifying the genes of *Streptomyces xanthochromogenes* strain RIA 1098 through the use of multistep random UV mutagenesis (Dzhavakhiya et al., 2015). This technique has been extensively employed in the pharmaceutical and chemical industries to modify the genomes of microorganisms in order to regulate metabolic pathways and generate a diverse range of drugs and enzymes (De Carvalho, 2017). In another approach, two distinct strategies are used, namely structure-guided engineering and directed evolution, have been recently employed to address the primary problems of low activity and poor stability of wild-type carbonyl reductase. First, using structure-guided engineering, a twofold mutant was created by complexing the crystal structure of the wild-type carbonyl reductase enzyme and NADH. The specific activity increased from 38.8 to 168.0 U mg^{-1}. The resultant double mutant was then employed as a template to construct a triple mutant carbonyl reductase by directed evolution, with thermostability and substrate tolerance 6.2 °C and 4.7-fold higher than the wild type (Li et al., 2017). Yin et al. (2021) used a novel mutation technique, atmospheric and room-temperature plasma (ARTP), which produces a plasma jet containing various highly uniformly activated chemical particles to mutate the DNA of microorganisms. This technique is used to increase the yield of transglutaminase in *Streptomyces mobaraensis* by 2.91-fold compared to two-stage fermentation strategy.

4.7 CHALLENGES AND FUTURE PERSPECTIVES

- Utilization of agro-industrial residues and industrial effluents as a low-cost substrate that can reduce the cost of enzyme.
- Isolation of novel strains for continuous production of stable enzyme are required to minimize production time and cost.
- Physiological conditions like pH, temperature, ionic strength and so forth should be maintained throughout the process for better catalytic activity.

- Genetic engineering and nanotechnology are the novel technologies adopted for the enhancement of production rate of enzymes.
- For industries the utilization of immobilized biocatalyst is uninterruptedly expanding with increased awareness of sustainable production processes. This approach gives an opportunity to modify ways of operation, employ the sustained use of enzymes and upgrade the stability of enzymes resulting in better activity. As there is not one absolute method for immobilization and each method comes with some drawback, developments are essential in the case of introducing new immobilization techniques and pioneering materials.
- During submerged fermentation of *Streptomyces*, the variable pellet size causes a disparity in the distribution of nutrients and oxygen.
- The formation of coarse biomass and large pellets in submerged fermentation of *Streptomyces* augments the mass transfer issue which restricts oxygen and nutrient supply to inner cells.
- As *Streptomyces* have complex secondary metabolites and intricate regulatory mechanisms, it is challenging to purify and characterize their enzymes. Insufficient genetic data adds to that.
- Although many technical challenges related to the production of microbes at an industrial level have been well attended, there is still a gap in our understanding of the technological conditions related to scale-up mostly because of insufficient infrastructure. Research is being done over countless fermentation systems to close this gap. Scaling up of *Streptomyces* has a huge industrial prospect because of their diversity, functions and metabolites, combined with genetic engineering which can provide novel solutions to modern bioprocess-related problems.

4.8 CONCLUSION

Extracellular enzymes from *Streptomyces* provide vast application in different industries such as household care, food, animal feed, technical industries, fine chemicals and pharma. Conventional approaches are still a part of industries in developing nations for products manufacturing that involve the incorporation of hazardous and life-threatening chemicals. These substances harm workers engaged in processing or contaminate the environment when they are released as industrial effluents. The incorporation of microbial enzymes can be a replacement to hazardous chemicals from different industrial sectors as they have unique properties such as high specificity, fast action and biodegradability that ensure the processes to run under milder reaction conditions, with improved yields and low waste generation. Enzymes obtained from *Streptomyces* spp. are widely utilized in various industries for their wide pH and temperature range. Application of isolation and culturing techniques for digging novel *Streptomyces* strains from unexplored habitats is a strategy to screen potential and efficient bioactive products secreted by them as a natural reserve biocatalyst. The adoption of synthetic biology and genetic engineering techniques has provided enough room for strain improvement and high-level protein production, which make the process easy and efficient for large-scale production to make better products. The implementation of industrial biotechnology and enzyme technology in different industrial sectors can increase the demand for new enzymes. This expansion of the application of enzymes is motivated not only due to growing environmental concerns but also an economic and quality effective approach.

REFERENCES

Adrio, J. L., & Demain, A. L. (2014). Microbial enzymes: tools for biotechnological processes. *Biomolecules*, *4*(1), 117–139. https://doi.org/10.3390/biom4010117

Afzali, E., Eslaminejad, T., Amirheidari, B., & Ansari, M. (2021). Cell immobilization of Streptomyces griseobrunneus by microcrystalline cellulose for production of cyclodextrin glucanotransferase enzyme. *Cogent Engineering*, *8*(1). https://doi.org/10.1080/23311916.2020.1868145

Ahorsu, R., Medina, F., & Constantí, M. (2018). Significance and challenges of biomass as a suitable feedstock for bioenergy and biochemical production: A review. *Energies*, *11*(12). https://doi.org/10.3390/en11123366

Alam, K., Mazumder, A., Sikdar, S., Zhao, Y.-M., Hao, J., Song, C., Wang, Y., Sarkar, R., Islam, S., Zhang, Y., & Li, A. (2022). Streptomyces: The biofactory of secondary metabolites. *Frontiers in Microbiology*, *13*, 968053. https://doi.org/10.3389/fmicb.2022.968053

Alokika, & Singh, B. (2019). Production, characteristics, and biotechnological applications of microbial xylanases. In *Applied Microbiology and Biotechnology* (Vol. 103, Issues 21–22, pp. 8763–8784). Springer Verlag. https://doi.org/10.1007/s00253-019-10108-6

Althuri, A., & Venkata Mohan, S. (2019). Single pot bioprocessing for ethanol production from biogenic municipal solid waste. *Bioresource Technology*, *283*, 159–167. https://doi.org/10.1016/j.biortech.2019.03.055

Anisha, G. S., & Prema, P. (2008). Cell immobilization technique for the enhanced production of a -galactosidase by Streptomyces griseoloalbus. *Bioresource Technology*, *99*(9), 3325–3330. https://doi.org/10.1016/j.biortech.2007.08.023

Barbuto Ferraiuolo, S., Cammarota, M., Schiraldi, C., & Restaino, O. F. (2021). Streptomycetes as platform for biotechnological production processes of drugs. *Applied Microbiology and Biotechnology*, *105*, 551–568.

Baskaran, B., & Muthukumarasamy, A. (2017). *Isolation, characterisation and enzymatic activity of Streptomyces sp . and its pH control during fermentation process. 11*(4), 114–118. https://doi.org/10.1049/iet-syb.2016.0048

Beatriz, B.; Paula, G.; Batista-Viera, F. *Immobilization of enzymes: A literature survey*; Springer Media: New York, 2013; Vol. 78.

Benítez-Mateos, A. I., Roura Padrosa, D., & Paradisi, F. (2022). Multistep enzyme cascades as a route towards green and sustainable pharmaceutical syntheses. *Nature chemistry*, *14*(5), 489–499.

Bhan, C., & Singh, J. (2020). Role of microbial lipases in transesterification process for biodiesel production. *Environmental Sustainability*, *3*, 257–266.

Bolourian, A., and Mojtahedi, Z. (2018). Streptomyces, shared microbiome member of soil and gut, as 'old friends' against colon cancer. *FEMS Microbiol. Ecol. 94*:fiy120. doi: 10.1093/femsec/fiy120

Brady, D.; Jordaan, J. Advances in enzyme immobilisation. *Biotechnol. Lett.* 2009, *31*, 1639–1650.

Brito-Cunha, C. C. D. Q., De Campos, I. T. N., De Faria, F. P., & Bataus, L. A. M. (2013). Screening and Xylanase Production by Streptomyces sp. Grown on Lignocellulosic Wastes. *Applied Biochemistry and Biotechnology*, *170*(3), 598–608. https://doi.org/10.1007/s12010-013-0193-3

Burke, F. (2008). Scale up and scale down of fermentation processes. *Practical Fermentation Technology*, 231–269.

Cao, L.; van Langen, L.; Sheldon, R. A. Immobilised enzymes: Carrier-bound or carrier-free. Curr. Opin. Biotechnol. 2003, *14*, 387– 394.

Cavaco-Paulo, A., Gübitz, G. M., & Textile Institute (Manchester, England) (Eds.). (2003). *Textile processing with enzymes*. CRC ; Woodhead.

Celaya-Herrera, S., Casados-Vázquez, L. E., Valdez-Vazquez, I., Barona-Gómez, F., Bideshi, D. K., & Barboza-Corona, J. E. (2021). A Cellulolytic Streptomyces Sp. Isolated from a Highly Oligotrophic Niche Shows Potential for Hydrolyzing Agricultural Wastes. *BioEnergy Research*, *14*(1), 333–343. https://doi.org/10.1007/s12155-020-10174-z

Choi, J. M., Han, S. S., & Kim, H. S. (2015). Industrial applications of enzyme biocatalysis: Current status and future aspects. *Biotechnology Advances*, *33*(7), 1443–1454. https://doi.org/10.1016/j.biotechadv.2015.02.014

Chowdhury, S., Dey, S., Di Girolamo, L., Smith, K. R., Pillarisetti, A., & Lyapustin, A. (2019). Tracking ambient PM2.5 build-up in Delhi national capital region during the dry season over 15 years using a high-resolution (1 km) satellite aerosol dataset. *Atmospheric Environment*, *204*, 142–150. https://doi.org/10.1016/j.atmosenv.2019.02.029

Christopher, L. P., Yao, B., & Ji, Y. (2014). Lignin Biodegradation with Laccase-Mediator Systems. *Frontiers in Energy Research*, *2*. https://doi.org/10.3389/fenrg.2014.00012

Claessen, D., de Jong, W., Dijkhuizen, L., & Wösten, H. A. B. (2006). Regulation of Streptomyces development: reach for the sky! In *Trends in Microbiology* (Vol. 14, Issue 7, pp. 313–319). https://doi.org/10.1016/j.tim.2006.05.008

Costa, S.; Azevedo, H.; Reis, R. Enzyme immobilization in biodegradable polymers for biomedical applications. *Biodegradable Systems in Tissue Engineering and Regenerative Medicine* 2004, *80*, 109–112.

Costa, S., Dedola, D., Pellizzari, S., Blo, R., Rugiero, I., Pedrini, P., & Tamburini, E. (2017). Lignin Biodegradation in Pulp-and-Paper Mill Wastewater by Selected White Rot Fungi. *Water*, *9*(12), 935. https://doi.org/10.3390/w9120935

Daraei, M. (2016). Evaluation and prediction of wood properties in pulp and paper production: Estimation of wood chemical properties and pulping results using Near Infrared Spectroscopy: A case study in BillerudKorsnäs in Gävle.

Dawson, T. P., Jackson, S. T., House, J. I., Prentice, I. C., & Mace, G. M. (2011). Beyond Predictions: Biodiversity Conservation in a Changing Climate. *Science*, *332*(6025), 53–58. https://doi.org/10.1126/science.1200303

De Carvalho, C. C. C. R. (2017). Whole cell biocatalysts: Essential workers from Nature to the industry. *Microbial Biotechnology*, *10*(2), 250–263. https://doi.org/10.1111/1751-7915.12363

de Lima Procópio, R. E., da Silva, I. R., Martins, M. K., de Azevedo, J. L., & de Araújo, J. M. (2012). Antibiotics produced by Streptomyces. *Brazilian Journal of Infectious Diseases*, *16*(5), 466–471. https://doi.org/10.1016/j.bjid.2012.08.014

de Mello, A. F. M., de Souza Vandenberghe, L. P., Herrmann, L. W., Letti, L. A. J., Burgos, W. J. M., Scapini, T., Manzoki, M.C., de Oliveira, P.Z. & Soccol, C. R. (2023). Strategies and engineering aspects on the scale-up of bioreactors for different bioprocesses. *Systems Microbiology and Biomanufacturing*, 1–21.

de Queiroz Brito Cunha, C.C., Gama, A.R., Cintra, L.C., Bataus, L.A.M., & Ulhoa, C.J. (2018). Improvement of bread making quality by supplementation with a recombinant xylanase produced by Pichia pastoris. *Plos one*, *13*(2), e0192996.

Dobson, L. (2008). *A quantitative study of the relationships between morphology, physiology and geldanamycin synthesis in submerged cultures of Streptomyces hygroscopicus var. geldanus* (Doctoral dissertation, Dublin City University).

Drumond Chequer, F. M., De Oliveira, G. A. R., Anastacio Ferraz, E. R., Carvalho, J., Boldrin Zanoni, M. V., & De Oliveir, D. P. (2013). Textile Dyes: Dyeing Process and Environmental Impact. In M. Gunay (Ed.), *Eco-Friendly Textile Dyeing and Finishing*. InTech. https://doi.org/10.5772/53659

Dubey, M. K., Zehra, A., Aamir, M., Meena, M., Ahirwal, L., Singh, S., Shukla, S., Upadhyay, R. S., Bueno-Mari, R., & Bajpai, V. K. (2017). Improvement Strategies, Cost Effective Production, and Potential Applications of Fungal Glucose Oxidase (GOD): Current Updates. *Frontiers in Microbiology*, *8*, 1032. https://doi.org/10.3389/fmicb.2017.01032

Dzhavakhiya, V. V., Voinova, T. M., Glagoleva, E. V., Petukhov, D. V., Ovchinnikov, A. I., Kartashov, M. I., Kuznetsov, B. B., & Skryabin, K. G. (2015). Strain Improvement of Streptomyces xanthochromogenes RIA 1098 for Enhanced Pravastatin Production at High Compactin Concentrations. *Indian Journal of Microbiology*, *55*(4), 440–446. https://doi.org/10.1007/s12088-015-0537-5

Dzhavakhiya, V., Savushkin, V., Ovchinnikov, A., Glagolev, V., Savelyeva, V., Popova, E., ... & Glagoleva, E. (2016). Scaling up a virginiamycin production by a high-yield Streptomyces virginiae VKM Ac-2738D strain using adsorbing resin addition and fed-batch fermentation under controlled conditions. *3 Biotech*, *6*(2), 240.

Edelvio, D. B. G., Leo, R. L. D., & Rita, D. C. M. D. M. (2018). Actinomycetes bioactive compounds: Biological control of fungi and phytopathogenic insect. *African Journal of Biotechnology*, *17*(17), 552–559. https://doi.org/10.5897/AJB2017.16323

El-Moslamy, S. H., Elnouby, M. S., Rezk, A. H., & El-Fakharany, E. M. (2023). Scaling-up strategies for controllable biosynthetic ZnO NPs using cell free-extract of endophytic Streptomyces albus: characterization, statistical optimization, and biomedical activities evaluation. *Scientific Reports*, *13*(1), 3200.

Fernández-Lorente, G.; Terreni, M.; Mateo, C.; Bastida, A.; Fernández-Lafuente, R.; Dalmases, P.; Huguet, J.; Guisán, J. M. Modulation of lipase properties in macro-aqueous systems by controlled enzyme immobilization: Enantioselective hydrolysis of a chiral ester by immobilized pseudomonas lipase. *Enzyme and Microbial Technology* 2001, 28, 389–396.

Fisher, A. K., Freedman, B. G., Bevan, D. R., & Senger, R. S. (2014). A review of metabolic and enzymatic engineering strategies for designing and optimizing performance of microbial cell factories. *Computational and Structural Biotechnology Journal*, *11*(18), 91–99. https://doi.org/10.1016/j.csbj.2014.08.010

Gao, S.; Wang, Y.; Diao, X.; Luo, G.; Dai, Y. Effect of pore diameter and cross-linking method on the immobilization efficiency of candida rugosa lipase in SBA-15. *Bioresource Technology* 2010, *101*, 3830–3837.

Granozzi, C., Billetta, R., Pass, R., No, A., Sollazzo, M., & Puglia, A. M. (1990). A breakdown in macromolecular synthesis preceding differentiation in Streptomyces coelicolor A3(2). In *Journal of General Microbiology* (Vol. 136).

Gulzar, T., Farooq, T., Kiran, S., Ahmad, I., & Hameed, A. (2019). Green chemistry in the wet processing of textiles. In *The Impact and Prospects of Green Chemistry for Textile Technology* (pp. 1–20). Elsevier. https://doi.org/10.1016/B978-0-08-102491-1.00001-0

Hassan ME, Yang Q, Xiao Z, Liu L, Wang N, Cui X, Yang L. Impact of immobilization technology in industrial and pharmaceutical applications. *3 Biotech*. 2019 Dec;9(12):440. doi:10.1007/s13205-019-1969-0. Epub 2019 Nov 8. PMID: 31750038; PMCID: PMC6841786.

Hodgson, D. A. (2000). *Primary metabolism and its control in streptomycetes: a most unusual group of bacteria.*

Hwang, K. S., Kim, H. U., Charusanti, P., Palsson, B. Ø., & Lee, S. Y. (2014). Systems biology and biotechnology of Streptomyces species for the production of secondary metabolites. *Biotechnology Advances*, *32*(2), 255–268.

Jesionowski, T.; Zdarta, J.; Krajewska, B. Enzyme immobilization by adsorption: A review. *Adsorption*. 2014, *20*, 801–821.

Jiang, J., Sun, Y. F., Tang, X., He, C. N., Shao, Y. L., Tang, Y. J., & Zhou, W. W. (2018). Alkaline pH shock enhanced production of validamycin A in fermentation of Streptomyces hygroscopicus. *Bioresource Technology*, *249*, 234–240.

Jørgensen, O. B., Karlsen, L. G., Nielsen, N. B., Pedersen, S., & Rugh, S. (1988). A New Immobilized Glucose Isomerase with High Productivity Produced by a Strain ofStreptomyces murinus. *Starch-Stärke*, *40(8)*, 307–313. doi:10.1002/star.19880400809

Ju, Y., Son, K., Jin, C., Hwang, B. S., Park, D.-J., & Kim, C.-J. (2018). Statistical optimization of culture medium for improved production of antimicrobial compound by Streptomyces rimosus AG-P1441. *Food Science and Biotechnology*, *27*, 581–590. https://doi.org/10.1007/s10068-017-0257-1

Kabir, S. M. F., Chakraborty, S., Hoque, S. M. A., & Mathur, K. (2019). Sustainability Assessment of Cotton-Based Textile Wet Processing. *Clean Technologies*, *1*(1), 232–246. https://doi.org/10.3390/cleantechnol1010016

Kalantzi, S., Kekos, D., & Mamma, D. (2019). Bioscouring of cotton fabrics by multienzyme combinations: Application of Box–Behnken design and desirability function. *Cellulose*, *26*(4), 2771–2790. https://doi.org/10.1007/s10570-019-02272-9

Kalia, S., & Sheoran, R. (2011). Modification of Ramie Fibers Using Microwave-Assisted Grafting and Cellulase Enzyme–Assisted Biopolishing: A Comparative Study of Morphology, Thermal Stability, and Crystallinity. *International Journal of Polymer Analysis and Characterization*, *16*(5), 307–318. https://doi.org/10.1080/1023666X.2011.587946

Kapoor, M., Kapoor, R. K., & Kuhad, R. C. (2007). Differential and synergistic effects of xylanase and laccase mediator system (LMS) in bleaching of soda and waste pulps. *Journal of Applied Microbiology*, *103*(2), 305–317. https://doi.org/10.1111/j.1365-2672.2006.03251.x

Kashiwagi, N., Ogino, C., & Kondo, A. (2017). Production of chemicals and proteins using biomass-derived substrates from a Streptomyces host. *Bioresource Technology*, *245*, 1655–1663.

Khattab, A. I., Babiker, E. H., & Saeed, H. A. (2016). Streptomyces : isolation, optimization of culture conditions and extraction of secondary metabolites Streptomyces : isolation , optimization of culture conditions and extraction of secondary metabolites. *International Current Pharmaceutical Journal*, *5*(3), 27–32. https://doi.org/10.3329/icpj.v5i3.26695

Kumar, A., & Chandra, R. (2020). Ligninolytic enzymes and its mechanisms for degradation of lignocellulosic waste in environment. *Heliyon*, *6*(2), e03170. https://doi.org/10.1016/j.heliyon.2020.e03170

Kohli, P., & Gupta, R. (2015). Alkaline pectinases: A review. *Biocatalysis and Agricultural Biotechnology*, *4*(3), 279–285. https://doi.org/10.1016/j.bcab.2015.07.001

Kumar, D., Bhardwaj, R., Jassal, S., Goyal, T., Khullar, A., & Gupta, N. (2021b). Application of enzymes for an eco-friendly approach to textile processing. *Environmental Science and Pollution Research*, *30*(28), 71838–71848. https://doi.org/10.1007/s11356-021-16764-4

Kumar, M., Kumar, P., Das, P., Solanki, R., & Kapur, M. K. (2020). Potential applications of extracellular enzymes from Streptomyces spp. in various industries. *Archives of Microbiology*, *202*(7), 1597–1615. https://doi.org/10.1007/s00203-020-01898-9

Kumar, P., Khusboo, Rajput. D., & Dubey, K. K. (2023). Insights into the mechanism of mycelium tr ansforma tion of Streptomyces toxytricini into pellet. *FEMS Microbes*, *4*. https://doi.org/10.1093/femsmc/xtad017

Kumar, R., Hu, F., Hubbell, C. A., Ragauskas, A. J., & Wyman, C. E. (2013). Comparison of laboratory delignification methods, their selectivity, and impacts on physiochemical characteristics of cellulosic biomass. *Bioresource Technology*, *130*, 372–381. https://doi.org/10.1016/j.biortech.2012.12.028

Kumar, V., Ahluwalia, V., Saran, S., Kumar, J., Patel, A. K., & Singhania, R. R. (2021a). Recent developments on solid-state fermentation for production of microbial secondary metabolites: Challenges and solutions. *Bioresource Technology*, *323*, 124566.

Kurzbaum, E.; Raizner, Y.; Kuc, M. E.; Kulikov, A.; Hakimi, B.; Kruh, L. I.; Menashe, O. Phenol biodegradation by bacterial cultures encapsulated in 3D microfiltration-membrane capsules. *Environmental Technology* 2020, 41, 2875–2883.

Li, M., Zhang, Z.-J., Kong, X.-D., Yu, H.-L., Zhou, J., & Xu, J.-H. (2017). Engineering Streptomyces coelicolor Carbonyl Reductase for Efficient Atorvastatin Precursor Synthesis. *Applied and Environmental Microbiology*, *83*(12), e00603–e00617. https://doi.org/10.1128/AEM.00603-17

Lim, J. H., Lee, C. R., Dhakshnamoorthy, V., Park, J. S., & Hong, S. K. (2015). Molecular characterization of streptomyces coelicolor a(3) sco6548 as a cellulose 1,4-ß-cellobiosidase. *FEMS Microbiology Letters*, *363*(3). https://doi.org/10.1093/femsle/fnv245

L Microbiology, Department of Spirit and Yeast Technology Wolczanska 171/173, odz University of Technology, Faculty of Biotechnology and Food Sciences, Institute of Fermentation Technology and Microbiology, Department of Spirit and Yeast Technology Wolczanska 171/173, PL-90-924 Lodz, Poland, Robak, K., Balcerek, M., & Lodz University of Technology, Faculty of Biotechnology and Food

Sciences, Institute of Fermentation Technology and Microbiology, Department of Spirit and Yeast Technology Wolczanska 171/173, PL-90-924 Lodz, Poland. (2018). Review of Second-Generation Bioethanol Production from Residual Biomass. *Food Technology and Biotechnology*, *56*(2). https://doi.org/10.17113/ftb.56.02.18.5428

Luo, L., Cai, J., Wang, C., Lin, J., Du, X., Zhou, A., & Xiang, M. (2016). Purification and characterization of an alkaliphilic endo-xylanase from *Streptomyces althioticus* LMZM and utilization in the pulp paper industry. *Journal of Chemical Technology & Biotechnology*, *91*(4), 1093–1098. https://doi.org/10.1002/jctb.4690

Manteca, A., Fernandez, M., & Sanchez, J. (2005). A death round affecting a young compartmentalized mycelium precedes aerial mycelium dismantling in confluent surface cultures of Streptomyces antibioticus. *Microbiology*, *151*(11), 3689–3697.

Manteca, A., Alvarez, R., Salazar, N., Yague, P., & Sanchez, J. (2008). Mycelium differentiation and antibiotic production in submerged cultures of Streptomyces coelicolor. *Applied and Environmental Microbiology*, *74*(12), 3877–3886.

Manteca, Á., & Yagüe, P. (2018). Streptomyces differentiation in liquid cultures as a trigger of secondary metabolism. *Antibiotics*, *7*(2), 41.

Marrazza, G. Piezoelectric biosensors for organophosphate and carbamate pesticides: A review. *Biosensors*. 2014, *4*, 301–317.

Mate, D. M., & Alcalde, M. (2017). Laccase: A multi-purpose biocatalyst at the forefront of biotechnology. *Microbial Biotechnology*, *10*(6), 1457–1467. https://doi.org/10.1111/1751-7915.12422

Meyer, H. P., Minas, W., & Schmidhalter, D. (2017). Industrial-scale fermentation. *Industrial Biotechnology: Products and Processes*, 1–53.

Mihajlovski, K., Buntić, A., Milić, M., Rajilić-Stojanović, M., & Dimitrijević-Branković, S. (2021). From Agricultural Waste to Biofuel: Enzymatic Potential of a Bacterial Isolate Streptomyces fulvissimus CKS7 for Bioethanol Production. *Waste and Biomass Valorization*, *12*(1), 165–174. https://doi.org/10.1007/s12649-020-00960-3

Moers, K., Celus, I., Brijs, K., Courtin, C. M., & Delcour, J. A. (2005). Endoxylanase substrate selectivity determines degradation of wheat water-extractable and water-unextractable arabinoxylan. *Carbohydrate Research*, *340*(7), 1319–1327. https://doi.org/10.1016/j.carres.2005.02.031

Mohamad, N.R., Marzuki, N.H., Buang, N.A., Huyop, F., Wahab, R.A. (2015 Mar). An overview of technologies for immobilization of enzymes and surface analysis techniques for immobilized enzymes. *Biotechnol Biotechnol Equip*. 4;*29*(2):205–220. doi: 10.1080/13102818.2015.1008192. Epub 2015 Feb 17. PMID: 26019635; PMCID: PMC4434042.

Murayama, K., Kano, K., Matsumoto, Y., & Sugimori, D. (2013). Crystal structure of phospholipase A1 from Streptomyces albidoflavus NA297. *Journal of Structural Biology*, *182*(2), 192–196. https://doi.org/10.1016/j.jsb.2013.02.003

Muthusamy, S., Selvan, S. T., Arunachalam, P., & Grasian, I. (2019). Bioconversion and bioethanol production from agro-residues through fermentation process using mangrove-associated actinobacterium *Streptomyces olivaceus* (MSU3). *Biofuels*, *10*(2), 167–179. https://doi.org/10.1080/17597269.2017.1309853

Nakamura. (2006). Effects of Reducing Agents and Baking Conditions on Potassium Bromate Residues in Bread. *Cereal Foods World*. https://doi.org/10.1094/CFW-51-0069

Nguyen, C. T., Dhakal, D., Pham, V. T. T., Nguyen, H. T., and Sohng, J.-K. (2020). Recent advances in strategies for activation and discovery/characterization of cryptic biosynthetic gene clusters in Streptomyces. *Microorganisms* 8:616. doi: 10.3390/microorganisms8040616

Nguyen, H. H. & Kim, M. An overview of techniques in enzyme immobilization. *Appl. Sci. Converg. Technol.* 2017, 26, 157–163.

Nicolopoulou-Stamati, P., Maipas, S., Kotampasi, C., Stamatis, P., & Hens, L. (2016). Chemical pesticides and human health: the urgent need for a new concept in agriculture. *Frontiers in Public Health*, *4*. https://doi.org/10.3389/fpubh.2016.00148

Niladevi, K. N., & Prema, P. (2008). Immobilization of laccase from Streptomyces psammoticus and its application in phenol removal using packed bed reactor. *World Journal of Microbiology and Biotechnology*, *24*(7), 1215–1222. https://doi.org/10.1007/s11274-007-9598-x

Özkara, A., Akyil, D., & Konuk, M. (2016). Pesticides, Environmental Pollution, and Health. In M. Larramendy & S. Soloneski (Eds.), *Environmental Health Risk—Hazardous Factors to Living Species*. InTech. https://doi.org/10.5772/63094

Pham, V. T. T., Nguyen, C. T., Dhakal, D., Nguyen, H. T., Kim, T.-S., and Sohng, J. K. (2021). Recent advances in the heterologous biosynthesis of natural products from Streptomyces. *Applied Sciences* 11:1851.

Raveendran, S., Parameswaran, B., Ummalyma, S. B., Centre for Biofuels, National Institute for Interdisciplinary Science and Technology, CSIR, Trivandrum-695 019, India, Centre for Biofuels, National Institute for Interdisciplinary Science and Technology, CSIR, Trivandrum-695 019, India, Institute of Bioresources and Sustainable Development, Takyelpat, Imphal -795 001, India, Abraham, A., Centre for Biofuels, National Institute for Interdisciplinary Science and Technology, CSIR, Trivandrum-695 019, India, Mathew, A. K., Centre for Biofuels, National Institute for Interdisciplinary Science and Technology, CSIR, Trivandrum-695 019, India, Madhavan, A., Rajiv Gandhi Centre for Biotechnology, Jagathy, Trivandrum – 695 014, India, Rebello, S., Communicable Disease Research Laboratory, St. Joseph's College, Irinjalakuda, India, Pandey, A., & Centre for Biofuels, National Institute for Interdisciplinary Science and Technology, CSIR, Trivandrum-695 019, Indi. (2018). Applications of microbial enzymes in food industry. *Food Technology and Biotechnology*, *56*(1). https://doi.org/10.17113/ftb.56.01.18.5491

Ravindran, R., & Jaiswal, A. (2016). Microbial enzyme production using lignocellulosic food industry wastes as feedstock: a review. *Bioengineering*, *3*(4), 30. https://doi.org/10.3390/bioengineering3040030

Ribeiro, R. M. M. G. P., Esperança, M. N., Sousa, A. P. A., Neto, Á. B., & Cerri, M. O. (2021). Individual effect of shear rate and oxygen transfer on clavulanic acid production by Streptomyces clavuligerus. *Bioprocess and Biosystems Engineering*, *44*, 1721–1732. https://doi.org/10.1007/s00449-021-02555-1

Robak, K., Balcerek, M., Lodz University of Technology, Faculty of Biotechnology and Food Sciences, Institute of Fermentation Technology and Microbiology, Department of Spirit and Yeast Technology Wolczanska 171/173, PL-90-924 Lodz, Poland, & Lodz University of Technology, Faculty of Biotechnology and Food Sciences, Institute of Fermentation Technology and Microbiology, Department of Spirit and Yeast Technology Wolczanska 171/173, PL-90-924 Lodz, Poland. (2018). Review of second-generation bioethanol production from residual biomass. *Food Technology and Biotechnology*, *56*(2). https://doi.org/10.17113/ftb.56.02.18.5428

Rother, C.; Nidetzky, B. Enzyme immobilization by microencapsulation: Methods, materials, and technological applications. *Encyclopedia of Industrial Biotechnology* 2014, 7, 1–21.

Rulli, M. C., Bellomi, D., Cazzoli, A., De Carolis, G., & D'Odorico, P. (2016). The water-land-food nexus of first-generation biofuels. *Scientific Reports*, *6*(1), 22521. https://doi.org/10.1038/srep22521

Salwan, R., & Sharma, V. (2020). Bioactive compounds of Streptomyces: biosynthesis to applications. *Studies in Natural Products Chemistry*, *64*, 467–491.

Salwan, R., Sharma, R., & Sharma, V. (2023). Characterization and immobilization of extracellular broad pH active multicopper oxidase from Streptomyces flavomacrosporus for dye-discoloration. *Biocatalysis and Agricultural Biotechnology*, 102955.

Sangale, M. K. (2012). A Review on Biodegradation of Polythene: The Microbial Approach. *Journal of Bioremediation and Biodegradation*, *03*(10). https://doi.org/10.4172/2155-6199.1000164

Sarkar, P., Meghvanshi, M., & Singh, R. (2011). Microbial Consortium: A New Approach in Effective Degradation of Organic Kitchen Wastes. *International Journal of Environmental Science and Development*, 170–174. https://doi.org/10.7763/IJESD.2011.V2.118

Selvaraj, J. N., Ganapathi, U., Vincent, S. G. P., Ramamoorthy, S., & Thavasimuthu, C. (2023). Statistical optimization of media components for antibiotic production in Streptomyces sp . CMSTAAHAL-3. *Electronic Journal of Biotechnology*, *65*, 1–13. https://doi.org/10.1016/j.ejbt.2023.03.005

Shao, H., Chen, M., Fei, X., Zhang, R., Zhong, Y., Ni, W., Tao, X., He, X., Zhang, E., Yong, B., & Tan, X. (2019). Complete genome sequence and characterization of a polyethylene biodegradation strain, streptomyces albogriseolus LBX-2. *Microorganisms*, 7(10). https://doi.org/10.3390/microorganisms7100379

Sharma, A., Tewari, R., Rana, S. S., Soni, R., & Soni, S. K. (2016). Cellulases: Classification, Methods of Determination and Industrial Applications. In *Applied Biochemistry and Biotechnology* (Vol. 179, Issue 8, pp. 1346–1380). Humana Press Inc. https://doi.org/10.1007/s12010-016-2070-3

Sharma, B., Dangi, A. K., & Shukla, P. (2018). Contemporary enzyme based technologies for bioremediation: a review. *Journal of Environmental Management*, *210*, 10–22.

Sharma, D., Bhardwaj, K. K., & Gupta, R. (2022). Immobilization and applications of esterases. *Biocatalysis and Biotransformation*, *40*(3), 153–168.

Sharma, V., Kaur, R., and Salwan, R. (2021). Streptomyces: Host for refactoring of diverse bioactive secondary metabolites. *3 Biotech 11*. 340. doi: 10.1007/s13205- 021-02872-y

Sheldon, R. A., & Pereira, P. C. (2017). Biocatalysis engineering: The big picture. *Chemical Society Reviews*, *46*(10), 2678–2691. https://doi.org/10.1039/c6cs00854b

Shen, J. (2019). Enzymatic treatment of wool and silk fibers. In *Advances in Textile Biotechnology* (pp. 77–105). Elsevier. https://doi.org/10.1016/B978-0-08-102632-8.00005-0

Shivalee, A., Lingappa, K., & Mahesh, D. (2018). Influence of bioprocess variables on the production of extracellular chitinase under submerged fermentation by Streptomyces pratensis strain KLSL55. *Journal of Genetic Engineering and Biotechnology*, *16*(2), 421–426. https://doi.org/10.1016/j.jgeb.2017.12.006

Shrestha, S., Chio, C., Khatiwada, J. R., Kognou, A. L. M., Chen, X., & Qin, W. (2023). Optimization of cultural conditions for pectinase production by Streptomyces sp. and characterization of partially purified enzymes. *Microbial Physiology*, *33*(1), 12–26. https://doi.org/10.1159/000528257

Singh Jadaun, J. (2018). Pectinase: a useful tool in fruit processing industries. *Nutrition & Food Science International Journal*, *5*(5). https://doi.org/10.19080/NFSIJ.2018.05.555673

Singh, S., and Khajuria, R. (2018). *New and Future Developments in Microbial Biotechnology and Bioengineering*. Amsterdam: Elsevier.

Singh, S., Singh, N., Kumar, V., Datta, S., Wani, A. B., Singh, D., Singh, K., & Singh, J. (2016). Toxicity, monitoring and biodegradation of the fungicide carbendazim. *Environmental Chemistry Letters*, *14*(3), 317–329. https://doi.org/10.1007/s10311-016-0566-2

Singh, V., Haque, S., Niwas, R., Srivastava, A., Pasupuleti, M., & Tripathi, C. (2017). Strategies for fermentation medium optimization: an in-depth review. *Frontiers in Microbiology*, *7*, 2087.

Talamantes, D., Biabini, N., Dang, H., Abdoun, K., & Berlemont, R. (2016). Natural diversity of cellulases, xylanases, and chitinases in bacteria. *Biotechnology for Biofuels*, *9*(1), 133. https://doi.org/10.1186/s13068-016-0538-6

Thomas, L., Joseph, A., Singhania, R. R., Patel, A. K., & Pandey, A. (2016). Industrial Enzymes: Xylanases. In *Current Developments in Biotechnology and Bioengineering: Production, Isolation and Purification of Industrial Products* (pp. 127–148). Elsevier Inc. https://doi.org/10.1016/B978-0-444-63662-1.00006-3

van Dissel, D., Claessen, D., & van Wezel, G. P. (2014). Morphogenesis of Streptomyces in submerged cultures. *Advances in Applied Microbiology*, *89*, 1–45.

Vurukonda, S. S. K. P., Giovanardi, D., & Stefani, E. (2018). Plant Growth Promoting and Biocontrol Activity of Streptomyces spp. As Endophytes. *International Journal of Molecular Sciences*, *19*(4), 952. https://doi.org/10.3390/ijms19040952

Walia, A., Guleria, S., Mehta, P., Chauhan, A., & Parkash, J. (2017). Microbial xylanases and their industrial application in pulp and paper biobleaching: A review. *3 Biotech*, *7*(1), 11. https://doi.org/10.1007/s13205-016-0584-6

Wang, L., Ravichandran, V., Yin, Y., Yin, J., and Zhang, Y. (2019). Natural products from mammalian gut microbiota. *Trends Biotechnology*, 37, 492–504.

Wang X, Li D, Qu M, Durrani R, Yang B, Wang Y. Immobilized MAS1 lipase showed high esterification activity in the production of triacylglycerols with n-3 polyunsaturated fatty acids. *Food Chemistry* 2017b;216:260–267. doi: 10.1016/j.foodchem.2016.08.041

Wang, X., Qin, X., Li, D., Yang, B., & Wang, Y. (2017a). One-step synthesis of high-yield biodiesel from waste cooking oils by a novel and highly methanol-tolerant immobilized lipase. *Bioresource Technology*, *235*, 18–24.

Wink, J., Mohammadipanah, F., & Hamedi, J. (2017). Biology and biotechnology of actinobacteria. In *Biology and Biotechnology of Actinobacteria*. ISBN 978-3-319-86840-0, 1st Edition VII, 395.

Won, K.; Kim, S.; Kim, K. J.; Park, H. W.; Moon, S. J. Optimization of lipase Entrapment in Ca-alginate gel beads. *Process Biochemistry* 2005, 40, 2149–2154.

Woolridge, E. (2014). Mixed enzyme systems for delignification of lignocellulosic biomass. *Catalysts*, *4*(1), 1–35. https://doi.org/10.3390/catal4010001

Wong, L. S.; Thirlway, J.; Micklefield, J. Direct site-selective covalent protein immobilization catalyzed by a phosphopantetheinyl transferase. *Journal of the American Chemical Society* 2008, 130, 12456–12464.

Yagüe, P., López-García, M. T., Rioseras, B., Sánchez, J., & Manteca, Á. (2013). Pre-sporulation stages of Streptomyces differentiation: State-of-the-art and future perspectives. In *FEMS Microbiology Letters* (Vol. 342, Issue 2, pp. 79–88). https://doi.org/10.1111/1574-6968.12128

Yamaguchi, T., Wei, M., Hagihara, N., Omori, M., Wanibuchi, H., & Fukushima, S. (2008). Lack of mutagenic and toxic effects of low dose potassium bromate on kidneys in the Big Blue rat. *Mutation Research/Genetic Toxicology and Environmental Mutagenesis*, *652*(1), 1–11. https://doi.org/10.1016/j.mrgentox.2007.11.004

Yeboah, P. J., Ibrahim, S. A., & Krastonov, A. (2023). A review of fermentation and the nutritional requirements for effective growth media for lactic acid bacteria. *Food Science and Applied Biotechnology*, *6*(2), 215–240.

Yin, X., Li, Y., Zhou, J., Rao, S., Du, G., Chen, J., & Liu, S. (2021). Enhanced production of transglutaminase in *Streptomyces mobaraensis* through random mutagenesis and site-directed genetic modification. *Journal of Agricultural and Food Chemistry*, *69*(10), 3144–3153. https://doi.org/10.1021/acs.jafc.1c00645

Zainith, S., Purchase, D., Saratale, G. D., Ferreira, L. F. R., Bilal, M., & Bharagava, R. N. (2019). Isolation and characterization of lignin-degrading bacterium Bacillus aryabhattai from pulp and paper mill wastewater and evaluation of its lignin-degrading potential. *3 Biotech*, 9(3), 92. https://doi.org/10.1007/s13205-019-1631-x

Zhang, Y., Hu, P., Muhammad, Y., Tang, Y., Shao, S., Gao, Z., et al. (2021). High-density immobilization of laccase on hollow nano-sphere NH2-MIL88(Fe) host with interfacial defects to improve enzyme activity and stability for Remazol brilliant blue R decolorization. *Chemical Engineering Journal* 405, 127003. doi:10.1016/j.cej.2020.127003

Zheng, Y., Wang, Y., Pan, J., Zhang, J., Dai, Y., & Chen, K. (2017). Semi-continuous production of high-activity pectinases by immobilized Rhizopus oryzae using tobacco wastewater as substrate and their utilization in the hydrolysis of pectin-containing lignocellulosic biomass at high solid content. *Bioresource Technology*, *241*, 1138–1144. https://doi.org/10.1016/j.biortech.2017.06.066

Zhou, Y., Han, L. R., He, H. W., Sang, B., Yu, D. L., Feng, J. T., & Zhang, X. (2018). Effects of agitation, aeration and temperature on production of a novel glycoprotein GP-1 by Streptomyces kanasenisi ZX01 and scale-up based on volumetric oxygen transfer coefficient. *Molecules*, *23*(1), 125.

Zhu, F., Lavine, L., O'Neal, S., Lavine, M., Foss, C., & Walsh, D. (2016). Insecticide resistance and management strategies in urban ecosystems. *Insects*, 7(1), 2. https://doi.org/10.3390/insects7010002

Chapter 5

Production of medicinal compound from streptomycetes

Sandeep Rathor, Ananya Tanwar, and Sakshi Verma
MM College of Pharmacy, Maharishi Markandeshwar (Deemed to be University) Mullana, Ambala, India

Pritam Singh
Chandigarh College of Pharmacy, Chadigarh Group of Colleges, Greater Mohali, India

5.1 INTRODUCTION

Gram-positive, aerobic soil bacteria called *Streptomyces* species are members of the *Actinobacteria* family. They develop filamentously from a single spore. Their filaments eventually form a network of branched filaments known as substrate mycelium as a result of tip expansion and branching (1). Under certain environmental conditions, such as nutrient constraint, streptomycetes transition from the vegetative phase (i.e., substrate mycelium) to the reproductive sporulation phase (i.e., aerial multinucleated mycelium). However, they have a linear chromosome that ranges in size from 8 to 10 MB, depending on the species; a high GC content; and many linear or circular plasmids. The presence of biosynthetic gene clusters that encode enzymes involved in the synthesis of secondary metabolites with a range of chemotypes, such as polyketides, lactams, nonribosomal peptides, and terpenes, is one of the genome's distinctive features in the *Streptomyces* species (2). Along with morphological differentiation (e.g., production of the aerial multinucleated mycelium), many secondary metabolites are produced during the shifting phase from the substrate mycelium to sporulation (3). Species of *Streptomyces* have been a major source of pharmaceuticals, particularly antibiotics (Figure 5.1). Many antibiotics were identified from different *Streptomyces* species and were put into clinical use between the late 1940s and the 1960s, which is also referred to as the "golden age of antibiotic discovery" (4). While the percentage of recently identified antibiotics that have been isolated from *Streptomyces* species has decreased to roughly 20–30%, new genome sequencing data show that this genus is capable of producing far more bioactive secondary metabolites than previously thought, many of which are still unclear. These secondary metabolites are used in medicine for a variety of purposes, such as immunosuppressants (rapamycin), antifungals (amphotericin B), anticancer drugs (doxorubicin),

DOI: 10.1201/9781032694634-5

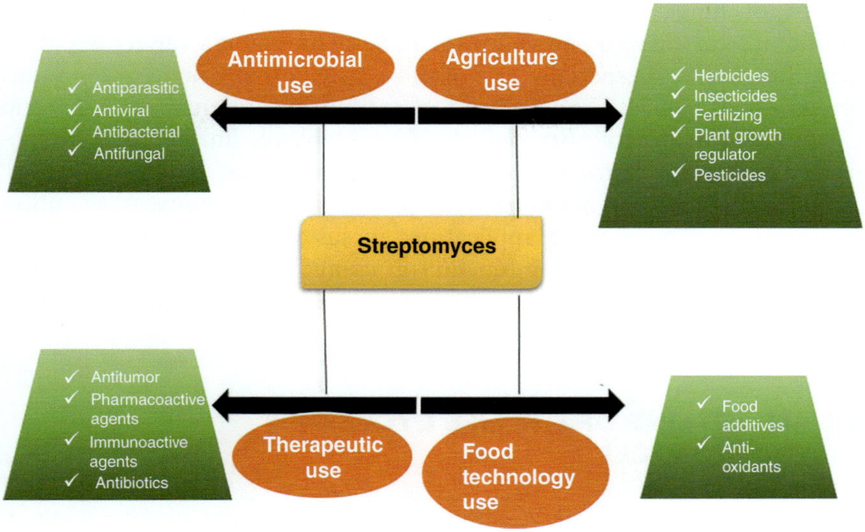

Figure 5.1 *Streptomyces* virtues in various fields.

antiparasitics (ivermectin), and antibiotics (daptomycin, for example) (5). The *Streptomyces* species is receiving a lot of attention because of the current, urgent need to find new antibacterial compounds due to the rapid rise in Gram-negative pathogens like *Acinetobacter baumannii*, *Pseudomonas aeruginosa*, and *Klebsiella pneumonia*, as well as antibiotic-resistant microbial pathogens like methicillin-resistant *Staphylococcus aureus* (6, 7). Crucially, compared to other antibiotic-producing microorganisms, *Streptomyces* species are more amenable to the application of high-throughput techniques (e.g., the combination of genome mining and mass spectrometry) and gene manipulation to maximize their potential of producing potent antibiotics. This is because of the vast amount of information that has been collected on the species. Fungal pathogens are a major global health concern for people and animals, particularly in tropical and subtropical areas. Even in cases when optimal cultivation, harvesting, storage, and handling practices are employed, fungi and their toxins are inherent pollutants of food and feed. On the other hand, fungicidal medications and other chemotherapeutic treatments are posing a risk of residue and side effects to humans. Therefore, the current active search for novel naturally occurring pharmacologically active agents has resulted in the discovery of numerous effective medications that may be used to treat a variety of fungal and mycotoxin-related diseases without causing environmental pollution or the emergence of pathogens resistant to fungicides.

To find fewer chemically dependent alternatives for protecting people and animals, thorough research is required. The biocontrol of fungal infections is a common application of microbial antagonists. These microorganisms consist of fungi, algae, and bacteria, including *Streptomyces* spp. (8). Actinomycetes are well known for their ability to inhibit a variety of pathogenic fungi. This is especially true of those species that belong to the genus *Streptomyces* (9–12). The generation of antifungal compounds (13–17) and extracellular hydrolytic enzymes (18–20) are typically linked to the antagonistic activity of *Streptomyces* toward fungal pathogens.

The potential use of actinobacteria, particularly *Streptomyces*, as antibiotics and their prospective applications in sustainable agriculture with a particular emphasis on their function in compost will be covered in this chapter. We'll look at the ways by which they improve the quality of compost and the composting process.

5.2 AN OUTLINE OF THE LIFE CYCLE OF *STREPTOMYCES*

Within the Bacteria domain, Actinobacteria is a sizable phylum that is extensively found in both terrestrial and aquatic environments (21). These filamentous, Gram-positive bacteria are distinguished by their high DNA content of guanine and cytosine (G + C) and their ability to generate both substrate and aerial mycelium (22). There are 374 genera in the phylum, with *Streptomyces* being the most prevalent. *Actinomyces*, *Kitasatospora*, *Micromonospora*, *Nocardia*, *Micrococcus*, *Arthrobacter*, and *Rhodococcus* are examples of non-*Streptomyces* actinobacteria that are also present but are less common in natural settings and require specialized techniques for their isolation, preservation, and cultivation (23). The life cycle of *Streptomyces* species starts with the appearance of one or two germ tubes, which triggers spore germination. The term "vegetative mycelium" (MV) refers to the structure of hyphae that resembles a mesh formed by emerging shoots that appear at points below the tip (stage 1). After the depletion of nutrients, the vegetative mycelium differentiates and becomes aerial hyphae (stage 2). Finally, stage 3 sees the aerial hyphae develop into long chains of spores. All three stages of this process can be manipulated and controlled through the regulation of the whi gene family, the bld gene family, and the polariscope (Figure 5.2) (24).

The life cycle of *Streptomyces* is characterized by compartmentalized hyphae formation, which starts with spore germination in solid cultures or mycelial fragmentation in liquid cultures. The vegetative and reproductive phases of the life cycle can be separated into these two categories (25), with some variations in the medium used in solid and liquid cultures. Spore germination and vegetative mycelium development, which extend deeply into solid medium culture, mark the beginning of the cycle of life in solid cultures (26). After a while, there is a growth in the aerial mycelium and the generation of antibiotics due to programmed cell death. Liquid cultures of *Streptomyces* start with compartmentalized hyphae formation (MI), which is followed by programmed cell death and the production of secondary metabolite-producing mycelia. This is in contrast to solid cultures, where

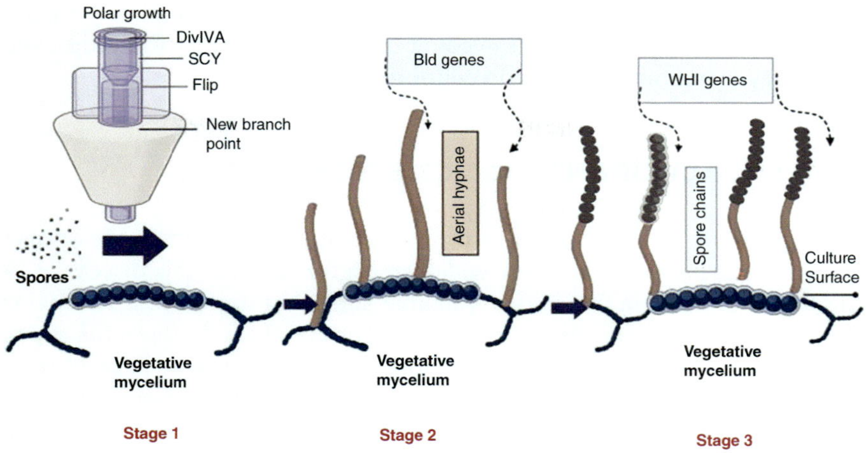

Figure 5.2 *Streptomyces* life cycle.

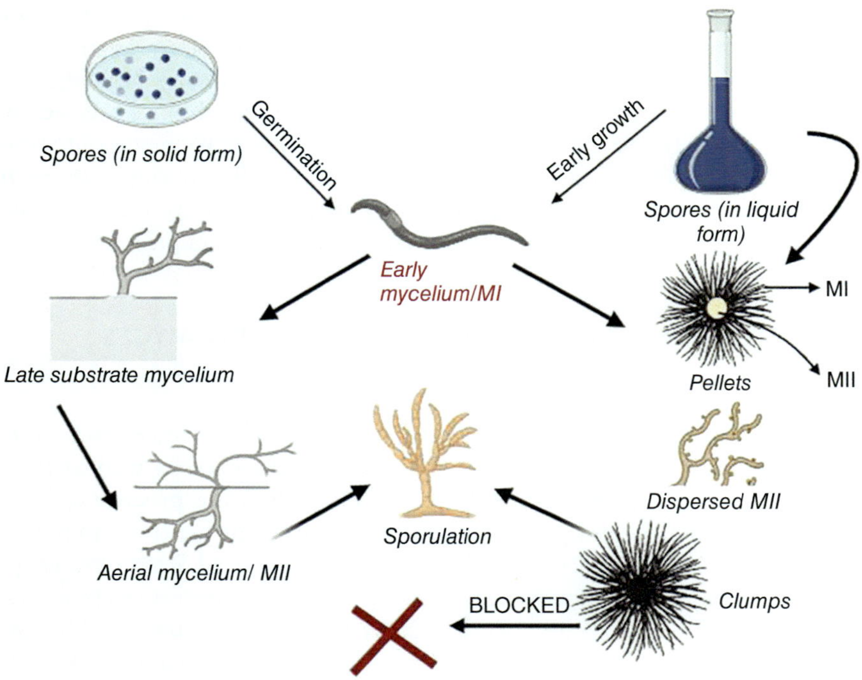

Figure 5.3 Overall *Streptomyces* life cycle in both liquid and solid cultures.

spore germination is the first stage. Aerial mycelium and spore production may be inhibited in some *Streptomyces* strains. The generation of secondary metabolites is caused by expressed proteins during the second stage of mycelial growth (MII), according to *Streptomyces* coelicolor proteomics research (Figure 5.3). Notwithstanding these results, the mechanisms underlying

Streptomyces differentiation in liquid media remain incompletely understood, and investigations into the connection between pellet formation and secondary metabolite production continue.

5.3 A SOURCE OF ANTIBIOTICS: STREPTOMYCES

Eighty percent of the antibiotics used today are derived from the species *Streptomyces*, which has been widely recognized since the discovery of the first antibiotic in 1942 as a valuable source of antibiotics. Following the isolation of streptothricin and streptomycin from *S. lavendulae* and *S. griseus*, respectively, the first antibiotic, actinomycin, was derived from *Streptomyces* antibiotics (27–29). As noted by (30, 31), several antibiotics, such as cephalosporins, tetracycline, nystatin, lincomycin, vancomycin, rifamycin, kanamycin, and daptomycin, have been produced from several *Streptomyces* spp. Because *Streptomyces* is known to emit antibiotics in its telluric environment to inhibit rivals, novel compounds of economic interest with marketable yields have been discovered (32). As in soil, interactions between biotic and abiotic factors are necessary for *Streptomyces* strains to produce more compounds and expand their chemical diversity. Nonetheless, recent research has demonstrated that the creation of antibiotics is typically seen when closely related strains coexist or share secondary metabolite biosynthesis pathways, not as a means of protection against toxins from rival strains. Furthermore, *Streptomyces*' ability to produce antibiotics can be reduced by food shortages and the presence of competitors (33).

5.4 EXTRACELLULAR ENZYMES FROM *STREPTOMYCES* SPECIES ARE USED IN INDUSTRY

The industrial sector has a great demand for microbial-derived enzymes due to their remarkable stability, productivity, and availability, as well as their cost-effectiveness and ecologically friendly manufacturing processes. Their ecological soundness and practicality further justify their use from an industrial standpoint (Figure 5.4) (34). A class of bacteria called *Streptomyces* sp. is well-known for its distinctive features and capacity to generate a large variety of extracellular enzymes (35). The latter has numerous uses in different industrial domains (36). The most widely utilized enzymes that *Streptomyces* produces include lipase, cellulase, xylanase, amylase, protease, and chitinase.

5.4.1 The enzyme amylase

The enzyme amylase (EC 3.2.1.1) breaks down starch into more digestible carbohydrates like glucose and maltose. These simpler sugars are vital starting materials for a variety of fermentative bioconversions, such as the

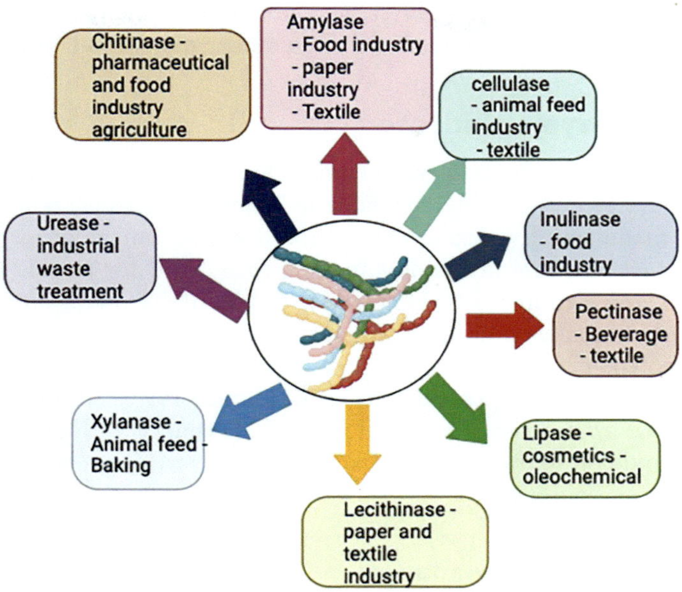

Figure 5.4 The variety and roles of the extracellular enzymes that *Streptomyces* species manufacture.

fermentation of lactic acid, citric acid, and ethanol. Because of their many uses in the production of paper, textiles, beer, food processing, detergents, tanning, papermaking, and biofuels, amylases are considered one of the most important industrial enzymes (37, 38). Research and development are still being done on this enzyme activity (amylase) to identify novel species with substantial physico-chemical potential (39). For the synthesis of amylase, several *Streptomyces* species have been employed (40–45). The starch processing business (46), particularly raw starch processing (47), has used *Streptomyces* amylase in medicine.

5.4.2 The protease enzymes

The production of proteases from *Streptomyces* has been the subject of numerous research studies, which have documented their extensive use in a variety of industrial applications (Figure 5.4). Because of their adaptability, these fields include food processing, cosmetic manufacturing, detergency, pharmaceuticals, chemical applications, and waste management (48). Additionally, they are employed in the dehairing stage of the leather processing industries, which eliminates the need for chemical application (49). The majority of proteases (EC 3.4) have been demonstrated to have a high level of resistance to a variety of environmental difficulties, such as extremes in temperature, salt, and pH (50). Furthermore, it has been discovered that proteases produced by *Streptomyces* spp. are useful in the recycling of agro-industrial wastes, including plant debris, feathers, nails, and hair (51).

5.4.3 The enzyme chitinase

Because of its capacity to degrade chitin, chitinase (EC 3.2.1.14) is a kind of enzyme that is extremely beneficial in industrial settings. Some actinobacteria manufacture chitinases that exhibit extraordinary properties like pH tolerance and thermostability, making them ideal for a wide range of industrial applications (52). Chitinases, produced by *Streptomyces* spp., are utilized to make chitin derivatives, such as chitosan, a naturally occurring biopolymer having uses in the food, agriculture, and medical sectors (53–55). Chitinases are produced by several *Streptomyces* strains, including *Streptomyces thermoviolaceus*. Chitobiose, a substance with potential uses as an antioxidant in the food and biomedical industries, has been extracted using these chitinases (56, 57).

5.4.4 The enzyme xylanases

Xylan, a complex polysaccharide that is primarily employed in the pulp and biobleaching industries to improve pulp quality, is the main component of hemicelluloses (58). Because of their wide-ranging applications in the food processing, paper, pharmaceutical, and cellulose manufacturing industries, xylanases (EC 3.2.1.8) are in great demand (59–61). They are also used in the baking industry, probiotics, and beverage production (62). In addition to producing xylanases, which are used in the pulp and paper sector (63, 64), *Streptomyces* sp. may break down a variety of agricultural leftovers, including oil cake and straw waste, which increases the production of biogas (65).

5.4.5 The enzyme lipases

Lipases (EC 3.1.1.3) are utilized in many different industries; they are used in the synthesis of taste compounds by esterification in food, cosmetics, fragrance, and medicine (67) and the generation of biodiesel (66) through lipase-catalyzed transesterification. In addition to their employment in tea processing, lipases are also employed in plastic degradation, agrochemical synthesis, waxes, biopolymers, lipid hydrolysis for bioremediation, and detergent composition for hydrolyzing greasy and oily stains. Lipases produced from streptomyces are also used as biosensors in probe technologies (68, 69). Moreover, they are used as degreasing agents in the textile, leather, and paper industries for pitch treatment, dewaxing, de-sizing, and scouring. Treating oily wastewater is one of lipases' other significant uses (70). It is known that lipases, which hydrolyze triglyceride ester linkages into glycerol and fatty acids, are produced by streptomyces exfoliates (71). Numerous possible uses for these lipases exist (72), such as fat catalysis, cosmetics, diagnostics, and detergents (73).

5.4.6 The enzyme cellulases

Primary products of cellulose breakdown include glucose, cellobiose, and cello-oligosaccharides; this process is carried out by cellulases, also referred to as glucan 1,4-β-glucosidase (EC 3.2.1.74) (74). These enzymes are essential for biomass hydrolysis (76) and function together to break down cellulose (75). As a result, they have been the subject of many investigations. Since cellulases can convert cellulosic biomass into glucose, they are highly sought after in the worldwide commercial market because of their numerous uses in a variety of industries (77, 78). Finding enzymes that are extremely stable and able to work under harsh pH and temperature settings is essential for the application of cellulases in several industries, including leather and detergents (79). Highly thermostable cellulases are produced by some *Streptomyces* species, including *Streptomyces ruber* and *Streptomyces rutgersensis* (80). Numerous sectors use cellulases extensively, including food, wine and beer, agriculture, textiles, animal feed, detergent, pulp and paper, and research and development. Apart from the enzymes mentioned before, *Streptomyces* is well known for its capacity to generate a wide range of additional enzymes that are highly consequential in other domains. For example, some strains of *Streptomyces* can produce azoreductase, which is needed for the biodegradation of azo colors in the paper and textile sectors (81). Another enzyme that *Streptomyces* produces, peroxidase, is used in food processing, pulp and paper bleaching, and environmental bioremediation (82). The fruit and vegetable processing industries use pectinase, an enzyme generated by *Streptomyces* strains (83). Another enzyme from *Streptomyces* that is involved in organic pollution is phenazine-1-carboxylic acid dioxygenase (84). Lastly, transglutaminase, an enzyme utilized in food processing to enhance the texture and quality of meat products, is known to be produced by *Streptomyces* (85). *Streptomyces* is a valuable organism for industrial and biotechnological applications because of the large spectrum of enzymes it produces.

5.5 THE SIGNIFICANT FUNCTION OF ACTINOMYCETES

Bacteria known as actinomycetes are useful for biotechnology and are frequently used to produce secondary metabolites (86). For many years, a major focus of study by numerous organizations worldwide has been the screening, isolation, and characterization of promising strains of actinomycetes producing potential secondary metabolites (87, 88). The main genera of Actinomycetes that produce biomolecules with commercial significance include *Streptomyces*, *Saccharopolyspora*, *Amycolatopsis*, *Micromonospora*, and *Actinoplanes* (89). Actinomycetes, particularly *Streptomyces* species, are widely acknowledged as significant industrial microorganisms due to their abundance of beneficial bioactive natural compounds that have potential uses (90–93) and are prolific producers of secondary metabolites, many of which have commercial importance as antibiotics, antiparasitics, and antifungal agents, herbicides, pesticides, anticancer or immunosuppressive agents as well as industrially important enzymes (94, 95). Of all known drugs, 70% have been isolated from Actinomycetes bacteria out of which 75% and 60% are used in medicine and agriculture respectively (96, 97). Secondary metabolites, to a large extent, are species-specific and in contrast to primary metabolites, often accumulate in substantial quantity, which is a key factor in their commercial importance (98, 99). Actinomycetes are of tremendous economic importance as the secondary metabolites produced by them include antibiotics, other medicinals, toxins, pesticides, and animal and plant growth factors. The best known of the secondary metabolites produced by Actinomycetes are the antibiotics (Table 5.1).

Table 5.1 Some clinically important antibiotics from actinomycetes

Antibiotics	*Produced by*	*Activity*	*Reference*
Sagamicin	*Micromonospora sagamiensis*	Antibacterial	(100)
Amiclenomycin	*Streptomyces lavendulae*	Antibacterial	(101)
Methylenomycin	*Streptomyces violaceoruber*	Antibacterial	(102)
Roseoflavin	*Streptomyces davawensis*	Antibacterial	(103, 104)
Minosaminomycin	*Streptomyces sp.*	Antibacterial	(105)
Libramycin.	*Streptomyces sp*	Antifungal	(106)
Candihexin	*Streptomyces viridoflavus*	Antifungal	(107)
Nanaomycin	*Streptomyces rosa*	Antifungal	(108)
Purpuromycin	*Actinoplanes ianthinogenes*	Antifungal	(109)
Zorbonomycin	*Streptomyces bikiniensis*	Antifungal	(110)
Validamycin	*Streptomyces hygroscopicus*	Antifungal	(111)
Rosamicin	*Micromonospora rosaria*	Antibacterial	(112)
Rifamycin	*Micromonospora rifamycinica*	Antibacterial	(113, 114)
Platenomycin	*Streptomyces platensis*	Antibacterial	(115)
Azalomycin	*Streptomyces malaysiensis*	Antifungal	(116)
Streptimidone	*Streptomyces sp.*	Agricultura	(117)
Kinamycin	*Streptomyces murayamaensis*	Antibacterial	(118)
Kuwaitimycin	*Streptomyces kuwaitinensis*	Antibacterial	(119)
Sarkomycin.	*Streptomyces sp*	Antitumor	(120)
Salinomycin	*Streptomyces albus*	Antiparasite	(121)
Antimycin	*Streptomyces antibioticus*	Antifungal	(122)
Antimycin	*Streptomyces lucitanusus*	Antifungal	(123)

5.6 CONCLUSION

This analysis concludes by highlighting the important roles that *Streptomyces* play in human health and the wide range of applications. By taking the place of chemical fertilizers, improving soil health through bioremediation, and serving as biological pest controllers, streptomycetes are important to agriculture. Furthermore, their involvement in the synthesis of antimicrobials, antioxidants, and anticancer agents in the healthcare and pharmaceutical industries creates new opportunities for the synthesis of high-value molecules. Actinobacteria also support biotechnology by generating important enzymes that generate large profit margins for a range of businesses. Lastly, we can draw attention to recent developments in nanomedicine, where streptomycetes have the potential to help resolve issues related to the insoluble nature of some active components. By helping to produce active principle shuttle molecules that are unable to reach therapeutic areas through the construction of nanoparticle-based drug delivery systems, these microorganisms make this possible. Overall, these microbes are genuinely wonderful for the improvement of our world and human life, and their many benefits demand more study and investigation to fully realize their unrealized potential.

REFERENCES

1. Dyson PJ. Streptomyces: molecular biology and biotechnology. (No Title). 2011.
2. Nett M, Ikeda H, Moore BS. Genomic basis for natural product biosynthetic diversity in the actinomycetes. *Natural Product Reports*. 2009;26(11): 1362–1384.
3. Flärdh K, Buttner MJ. Streptomyces morphogenetics: dissecting differentiation in a filamentous bacterium. *Nature Reviews Microbiology*. 2009 Jan;7(1): 36–49.
4. Berdy J. Bioactive microbial metabolites. *The Journal of Antibiotics*. 2005 Jan;58(1):1–26.
5. Newman DJ, Cragg GM. Natural products as sources of new drugs over the last 25 years. *Journal of Natural Products*. 2007 Mar 23;70(3):461–477.
6. Bush K, Courvalin P, Dantas G, Davies J, Eisenstein B, Huovinen P, Jacoby GA, Kishony R, Kreiswirth BN, Kutter E, Lerner SA. Tackling antibiotic resistance. *Nature Reviews Microbiology*. 2011 Dec;9(12):894–896.
7. Fischbach MA, Walsh CT. Antibiotics for emerging pathogens. *Science*. 2009 Aug 28;325(5944):1089–1093.
8. Brimner TA, Boland GJ. A review of the non-target effects of fungi used to biologically control plant diseases. *Agriculture, Ecosystems & Environment*. 2003 Nov 1;100(1):3–16.
9. El-Tarabily KA, Soliman MH, Nassar AH, Al-Hassani HA, Sivasithamparam K, McKenna F, Hardy GS. Biological control of Sclerotinia minor using a

chitinolytic bacterium and actinomycetes. *Plant Pathology*. 2000 Oct;49(5): 573–583.

10. Errakhi R, Bouteau F, Lebrihi A, Barakate M. Evidences of biological control capacities of Streptomyces spp. against Sclerotium rolfsii responsible for damping-off disease in sugar beet (Beta vulgaris L.). *World Journal of Microbiology and Biotechnology*. 2007 Nov;23:1503–1509.
11. Xiao K, Kinkel LL, Samac DA. Biological control of Phytophthora root rots on alfalfa and soybean with Streptomyces. *Biological Control*. 2002 Mar 1;23(3): 285–295.
12. Joo GJ. Production of an anti-fungal substance for biological control of Phytophthora capsici causing phytophthora blight in red-peppers by Streptomyces halstedii. *Biotechnology Letters*. 2005 Feb;27:201–205.
13. Trejo-Estrada SR, Paszczynski A, Crawford DL. Antibiotics and enzymes produced by the biocontrol agent Streptomyces violaceusniger YCED-9. *Journal of Industrial Microbiology and Biotechnology*. 1998 Aug 1;21(1–2):81–90.
14. Ouhdouch Y, Barakate M, Finance C. Actinomycetes of Moroccan habitats: isolation and screening for antifungal activities. *European Journal of Soil Biology*. 2001 Apr 1;37(2):69–74.
15. Getha K, Vikineswary S. Antagonistic effects of Streptomyces violaceusniger strain G10 on Fusarium oxysporum f. sp. cubense race 4: indirect evidence for the role of antibiosis in the antagonistic process. *Journal of Industrial Microbiology and Biotechnology*. 2002 Jun;28:303–310.
16. Fguira LF, Fotso S, Ameur-Mehdi RB, Mellouli L, Laatsch H. Purification and structure elucidation of antifungal and antibacterial activities of newly isolated Streptomyces sp. strain US80. *Research in Microbiology*. 2005 Apr 1;156(3):341–347.
17. Taechowisan T, Lu C, Shen Y, Lumyong S. Secondary metabolites from endophytic Streptomyces aureofaciens CMUAc130 and their antifungal activity. *Microbiology*. 2005 May;151(5):1691–1695.
18. Valois D, Fayad K, Barasubiye T, Garon M, Dery C, Brzezinski R, Beaulieu C. Glucanolytic actinomycetes antagonistic to Phytophthora fragariae var. rubi, the causal agent of raspberry root rot. *Applied and Environmental Microbiology*. 1996 May;62(5):1630–1635.
19. Mahadevan B, Crawford DL. Properties of the chitinase of the antifungal biocontrol agent Streptomyces lydicus WYEC108. Enzyme and Microbial Technology. 1997 May 15;20(7):489–943.
20. Mukherjee G, Sen SK. Purification, characterization, and antifungal activity of chitinase from Streptomyces venezuelae P 10. *Current Microbiology*. 2006 Oct;53:265–269.
21. Barka EA, Vatsa P, Sanchez L, Gaveau-Vaillant N, Jacquard C, Klenk HP, Clément C, Ouhdouch Y, van Wezel GP. Taxonomy, physiology, and natural products of Actinobacteria. *Microbiology and Molecular Biology Reviews*. 2016 Mar;80(1):1–43.
22. Mawang CI, Azman AS, Fuad AS, Ahamad M. Actinobacteria: An eco-friendly and promising technology for the bioaugmentation of contaminants. *Biotechnology Reports*. 2021 Dec 1;32:e00679.
23. Lee LH, Zainal N, Azman AS, Mutalib NS, Hong K, Chan KG. Mumia flava gen. nov., sp. nov., an actinobacterium of the family Nocardioidaceae. *International Journal of Systematic and Evolutionary Microbiology*. 2014 May;64(Pt_5):1461–1467.

24. Xu Z, Ji L, Tang W, Guo L, Gao C, Chen X, Liu J, Hu G, Liu L. Metabolic engineering of Streptomyces to enhance the synthesis of valuable natural products. *Engineering Microbiology*. 2022 Jun 1;2(2):100022.
25. Lajtai-Szabó P, Hülber-Beyer É, Nemestóthy N, Bélafi-Bakó K. The role of physical support in secondary metabolite production by Streptomyces species. *Biochemical Engineering Journal*. 2022 Jul 1;185:108495.
26. Ser HL, Tan LT, Law JW, Chan KG, Duangjai A, Saokaew S, Pusparajah P, Ab Mutalib NS, Khan TM, Goh BH, Lee LH. Focused review: cytotoxic and antioxidant potentials of mangrove-derived Streptomyces. *Frontiers in Microbiology*. 2017 Nov 1;8:2065.
27. Waksman SA, Woodruff HB. Bacteriostatic and bactericidal substances produced by a soil Actinomyces. *Proceedings of the Society for Experimental Biology and Medicine*. 1940 Nov;45(2):609–614.
28. Waksman SA, Woodruff HB. Streptothricin, a new selective bacteriostatic and bactericidal agent, particularly active against Gram-negative bacteria. *Proceedings of the Society for Experimental Biology and Medicine*. 1942 Feb;49(2):207–210.
29. Cuozzo S, de LeBlanc AD, LeBlanc JG, Hoffmann N, Tortella GR. Streptomyces genus as a source of probiotics and its potential for its use in health. *Microbiological Research*. 2023 Jan 1;266:127248.
30. Procópio RE, Silva IR, Martins MK, Azevedo JL, Araújo JM. Antibiotics produced by Streptomyces. *Brazilian Journal of Infectious Diseases*. 2012;16: 466–471.
31. Quinn GA, Banat AM, Abdelhameed AM, Banat IM. Streptomyces from traditional medicine: Sources of new innovations in antibiotic discovery. *Journal of Medical Microbiology*. 2020 Aug;69(8):1040.
32. Zong G, Fu J, Zhang P, Zhang W, Xu Y, Cao G, Zhang R. Use of elicitors to enhance or activate the antibiotic production in Streptomyces. *Critical Reviews in Biotechnology*. 2022 Nov 17;42(8):1260–1283.
33. Westhoff S, Kloosterman AM, van Hoesel SF, van Wezel GP, Rozen DE. Competition sensing changes antibiotic production in streptomyces. *MBio*. 2021 Feb 23;12(1):10–128.
34. Ferreira RG, Azzoni AR, Freitas S. On the production cost of lignocellulose-degrading enzymes. *Biofuels, Bioproducts and Biorefining*. 2021 Jan;15(1): 85–99.
35. Sarmiento-Vizcaíno A, Martín J, Ortiz-López FJ, Reyes F, García LA, Blanco G. Natural products, including a new caboxamycin, from Streptomyces and other Actinobacteria isolated in Spain from storm clouds transported by Northern winds of Arctic origin. *Frontiers in Chemistry*. 2022 Nov 3;10:948795.
36. Singh SP, Upadhyay SK, editors. *Bioprospecting of Microorganism Based Industrial Molecules*. Wiley; 2021 Dec 13.
37. Waheeb MS, Elkhatib WF, Yassien MA, Hassouna NA. Production of Cellulase by Soil Isolated Streptomyces sp. *Archives of Pharmaceutical Sciences Ain Shams University*. 2021 Dec 1;5(2):225–233.
38. Mondal S, Mondal K, Halder SK, Thakur N, Mondal KC. Microbial Amylase: Old but still at the forefront of all major industrial enzymes. *Biocatalysis and Agricultural Biotechnology*. 2022 Oct 22:102509.
39. Lee LH, Azman AS, Zainal N, Eng SK, Mutalib NS, Yin WF, Chan KG. Microbacterium mangrovi sp. nov., an amylolytic actinobacterium isolated

from mangrove forest soil. *International Journal of Systematic and Evolutionary Microbiology*. 2014 Oct;64(Pt_10):3513–3519.
40. Al-Dhabi NA, Esmail GA, Ghilan AK, Arasu MV, Duraipandiyan V, Ponmurugan K. Isolation and purification of starch hydrolysing amylase from Streptomyces sp. Al-Dhabi-46 obtained from the Jazan region of Saudi Arabia with industrial applications. *Journal of King Saud University-Science*. 2020 Jan 1;32(1):1226–1232.
41. Boovaragamoorthy GM, Anbazhagan M, Piruthiviraj P, Pugazhendhi A, Kumar SS, Al-Dhabi NA, Ghilan AK, Arasu MV, Kaliannan T. Clinically important microbial diversity and its antibiotic resistance pattern towards various drugs. *Journal of Infection and Public Health*. 2019 Nov 1;12(6):783–788.
42. Gurusamy S, Kulanthaisamy MR, Hari DG, Veleeswaran A, Thulasinathan B, Muthuramalingam JB, Balasubramani R, Chang SW, Arasu MV, Al-Dhabi NA, Selvaraj A. Environmental friendly synthesis of TiO2-ZnO nanocomposite catalyst and silver nanomaterials for the enhanced production of biodiesel from Ulva lactuca seaweed and potential antimicrobial properties against the microbial pathogens. *Journal of Photochemistry and Photobiology B: Biology*. 2019 Apr 1;193:118–130.
43. Rajkumari J, Magdalane CM, Siddhardha B, Madhavan J, Ramalingam G, Al-Dhabi NA, Arasu MV, Ghilan AK, Duraipandiayan V, Kaviyarasu K. Synthesis of titanium oxide nanoparticles using Aloe barbadensis mill and evaluation of its antibiofilm potential against Pseudomonas aeruginosa PAO1. *Journal of Photochemistry and Photobiology B: Biology*. 2019 Dec 1;201:111667.
44. Roopan SM, Priya DD, Shanavas S, Acevedo R, Al-Dhabi NA, Arasu MV. CuO/C nanocomposite: Synthesis and optimization using sucrose as carbon source and its antifungal activity. *Materials Science and Engineering: C*. 2019 Aug 1;101:404–414.
45. Valsalam S, Agastian P, Esmail GA, Ghilan AK, Al-Dhabi NA, Arasu MV. Biosynthesis of silver and gold nanoparticles using Musa acuminata colla flower and its pharmaceutical activity against bacteria and anticancer efficacy. *Journal of Photochemistry and Photobiology B: Biology*. 2019 Dec 1;201:111670.
46. Mehta D, Satyanarayana T. Bacterial and archaeal α-amylases: diversity and amelioration of the desirable characteristics for industrial applications. *Frontiers in Microbiology*. 2016 Jul 28;7:1129.
47. Lakshmi SA, Shafreen RB, Priyanga A, Shiburaj S, Pandian SK. A highly divergent α-amylase from Streptomyces spp.: An evolutionary perspective. *International Journal of Biological Macromolecules*. 2020 Nov 15;163:2415–2428.
48. Ashraf M, Hussain N, Baqar Z, Kumar A, Ferreira LF, Iqbal HM. Bioprospecting microbial proteases in various industries/sectors. In *Microbial Biomolecules* 2023 Jan 1 (pp. 301–324). Academic Press.
49. Al-Dhabi NA, Esmail GA, Ghilan AK, Arasu MV, Duraipandiyan V, Ponmurugan K. Characterization and fermentation optimization of novel thermo stable alkaline protease from Streptomyces sp. Al-Dhabi-82 from the Saudi Arabian environment for eco-friendly and industrial applications. *Journal of King Saud University-Science*. 2020 Jan 1;32(1):1258–1264.
50. Rathnan RK, Ambili M. Cellulase enzyme production by Streptomyces sp using fruit waste as substrate. *Australian Journal of Basic and Applied Sciences*. 2011;5(12):1114–1118.

51. Bentley SD, Chater KF, Cerdeño-Tárraga AM, Challis GL, Thomson NR, James KD, Harris DE, Quail MA, Kieser H, Harper D, Bateman A. Complete genome sequence of the model actinomycete Streptomyces coelicolor A3 (2). *Nature*. 2002 May 9;417(6885):141–147.
52. Kulkarni N, Gadre RV. Production and properties of an alkaline, thermophilic lipase from Pseudomonas fluorescens NS2W. *Journal of Industrial Microbiology and Biotechnology*. 2002 Jun;28:344–348.
53. El-Araby A, El Ghadraoui L, Errachidi F. Usage of biological chitosan against the contamination of post-harvest treatment of strawberries by aspergillus niger. *Frontiers in Sustainable Food Systems*. 2022 Aug 8;6:881434.
54. El-Araby A, El Ghadraoui L, Errachidi F. Physicochemical properties and functional characteristics of ecologically extracted shrimp chitosans with different organic acids during demineralization step. *Molecules*. 2022 Nov 28;27(23):8285.
55. Khan FI, Bisetty K, Singh S, Permaul K, Hassan MI. Chitinase from Thermomyces lanuginosus SSBP and its biotechnological applications. *Extremophiles*. 2015 Nov;19:1055–1066.
56. Bhattacharya D, Nagpure A, Gupta RK. Bacterial chitinases: properties and potential. *Critical Reviews in Biotechnology*. 2007 Jan 1;27(1):21–28.
57. Setyahadi S, Putri SF. Screening microbe producing chitinase for inhibiting Ganoderma boninense. *InIOP Conference Series: Earth and Environmental Science* 2019 Feb 1 (Vol. 230, No. 1, p. 012100). IOP Publishing.
58. Priya BS, Stalin T, Selvam K. Efficient utilization of xylanase and lipase producing thermophilic marine actinomycetes (Streptomyces albus and Streptomyces hygroscopicus) in the production of ecofriendly alternative energy from waste. *African Journal of Biotechnology*. 2012;11(78):14320–14325.
59. Pinheiro VE, Ferreira JA, Betini JH, Kamimura ES, Polizeli ML. Utilizing a novel fungal enzymatic cocktail as an eco-friendly alternative for cellulose pulp biobleaching. *BioResources*. 2021 Nov 1;16(4):7509.
60. Sharma P, Dutta J, Thakur D. Future prospects of actinobacteria in health and industry. In *New and Future Developments in Microbial Biotechnology and Bioengineering* 2018 Jan 1 (pp. 305–324). Elsevier.
61. Walia A, Guleria S, Mehta P, Chauhan A, Parkash J. Microbial xylanases and their industrial application in pulp and paper biobleaching: a review. *3 Biotech*. 2017 May;7:1–2.
62. de Souza HF, Borges LA, Gonçalves VD, Dos Santos JV, Bessa MS, Carosia MF, de Carvalho MV, Brandi IV, Kamimura ES. Recent advances in the application of xylanases in the food industry and production by actinobacteria: A review. *Food Research International*. 2022 Nov 8:112103.
63. Li Y, Zhang X, Lu C, Lu P, Yin C, Ye Z, Huang Z. Identification and Characterization of a Novel Endo-β-1, 4-Xylanase from Streptomyces sp. T7 and Its Application in Xylo-Oligosaccharide Production. *Molecules*. 2022 Apr 13;27(8):2516.
64. Kuhad RC, Singh A. *Biotechnology for environmental management and resource recovery*. New Delhi, India:: Springer; 2013 Mar 25.
65. Chakraborty S, Khopade A, Kokare C, Mahadik K, Chopade B. Isolation and characterization of novel α-amylase from marine Streptomyces sp. D1. *Journal of Molecular Catalysis B: Enzymatic*. 2009 Jun 1;58(1–4):17–23.

66. Thatoi H, Mohapatra S, Das SK, editors. *Bioprospecting of Enzymes in Industry, Healthcare and Sustainable Environment*. Springer Singapore; 2021 Jan 23.
67. Vivek K, Sandhia GS, Subramaniyan SJ. Extremophilic lipases for industrial applications: A general review. *Biotechnology Advances*. 2022 Nov 1;60:108002.
68. Bolina IC, Gomes RA, Mendes AA. Biolubricant production from several oleaginous feedstocks using lipases as catalysts: Current scenario and future perspectives. *Bioenergy Research*. 2021 Jan 5:1–9.
69. Chandra P, Enespa, Singh R, Arora PK. Microbial lipases and their industrial applications: a comprehensive review. *Microbial Cell Factories*. 2020 Dec;19:1–42.
70. Gao K, Chu W, Sun J, Mao X. Identification of an alkaline lipase capable of better enrichment of EPA than DHA due to fatty acids selectivity and regioselectivity. *Food Chemistry*. 2020 Nov 15;330:127225.
71. Aly MM, Tork S, Al-Garni SM, Nawar L. Production of lipase from genetically improved Streptomyces exfoliates LP10 isolated from oil-contaminated soil. *African Journal of Microbiology Research*. 2012 Feb 16;6(6):1125–1137.
72. Parameswaran B, Varjani S, Raveendran S. Green bio-processes. *Energy, Environment, and Sustainability*. ISBN 978-981-13-3263-0; XVI, 446, 2019.
73. Ninawe S, Lal R, Kuhad RC. Isolation of three xylanase-producing strains of actinomycetes and their identification using molecular methods. *Current Microbiology*. 2006 Sep;53:178–182.
74. Zukovski L, Fontana RC, Pauletti G, Camassola M, Dillon AJ. Fodder radish cake (Raphanus sativus L.) as an alternative biomass for the production of cellulases and xylanases in solid-state cultivation. *Brazilian Journal of Chemical Engineering*. 2017 Jul;34:709–714.
75. Romsaiyud A, Songkasiri W, Nopharatana A, Chaiprasert P. Combination effect of pH and acetate on enzymatic cellulose hydrolysis. *Journal of Environmental Sciences*. 2009 Jan 1;21(7):965–970.
76. Budihal SR, Agsar D, Patil SR. Enhanced production and application of acidothermophilic Streptomyces cellulase. *Bioresource Technology*. 2016 Jan 1;200:706–712.
77. Ejaz U, Sohail M, Ghanemi A. Cellulases: From bioactivity to a variety of industrial applications. *Biomimetics*. 2021 Jul 5;6(3):44.
78. Areeshi MY. Microbial cellulase production using fruit wastes and its applications in biofuels production. *International Journal of Food Microbiology*. 2022 Oct 2;378:109814.
79. Nawani N, Aigle B, Mandal A, Bodas M, Ghorbel S, Prakash D. Actinomycetes: Role in biotechnology and medicine. *BioMed Research International*. 2013 May 30;2013.
80. Kar S, Ray RC. Statistical optimization of α-amylase production by Streptomyces erumpens MTCC 7317 cells in calcium alginate beads using response surface methodology. *Polish Journal of Microbiology*. 2008;57(1):49.
81. Dong H, Guo T, Zhang W, Ying H, Wang P, Wang Y, Chen Y. Biochemical characterization of a novel azoreductase from Streptomyces sp.: application in eco-friendly decolorization of azo dye wastewater. *International Journal of Biological Macromolecules*. 2019 Nov 1;140:1037–1046.

82. Preethi PS, Vickram S, Das R, Hariharan NM, Rameshpathy M, Subbaiya R, Karmegam N, Kim W, Govarthanan M. Bioprospecting of novel peroxidase from Streptomyces coelicolor strain SPR7 for carcinogenic azo dyes decolorization. *Chemosphere*. 2023 Jan 1;310:136836.
83. Govindaraji PK, Vuppu S. Characterisation of pectin and optimization of pectinase enzyme from novel Streptomyces fumigatiscleroticus VIT-SP4 for drug delivery and concrete crack-healing applications: An eco-friendly approach. *Saudi Journal of Biological Sciences*. 2020 Dec 1;27(12):3529–3540.
84. Deng R-X, Yue S-J, Wang W, et al. Identification, biological evaluation, and improved biotransformation of a phenazine antioxidant using *Streptomyces lomondensis* S015 whole cells. *Process Biochemistry* 2023; 125: 154–161.
85. Ceresino EB, de Melo RR, Kuktaite R, Hedenqvist MS, Zucchi TD, Johansson E, Sato HH. Transglutaminase from newly isolated Streptomyces sp. CBMAI 1617: Production optimization, characterization and evaluation in wheat protein and dough systems. *Food Chemistry*. 2018 Feb 15;241:403–410.
86. Balagurunathan R, Radhakrishnan M, Somasundaram ST. L-Glutaminase producing actinomycetes from marine sediments–selective isolation, semi quantitative assay and characterization of potential strain. *Australian Journal of Basic and Applied Sciences* 2010;4(5):698–705.
87. Hacene H, Daoudi-Hamdad F, Bhatnagar T, Baratti JC, Lefebvre G. H107, a new aminoglycoside anti-Pseudomonas antibiotic produced by a new strain of Spirillospora. *Microbios*. 2000 Jan 1;102(402):69–77.
88. Laidi RF, Kansoh AL, Elshafei AM, Cheikh B. Taxonomy, identification and biological activities of a novel isolate of Streptomyces tendae. *Arab Journal of Biotechnology*. 2006;9(3):427–436.
89. Solanki A, Kim JD, Lee KB. Nanotechnology for regenerative medicine: nanomaterials for stem cell imaging. *Nanomedicine*. (2008) 3(4), 567–578.
90. Korn-Wendisch F, Kutzner HJ. The family Streptomycetaceae. *The Prokaryotes*. 1992(Ed. 2):921–995.
91. Williams ST. *Bergey's manual of systematic bacteriology*. Williams & Wilkins; 1989.
92. Goodfellow M, Williams ST. Ecology of actinomycetes. *Annual Review of Microbiology*. 1983 Oct;37(1):189–216.
93. Lakshmipathy DT, Kannabiran K. A morphological, biochemical and biological studies of halophilic Streptomyces sp. isolated from saltpan environment. *American Journal of Infectious Diseases*. 2009;5(3):200–206.
94. Takahashi Y, Omura S. Isolation of new actinomycete strains for the screening of new bioactive compounds. *The Journal of General and Applied Microbiology*. 2003;49(3):141–154.
95. Atta HM, Ahmad MS. Antimycin-A antibiotic biosynthesis produced by Streptomyces Sp. AZ-AR-262: taxonomy, fermentation, purification and biological activities. *Australian Journal of Basic and Applied Sciences*. 2009;3(1): 126–135.
96. Miyadoh S. Research on antibiotic screening in Japan over the last decade: a producing microorganism approach. *Actinomycetologica*. 1993 Dec 25;7(2): 100–106.
97. Tanaka Y, Omura S. Agroactive compounds of microbial origin. *Annual Review of Microbiology*. 1993 Oct;47(1):57–87.

98. Turner WB. Secondary metabolism with special reference to Actinomycetales. *InSociety for Applied Bacteriology Symposium Series* 1973 Jan (Vol. 2, pp. 209–217).
99. Candeias JA. A engenharia genética. *Revista de saúde pública*. 1991 Feb;25(1): 3–10.
100. Okachi R, Kawamoto I, Takasawa S, Yamamoto M, Sato S, Sato T, Nara T. A new antibiotic XK-62-2 (Sagamicin) I. Isolation, physicochemical and antibacterial properties. *The Journal of Antibiotics*. 1974;27(10):793–800.
101. Okami Y, Kitahara T, Hamada M, Naganawa H, Kondo S, Maeda K, Takeuchi T, Umezawa H. Studies on a new amino acid antibiotic, amiclenomycin. *The Journal of Antibiotics*. 1974;27(9):656–664.
102. Haneishi T, Kitahara N, Takiguchi Y, Arai M, Sugawara S. New antibiotics, methylenomycins A and B I. Producing organism, fermentation and isolation, biological activities and physical and chemical properties. *The Journal of Antibiotics*. 1974;27(6):386–392.
103. Matsui K, Juri N, Marukawa S, Kasai S. Roseoflavin formation by Streptomyces davawensis. *Agricultural and Biological Chemistry*. 1979;43(10):2189–2190.
104. Grill S, Busenbender S, Pfeiffer M, Köhler U, Mack M. The bifunctional flavokinase/flavin adenine dinucleotide synthetase from Streptomyces davawensis produces inactive flavin cofactors and is not involved in resistance to the antibiotic roseoflavin. *Journal of Bacteriology*. 2008 Mar 1;190(5):1546–1553.
105. Hamada M, Kondo S, Yokoyama T, Miura K, Iinuma K, Yamamoto H, Maeda K, Takeuchi T, Umezawa H. Minosaminomycin, a new antibiotic containing myo-inosamine. *The Journal of Antibiotics*. 1974;27(1):81–83.
106. Yahagi T, Sugawara H, Uramoto M, Suzuki S. A new antibiotic libramycin A. *The Journal of Antibiotics*. 1974;27(2):143–144.
107. Martin JF, McDaniel LE. Isolation, purification and properties of the hexaene macrolides candihexin I and candihexin II. *The Journal of Antibiotics*. 1974;27(8):610–619.
108. Omura S, Tanaka H, Koyama Y, Oiwa R, Katagiri M, Awaya J, Nagai T, Hata T. Nanaomycins A and B*, new antibiotics produced by a strain of Streptomyces. *The Journal of Antibiotics*. 1974;27(5):363–365.
109. Coronelli C, Pagani H, Bardone MR, Lancini GC. Purpuromycin, a new antibiotic isolated from Actinoplanes ianthinogenes N. sp. *The Journal of Antibiotics*. 1974;27(3):161–168.
110. Argoudelis AD, Bergy ME, Pyke TR. Zorbamycin and related antibiotics I. Production, isolation and characterization. *The Journal of Antibiotics*. 1971;24(8):543–557.
111. Wu H, Qu S, Lu C, Zheng H, Zhou X, Bai L, Deng Z. Genomic and transcriptomic insights into the thermo-regulated biosynthesis of validamycin in Streptomyces hygroscopicus 5008. *BMC Genomics*. 2012 Dec;13(1):1–4.
112. Anzai Y, Iizaka Y, Li W, Idemoto N, Tsukada SI, Koike K, Kinoshita K, Kato F. Production of rosamicin derivatives in Micromonospora rosaria by introduction of d-mycinose biosynthetic gene with Φ C31–derived integration vector pSET152. *Journal of Industrial Microbiology and Biotechnology*. 2009 Aug 1;36(8):1013.
113. Huang H, Lv J, Hu Y, Fang Z, Zhang K, Bao S. Micromonospora rifamycinica sp. nov., a novel actinomycete from mangrove sediment. *International Journal of Systematic and Evolutionary Microbiology*. 2008 Jan;58(1):17–20.

114. Huang H, Wu X, Yi S, Zhou Z, Zhu J, Fang Z, Yue J, Bao S. Rifamycin S and its geometric isomer produced by a newly found actinomycete, *Micromonospora rifamycinica. Antonie Van Leeuwenhoek*. 2009 Feb;95:143–148.
115. Furumai T, Seki Y, Takeda K, Kinumaki A, Suzuki M. An approach to the biosynthesis of macrolide antibiotic platenomycin. *The Journal of Antibiotics*. 1973 Nov 25;26(11):708–710.
116. Cheng J, Yang SH, Palaniyandi SA, Han JS, Yoon TM, Kim TJ, Suh JW. Azalomycin F complex is an antifungal substance produced by Streptomyces malaysiensis MJM1968 isolated from agricultural soil. *Journal of the Korean Society for Applied Biological Chemistry*. 2010 Oct;53:545–552.
117. Chatterjee S, Vijayakumar EK, Chatterjee S, Blumbach J, Ganguli BN. 5-Hydroxy-9-methylstreptimidone, a New Glutarimide from a Streptomyces sp. MIL Y-9065403. *The Journal of Antibiotics*. 1995 Mar 25;48(3):271–273.
118. Gould SJ, Hong ST, Carney JR. Cloning and heterologous expression of genes from the kinamycin biosynthetic pathway of Streptomyces murayamaensis. *The Journal of Antibiotics*. 1998 Jan 25;51(1):50–57.
119. Shimi IR, Dewedar A, Shoukry S. Kuwaitimycin, a new antibiotic. *The Journal of Antibiotics*. 1973 Oct 25;26(10):593–601.
120. Umezawa H. Sarkomycin, an anti-tumor substance produced by streptomyces. *Journal of Antibiotics Series A*. 1953;6:101.
121. Naidenova MM, Michaĭilov YG, Ivanova I. Oxadin effect on actinomycetes strains. *Antibiotiki i Khimioterapiia= Antibiotics and Chemoterapy [sic]*. 2001 Jan 1;46(1):7–9.
122. Xu LY, Quan XS, Wang C, Sheng HF, Zhou GX, Lin BR, Jiang RW, Yao XS. Antimycins A19 and A20, two new antimycins produced by marine actinomycete Streptomyces antibioticus H74-18. *The Journal of Antibiotics*. 2011 Oct;64(10):661–665.
123. Han Z, Xu Y, McConnell O, Liu L, Li Y, Qi S, Huang X, Qian P. Two antimycin A analogues from marine-derived actinomycete Streptomyces lusitanus. *Marine Drugs*. 2012 Mar 22;10(3):668–676.

Chapter 6

Streptomyces as a novel source of enzymes and approaches to enhance the production of enzymes

Rabab Anjum
IMS Engineering College, Ghaziabad, India

Santosh Kumar Mishra
Department of Life Sciences, Sharda School of Basic Science and Research, Sharda University, Greater Noida, India

D. Vasanth
Department of Biotechnology, National Institute of Technology, Raipur, India

Mohit Kumar
Department of Life Sciences, Sharda School of Basic Science and Research, Sharda University, Greater Noida, India

6.1 INTRODUCTION

Bioactive enzymes from *Streptomyces* are widely available and have a variety of uses in both industry and medicine (de Lima Procópio et al. 2012). These gram-positive, filamentous bacteria are members of the class Actinobacteria. They also generate spores. Terrestrial vertebrates first appeared around 382 million years ago when *Streptomyces* split from their closest related Kitasatospora during the late Devonian period (McDonald and Curriea 2017; Chater 2006). *Streptomyces* are non-motile bacteria that multiply by producing thread-like hyphae that puncture surfaces in search of food. *Streptomyces* produces aerial hyphae that divide when resources are few, resulting in spores that are resilient to adverse conditions and rapidly spread to other locations or nutrient sources (Chater 2016).

6.1.1 Bioactive compounds from *Streptomyces*

During the development stage, *Streptomyces* generates secondary metabolites, which are substances that aren't necessarily required for the organism's growth or reproduction but might provide it with an edge over rivals (Chater 2016). These metabolites support the vegetative bacterial cells

 DOI: 10.1201/9781032694634-6

through the sequestration of metals like iron (siderophores), pigmentation, competition inhibition (antibiotics), and enhanced interspecies communication with other species (Bibb 2005). This molecular diversity is made feasible because *Streptomyces* has a relatively big genome; it may be up to four times larger than the genomes of certain other bacteria (Chater 2016). The ability to produce active surface biomolecules, such as bioemulsifiers and biosurfactants, is quite high in *Streptomyces* species. *Streptomyces* strains were used to manufacture antidiabetic acarbose by microbial fermentation. Certain *Streptomyces* species have demonstrated the capacity to produce inhibitors of cholesterol production, such as pravastatin. Recently, ecologically acceptable "nanofactories" for the manufacture of nanoparticles have been found to be *Streptomyces* species. Certain *Streptomyces* species show promise in the synthesis of vitamin B12 (El-Naggar 2021).

Streptomyces has an innate capacity to produce a variety of metabolites with distinct biological functions. These metabolites can have a linear chain that leads to either complicated heterocyclic molecules or simple cyclic compounds, depending on their structure. Under various growing circumstances, the synthesis of xylanase and cellulases by *Streptomyces lividans* 1326 was investigated by Kluepfel et al. (1986). At 40 °C, *S. lividans* produced significant amounts of xylanxylanohydrolase (50 IU/mL) and appropriate levels of extracellular glucan-glucanohydrolase (1 IU/mL). It was discovered that the enzyme's stability was noticeably higher than that of the xylanases and cellulases made by other bacteria (Kluepfel et al. 1986). The production of extracellular enzymes such as DD-carboxypeptidase by genetically modified *Streptomyces lividans* TK24 in Lennox broth medium has also been reported (Erpicum et al. 1990).

6.1.2 Isolation of *Streptomyces* from natural sources

Due to their complex and powerful secondary metabolites, *Streptomyces* species have been contemplated as a warehouse of a diverse range of natural products. The complete process for production of enzymes using *Streptomyces* has been presented in Figure 6.1. They are the largest current producer of new and innovative secondary metabolites (Donald et al., 2022). The process that leads to the synthesis of secondary metabolites from *Streptomyces* includes enzyme inhibitors, antibiotics, antifungals, antibacterials, and anticancer agents (Manivasagan et al. 2014). The aerial species of *Streptomyces* are a more essential source of unique antibiotics in comparison with vegetative species. This is because of the adaptations to extreme conditions and unique compounds found in marine *Streptomyces*. These species produce unique bioactive compounds that can have diverse applications in medicine, biotechnology, and environmental science. Presence of extreme conditions in marine environments like high salinity and pressure lead to development of unique metabolic pathways and bioactive compounds, which make them valuable for bioprospecting.

Secondary metabolites are known to be produced during the lag phase and to reach a stationary phase; however, in the case of *Streptomyces*, secondary metabolites can also be produced during the exponential, stationary,

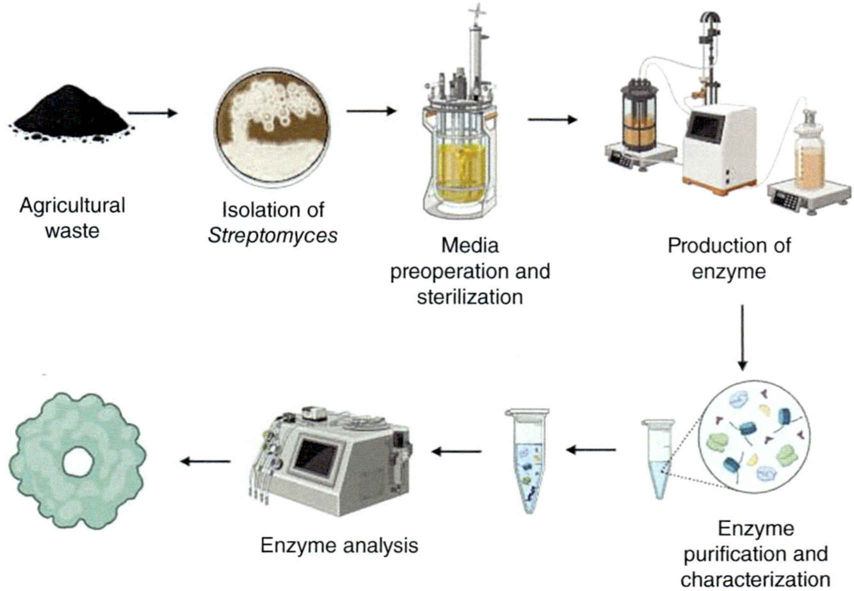

Figure 6.1 Process for production of enzyme from *Streptomyces*.

and death phases (Harir et al., 2018). Low molecular organic compounds, known as secondary metabolites, are synthesized by different bacteria and viruses. These substances do not play a direct role in the primary metabolic functions related to the growth and development of the organism. However, they have demonstrated significant importance for human health, the environment, and the economic well-being of society. These metabolites play an essential role in various aspects of science and industry. Their uses in various fields are given in the following:

6.1.2.1 Pharmaceuticals

Bacterial secondary metabolites have yielded numerous antibiotics and other drugs that are crucial for treating infectious diseases. The excessive use of antibiotics and other drugs like antifungal and antibacterial in human therapy and for farm animals have created a pathogenic bacterial resistance to multiple drugs (Romano-Armada et al. 2020). Therefore, secondary metabolites are continuously required to develop new antibiotics and other drugs to fight antibiotic-resistant bacteria and fungal and viral diseases. For example, penicillin (from *Penicillium* bacteria) and vancomycin (from *Streptomyces*) are some useful antibiotics which are currently used.

6.1.2.2 Biotechnology

Bacterial secondary metabolites are used in biotechnological processes for the production of enzymes, bioactive compounds, and other bio-based products. The enzymes produced are used as biocatalysts to synthesize chemicals in various industries like pharmaceuticals, food, and fine chemicals (Khushboo et al. 2022). Some *Streptomyces* species produce enzymes used in various biotechnological processes, such as amylase for starch degradation or proteases for protein hydrolysis. These metabolites are also used as immunosuppressants such as compounds like rapamycin from *Streptomyces hygroscopicus* are used in organ transplant procedures to suppress the immune system's response. Like all these there are many more uses of secondary metabolites in the vast field of biotechnology.

6.1.2.3 Bioremediation

Various living things, including plants, are employed in bioremediation to neutralize or detoxify toxins and poisons from the environment. Microorganisms produce metabolites that have the ability to metabolize and break down pollutants, converting them into less toxic forms. When compared to other bacteria, *Streptomyces* is a very desirable option for bioremediation because of its remarkable adaptability and tolerance to a wide range of environmental stressors (Romano-Armada et al. 2020).

6.1.2.4 Agriculture

Bacterial metabolites produce biopesticides, biofertilizers, and plant growth regulators. They to help manage plant diseases and reduce the demand for synthetic pesticides and fertilizers. Since the extensive use of synthetic pesticides has led to the emergence of bacteria resistant to pesticides, using biopesticides and biofertilizers made from *Streptomyces* species' secondary metabolites enhances the sustainability of agricultural output since they are ecologically benign. The genus *Streptomyces* is a serious candidate for the biological control of phytopathogenic microorganisms because of its vast biotechnological potential (Pacios-Michelena et al. 2021). To effectively control certain plant diseases, certain products based on *Streptomyces* species have been developed and evaluated. For example, to manage *Ceratocystis radicicola*, which causes black burns on date palms, *Streptomyces griseoviridis* strain K61 (MycostopR) has been employed (Pacios-Michelena et al. 2021).

6.1.2.5 Cosmetics and personal care

Some secondary metabolites have cosmetic and skincare applications, including antimicrobial properties and ingredients for antiaging products. So, the cosmetic industry and researchers have extensively explored the creation of bioactive cosmetics formulations to reduce the signs of aging and enhance skin beauty and health in response to the strong consumer demand for skincare products that have proven effectiveness (Saraiva et al. 2023).

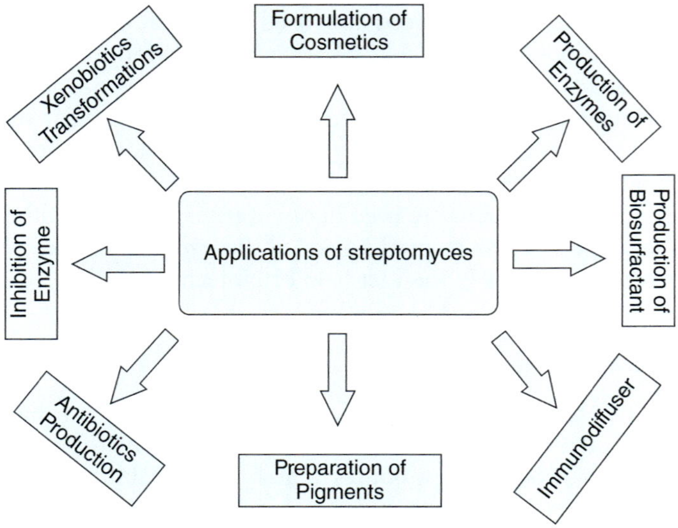

Figure 6.2 Pharmaceutical and biotechnological applications of *Streptomyces*.

For example, algae-derived secondary metabolites have been found to offer skin benefits, including shielding against UV radiation and decreasing skin laxity, uneven texture, and wrinkles. Similarly, some species of *Streptomyces* also produce secondary metabolites that have some antimicrobial and antioxidant properties and are used in the cosmetics industry. One of the examples of this is *Streptomyces chattanoogensis* THA- 663's metabolites that have an impressive antiaging activity on skin (Liu et al. 2022). Different areas where *Streptomyces* can be used for different application is summarized in Figure 6.2.

6.2 BIOACTIVITY OF *STREPTOMYCES*

6.2.1 Bioactive compounds

Streptomyces are responsible for generating approximately 70–80% of natural bioactive compounds renowned for their applications in pharmaceuticals and agrochemicals. Continuously, novel metabolites with diverse biological properties are extracted from diverse strains of *Streptomyces*. Since 1955, the *Streptomyces* genus has been the primary source of fresh antibiotics. They serve as the origin of antibacterial, antifungal, antitumor, antiviral, antiparasitic, insecticidal, pesticidal, and herbicidal agents, in addition to pharmacological compounds like immunomodulators, vasoactive substances, and neuroactive agents. The second most significant products of *Streptomyces* after antibiotics are enzymes, for example cellulases, proteases, pectinases, lipases, amylases, and so on.

6.2.2 Formulation of enzymes from *Streptomyces*

A wide array of enzymes with applications in microbial and biotechnological fields can be derived from various actinomycetes genera. Analyzing the genomic and protein sequencing data thoroughly, actinomycetes are perpetually examined and harnessed for the production of amylases, proteases, xylanases, cellulases, and various other enzymatic products. Some species of *Streptomyces* produce extracellular enzymes having the potential to remove the toxic chemicals that are being used in various industries (Kumar et al. 2020) (Table 6.1).

Table 6.1 Some examples of enzymes produced using actinobacteria and their industrial utilizations (Kumar et al. 2020)

Enzyme	*Industry*	*Application*	*Species used*
Amylase	Textile	Starch removal from woven fabrics	*Streptomyces erumpons*
	Baking	Softening of bread; volume	
	Paper and pulp	Deinking	
	Starch	Production of glucose; syrups	
Lipase	Detergent	Stain removal	*Streptomyces griseus*
	Baking	Stability of dough	
	Dairy	Flavoring of cheese	
	Textile	Cleaning, deinking	
Pectinase	Beverage	Clarification, mashing	*S. lydicus*
	Textile	Scouring	
Chitinase	Bioremediation	Chitin waste utilization	*Streptomyces griseus*
Penicillin amidase	Commercial significance	Industrial production of 6-aminopenicillanic acid	*Streptomyces clavuligerus*
Xylanase	Baking	Dough conditioning	*Streptomyces* spp.
	Animal feeds	Digestibility	
	pulp and paper	Boosting of bleach	

6.3 LARGE SCALE PRODUCTION OF ENZYMES FROM *STREPTOMYCES*

Streptomyces stands out as a resourceful genus with remarkable applications in biotechnology, particularly as a pivotal platform for drug production. Despite the complicated life cycle and metabolism of *Streptomyces*, existing review papers often concentrate on isolated aspects, such as isolating new strains, identification of metabolites, morphological studies, and genetic or metabolic investigations.

The study comprehensively outlines cutting-edge advancements in biotechnological manufacturing and bioconversion operations, specifically focusing on anticancer drugs, antibiotics, steroidal drugs, and immunosuppressant and anthelmintic agents produced by *Streptomyces*. Emphasis is placed on process development aspects, illuminating diverse approaches and technologies employed to enriched production yields. The investigation explores the impact of nutrients and oxygen on *Streptomyces* metabolism, introduces novel fed-batch fermentation tactics, innovative precursor argumentation methods, and distinct bioreactor designs. It also includes biotechnological strategies coupled with metabolic engineering and genetic tools targeted at enhancing strain performance.

The investigation extends to the utilization of whole, free, and restrained cells on unconventional supports for bioconversion processes of drugs. Thirty notable investigations from the past eight years are highlighted, providing a snapshot of contemporary progress in the field. The significant topics include the influence of carbon, nitrogen, oxygen, and nutrients on *Streptomyces* metabolism, advancements in fed-batch fermentation, pioneering techniques to precursor supplementation, and the design of specialized bioreactors.

Microbial secondary metabolites are widely used in a variety of industries, including food additives, insecticides, and pharmaceuticals. To formulate these compounds, diverse fermenting techniques are employed across numerous scales, spanning from laboratory settings to manufacturing at larger scales, each manifesting distinct advantages and shortcomings. There are four cultivation techniques:

- Traditional immobilization
- Solid-state fermentation
- Microparticle-enhanced cultivation
- Membrane gradostat reactor.

The focus is on the significance of physical support, both in terms of its presence and its impact on microbial processes. The genus *Streptomyces* was selected for its critical role in the generation of secondary metabolites and its sensitivity to cultivation conditions (Lajtai-Szabó et al. 2022). The methods discussed are explained and compared based on research findings linked to this genus. Instead of identifying a technique that works for everyone, the goal is to provide a critical evaluation of their current state and upcoming obstacles.

6.4 CONCLUSION

Based on the preceding explanation, it is possible to conclude that various *Streptomyces* species are responsible for producing a wide range of secondary metabolites that may be employed as antimicrobials, enzymes, and agricultural products. Several scientists have discovered a variety of cosmetic and personal applications. To manufacture these metabolites in large quantities from various *Streptomyces* species, it is necessary to optimize the media and growth conditions under which these bacteria may thrive. Fermentative parameter optimization may be achieved using a variety of strategies, some of which are discussed in this book chapter, such as response surface methodology connected with Plackett–Burman design, central composite design, artificial neural network linked with genetic algorithms, and others. Nowadays, artificial intelligence is employed to optimize suitable medium and culture conditions in order to improve the production of desired metabolites. Optimizing bioprocess parameters would further open up the possibility of large-scale synthesis of desired metabolites by cultivating the *Streptomyces* strain, including production at industrial scale. With the progress of the processes involved in optimization studies included in bioprocesses and scale-up approaches, it will play a critical role in realizing *Streptomyces*' full potential for solving present and future difficulties in medicine, agriculture, and industry.

REFERENCES

Bibb, Mervyn J. 2005. "Regulation of Secondary Metabolism in Streptomycetes." *Current Opinion in Microbiology*. Elsevier Ltd. https://doi.org/10.1016/j.mib.2005.02.016

Chater, Keith F. 2006. "Streptomyces Inside-out: A New Perspective on the Bacteria That Provide Us with Antibiotics." *Philosophical Transactions of the Royal Society B: Biological Sciences* 361 (1469): 761–768. https://doi.org/10.1098/rstb.2005.1758

Chater, Keith F. 2016. "Recent Advances in Understanding Streptomyces." *F1000Research* 5 (November): 2795. https://doi.org/10.12688/f1000research.9534.1

Donald, L., Pipite, A., Subramani, R., Owen, J., Keyzers, R.A. and Taufa, T., 2022. Streptomyces: Still the biggest producer of new natural secondary metabolites, a current perspective. *Microbiology Research*, *13*(3), 418–465. https://doi.org/10.3390/microbiolres13030031

El-Naggar, Noura El Ahmady 2021. "Streptomyces-Based Cell Factories for Production of Biomolecules and Bioactive Metabolites." In *Microbial Cell Factories Engineering for Production of Biomolecules*, 183–234. Elsevier. https://doi.org/10.1016/B978-0-12-821477-0.00011-8

Erpicum, Thomas, Benoît Granier, Marianne Delcour, V. Mauro Lenzini, Marline Nguyen Distèche, Jean Dusart, and Jean-Marie Frère. 1990. "Enzyme Production by Genetically Engineered Streptomyces Strains: Influence of Culture Conditions." *Biotechnology and Bioengineering* 35 (7): 719–726. https://doi.org/10.1002/BIT.260350709

Harir, M, H Bendif, ... M Bellahcene - Basic biology and, and undefined 2018. "Streptomyces Secondary Metabolites." *Books.Google.Com*. Accessed January

24, 2024. https://books.google.co.in/books?hl=en&lr=&id=fi-RDwAAQBAJ&oi=fnd&pg=PA99&dq=Streptomyces,+secondary+metabolites+can+also+be+produced+during+the+exponential,+stationary,+and+death+phases+&ots=YV-P0_lwsg&sig=pxR_lQlHHyMSKTqbD_VSixpwRc4

Khushboo, Punit Kumar, Kashyap K. Dubey, Zeba Usmani, Minaxi Sharma, and Vijai Kumar Gupta. 2022. "Biotechnological and Industrial Applications of Streptomyces Metabolites." *Biofuels, Bioproducts and Biorefining* 16 (1): 244–264. https://doi.org/10.1002/BBB.2294

Kluepfel, D., F. Shareck, F. Mondou, and R. Morosoli. 1986. "Characterization of Cellulase and Xylanase Activities of Streptomyces Lividans." *Applied Microbiology and Biotechnology* 24 (3): 230–234. https://doi.org/10.1007/BF00261542

Kumar, Munendra, Prateek Kumar, Payal Das, Renu Solanki, and Monisha Khanna Kapur. 2020. "Potential Applications of Extracellular Enzymes from Streptomyces Spp. in Various Industries." *Archives of Microbiology*. Springer. https://doi.org/10.1007/s00203-020-01898-9

Lajtai-Szabó, Piroska, Éva Hülber-Beyer, Nándor Nemestóthy, and Katalin Bélafi-Bakó. 2022. "The Role of Physical Support in Secondary Metabolite Production by Streptomyces Species." *Biochemical Engineering Journal* 185 (July): 108495. https://doi.org/10.1016/J.BEJ.2022.108495

Lima Procópio, Rudi Emerson, De, Ingrid Reis Da Silva, Mayra Kassawara Martins, João Lúcio De Azevedo, and Janete Magali De Araújo. 2012. "Antibiotics Produced by Streptomyces." *Brazilian Journal of Infectious Diseases*. Brazilian Society of Infectious Diseases. https://doi.org/10.1016/j.bjid.2012.08.014

Liu, Xiao Fang, Jun Xiao Wang, Xin Ai Chen, Yu Liu, and Yong Quan Li. 2022. "Activation and Characterization of Lanthomicins A–C by Promoter Engineering in Streptomyces Chattanoogensis L10." *Frontiers in Microbiology* 13 (May): 902990. https://doi.org/10.3389/FMICB.2022.902990/BIBTEX

Manivasagan, Panchanathan, Jayachandran Venkatesan, Kannan Sivakumar, and Se Kwon Kim. 2014. "Pharmaceutically Active Secondary Metabolites of Marine Actinobacteria." *Microbiological Research* 169 (4): 262–278. https://doi.org/10.1016/J.MICRES.2013.07.014

McDonald, Bradon R., and Cameron R. Curriea. 2017. "Lateral Gene Transfer Dynamics in the Ancient Bacterial Genus Streptomyces." *MBio* 8 (3). https://doi.org/10.1128/mBio.00644-17

Pacios-Michelena, Sandra, Cristobal N. Aguilar González, Olga B. Alvarez-Perez, Raul Rodriguez-Herrera, Mónica Chávez-González, Roberto Arredondo Valdés, Juan A. Ascacio Valdés, Mayela Govea Salas, and Anna Ilyina. 2021. "Application of Streptomyces Antimicrobial Compounds for the Control of Phytopathogens." *Frontiers in Sustainable Food Systems* 5 (September): 696518. https://doi.org/10.3389/FSUFS.2021.696518/BIBTEX

Romano-Armada, Neli, María Florencia Yañez-Yazlle, Verónica P. Irazusta, Verónica B. Rajal, and Norma B. Moraga. 2020. "Potential of Bioremediation and PGP Traits in Streptomyces as Strategies for Bio-Reclamation of Salt-Affected Soils for Agriculture." *Pathogens* 9 (2): 117. https://doi.org/10.3390/PATHOGENS9020117

Saraiva, Sofia M., Sónia P. Miguel, André R.T.S. Araujo, Márcio Rodrigues, Maximiano P. Ribeiro, and Paula Coutinho. 2023. "Cosmetic Industry: Natural Secondary Metabolites for Beauty and Aging." *Natural Secondary Metabolites: From Nature, Through Science, to Industry*, February, 853–891. https://doi.org/10.1007/978-3-031-18587-8_27/COVER

Chapter 7

Application of streptomycetes in medicine

Kanchan Chatterjee

Department of Zoology, Doon (PG) College of Agriculture Science and Technology Sealqui Dehradun, Uttarakhand, India

Mohit Kumar

Department of Life Sciences, Sharda University, Greater Noida, India

Neeraj Agarwal

Noida Institute of Engineering and Technology, Greater Noida, India

Santosh Kumar Mishra and Nupur

Department of Life Sciences, Sharda University, Greater Noida, India

7.1 INTRODUCTION

Due to increased resistance toward antibiotics and continuously rising problems of infections caused by different microorganisms including bacteria, fungus and viruses, new novel antimicrobial agents are in need to be discovered. Most antimicrobial agents or components are derived from natural products out of which actinomycetes bacteria are a rich source of antimicrobial components which they produced as secondary metabolites during their growth period. *Streptomycetes* are a group of over 500 bacterial microorganism species and a member of Actinobacteria that was first discovered in 1943 by Dr. Selman Waksman. Together with his coworker Dr. Albert Schatz, they isolated the first forms of antibiotic called streptomycin from a strain of *S. griseus*. Waksman coined the term 'antibiotic' and got a patent. He won the Nobel prize for medicine in 1952 (Wainwright 1991). The discovery of streptomycin marked the beginning of explorations and research that has unravelled the huge therapeutic potentials of *Streptomyces*. They are filamentous, spore-forming and Gram-positive bacteria and are the source of many antibiotics (Waksman and Lechevalier 1949; Sarmiento-Vizcaíno et al. 2018). When the sources are sufficient, *Streptomyces* grow by throwing hyphae in search of nutrients and when the sources are limited and they produce ariel hyphae and produce spores to combat with an unfavourable environment. They can act as endophyte or epiphyte (Nalini and Prakash 2017; Zhao et al. 2020). They have cosmopolitan distribution

DOI: 10.1201/9781032694634-7

mainly found in soil but also obtained from other sources, such as manure, deep sea, high mountains, traditional medicinal plants such as *Lychnophora ericoides* and so on. *Streptomyces* isolated from the plant *Vochysia divergens* in Brazil have been found to produce Brevianamide F and cyclo-(L-Pro-L-Phe), which are the chemicals effective against methicillin-resistant *Staphylococcus aureus* MRSA (Gos et al. 2017). *Heracleum souliei* in China is a plant used in traditional medicine to derive *Streptomyces*-based antibiotics like pluramycin (Zhang et al. 2014). At higher elevations, they can be segregated from soil ecosystems such as gardens and dumpsites, where their pH ranges from slightly acidic to alkaline and temperatures range from 29 to 33 °C. Many different kinds of antibiotics are isolated from *Streptomyces* species like cyclodipeptides, 2,3-dihydro-2,2-dimethyl-4(1H)-quinazolinone, cycloheximide, actiphenol, diketopiperazine antibiotics and nocardamine (Conti et al. 2016). *Streptomyces* have links with invertebrates in traditional medicine in India as well as other parts of world. It can be found in association with invertebrates and has been used in medicine from ancient times to cure a large number of diseases. Some parts of an ant's exoskeleton are associated with the *Streptomyces* sp. and are used to relieve tooth ache, wound, boil and scabies (Chakravorty, Ghosh, and Meyer-Rochow 2011; Seabrooks and Hu 2017). African *Tetraponera penzigi* ants are associated with the production of formicamycins, a powerful antibiotic used for inhibition of MRSA and vancomycin-resistant enterococci (Qin et al. 2017). Other invertebrates such as sponges have been used in traditional medicines for a long time and are reported to have symbiotic association with antibiotic-producing *Streptomyces sp.* Recently, a novel compound, *Hexadecamethyl octasiloxane*, has been isolated from *Streptomyces* sp. that exhibits antibacterial properties against *Streptococcus agalactiae* (Abdelaziz et al. 2024). Additionally, several researchers from different locations extract *Streptomyces* sp. that are effective against both Gram-positive and Gram-negative bacteria, as well as *Mycobacterium tuberculosis* (Balasubramanian et al. 2018). Moon milk from the surfaces of caves associated with the abundance of *Streptomyces* sp. which have the potential antimicrobial activity against a variety of bacteria (Rangseekaew and Pathom-Aree 2019) including multi-resistant *Rasamsonia argillacea*, which is the potent causative agent of invasive mycosis in chronic granulomatous disease and cystic fibrosis (Maciejewska et al. 2016). Use of soil in beauty treatments is well known to everyone and is widely used from ancient times in different parts of the world. Importance of the soil depends upon the property of the soil which is due to the microorganism present in the particular soil. It is observed that soil-derived *Streptomyces* sp. are predominant in traditional medicine for example. Boho clay used in traditional therapy contained a new species *Streptomyces* sp. myrophorea, which inhibited several strains of carbapenem-resistant *Acinetobactor* sp., vancomycin-resistant *Enterococcus* and *Pseudomonas* spp. (Terra et al. 2018). The identification of the potential life-saving secrets harbored by *Streptomycetes* has resulted in more and more research studies that seek to tap more useful antibiotics from these indispensable bacteria.

7.2 HISTORY OF STREPTOMYCETES

Streptomycetes are the largest group of actinobacteria and are found in both terrestrial and aquatic environments. They are aerobic, Gram-positive, non–acid-fast, non-motile, and have a complex life cycle. They are used to produce a wide variety of antibiotics including streptomycin, which was the first specific treatment for tuberculosis, and isoniazid, which is also used to treat tuberculosis today. Known as the miracle cure of the 1940s and 1950s, streptomycin was one of the first antibiotics to be used in cases of tuberculous meningitis and was considered to be a wonder drug at the time. However, by the 1960s, more and more strains of the *Mycobacterium tuberculosis* that were resistant to the drug were being discovered and so it became common practice to use a combination of two or more drugs to treat the condition. This was the point at which the use of streptomycin in the treatment of tuberculosis began to decline, particularly in more developed countries, and today there are many strains of *Mycobacterium tuberculosis* that are completely resistant to treatment with streptomycin.

Effective use of their medical applications is so prominent that the number of drugs has sixth place among the entire number of antibiotics. They are used to produce streptomycin, neomycin, gentamicin, and many other antibiotics which are prescribed for the treatment of various infections caused by microorganisms. Penicillin was the other source of antibiotics till 1940, but when the antibiotics isolated from streptomycetes were introduced, penicillin was bitterly defeated. Today, this class of antibiotics are 76% of the total antibiotics used in the world. Streptomycetes are known for their chemical diversity that leads to therapeutic potential, ease of use and consumer acceptance in the drug market. They are known to produce a large number of antibiotic compounds. Culturing and employing them for antibiotic production on an industrial scale is easy. The use of streptomycetes and the repair or restoring of their degenerated habitats are necessary for the sustainable development of society. Streptomycetes are greatly contributing to the development of various medical industries and treatment of disease through their antibiotic properties.

7.2.1 Some of the bioactive molecule isolated from *Streptomyces* sp.

Streptomycetes are a group of bacteria that produce many natural bioactive compounds with various biological activities (Figure 7.1). They are the source of over two-thirds of the clinically useful antibiotics of natural origin, such as erythromycin, kanamycin, neomycin, streptomycin, tetracycline,

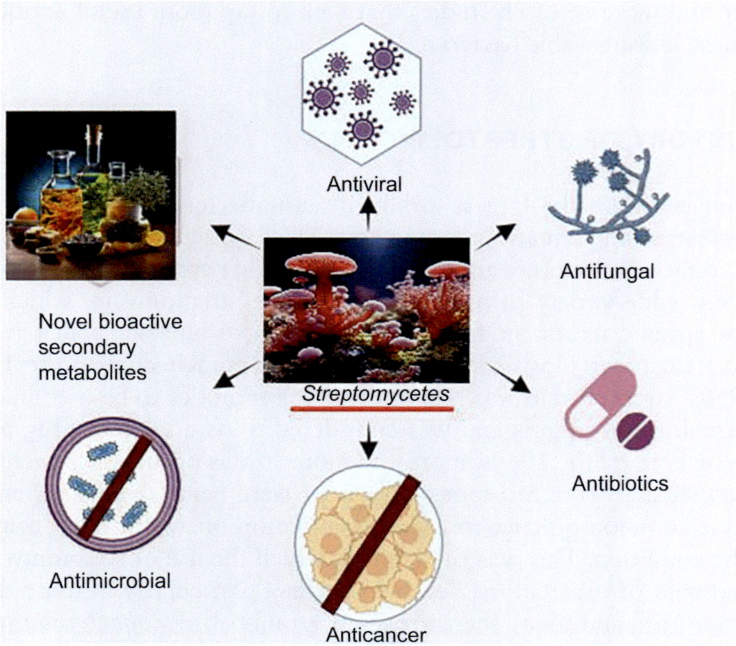

Figure 7.1 Bioactive compounds produced by streptomycetes.

vancomycin, clindamycin, lincomycin, chloramphenicol, cineromycin A+B, daptomycin (lipopeptide antibiotic) and fosfomycin (broad-spectrum antibiotic). They also produce other drugs, such as bialaphos (herbicide), bleomycin (anticancer), clavulanic acid (β-lactamase inhibitor), ivermectin (anti parasitic, anti-onchocerciasis and anti-lymphatic filariasis), nystatin (antifungal), rapamycin (antifungal, antitumour and immunosuppressive), saframycin(s) (anticancer) and so on (Figure 7.2). Streptomycetes are widely used in biotechnology and biotransformation processes to produce or modify drugs. Streptomycetes are therefore very important in medicine, as they offer a rich and diverse source of natural products with potential therapeutic applications.

7.2.2 Secondary metabolites of streptomycetes

Actinomycetes are the utmost source of different bioactive compounds which play a major role in the derivation of natural antibiotics in large proportion of those used in medicines, and the majority of these molecules are synthesized by *Streptomyces* genus. In recent years researchers applied modifications in the fermentation process to increase the production of metabolites (Latha, Sivaranjani, and Dhanasekaran 2017). Genome sequencing also plays a significant role in identifying the genes responsible

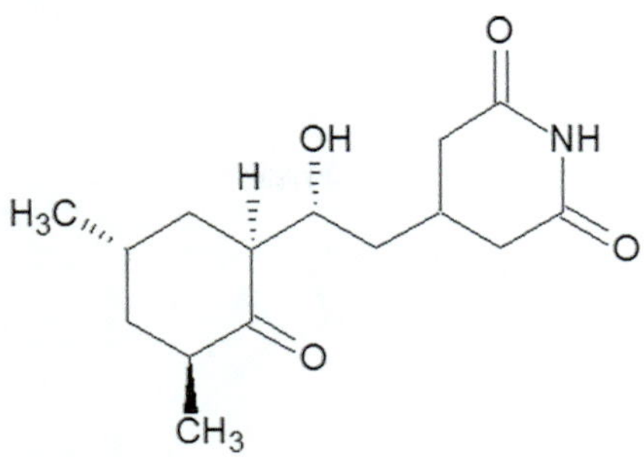

Cycloheximide

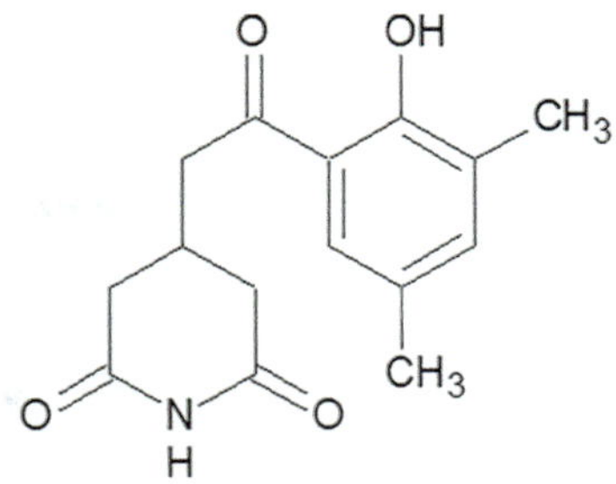

Actiphenol

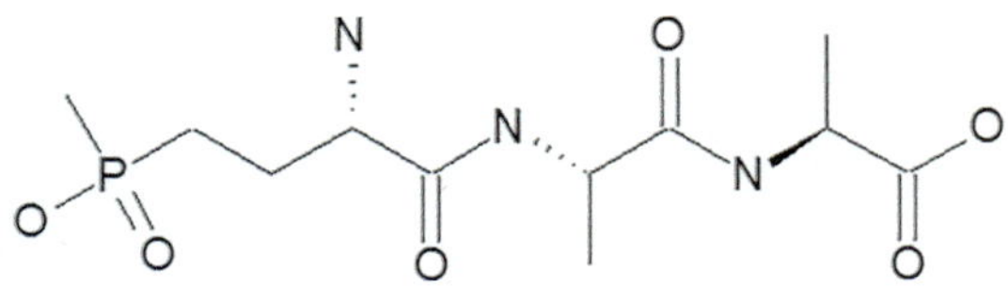

Bialaphus

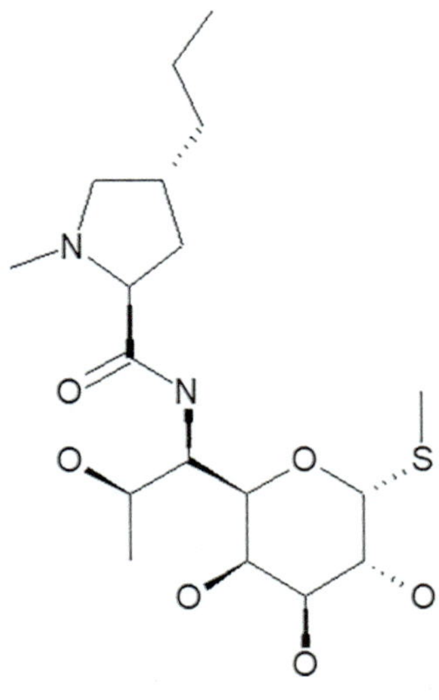

Lincomycin

Figure 7.2 Chemical structures of some compounds derived from streptomycetes.

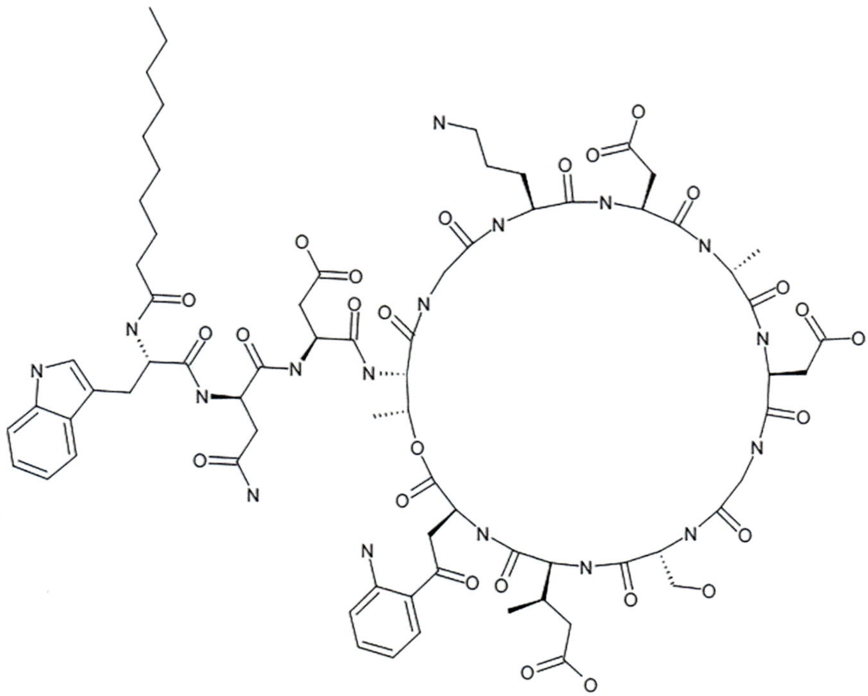

Daptomycin

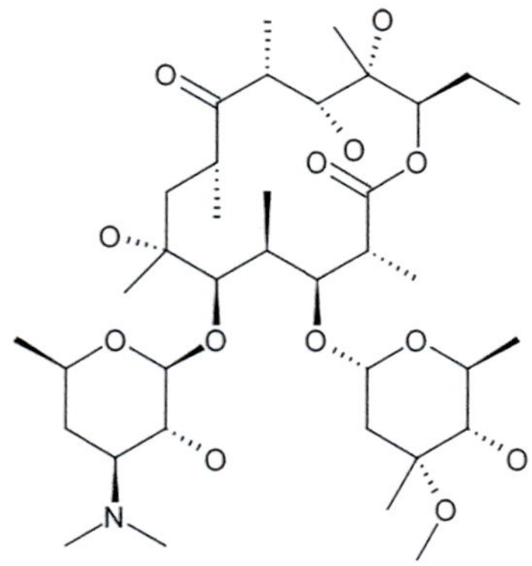

Erythromycin

Figure 7.2 (Continued)

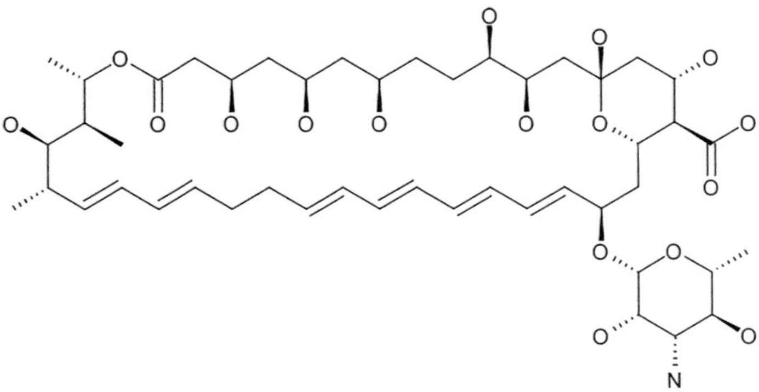

Nystatin

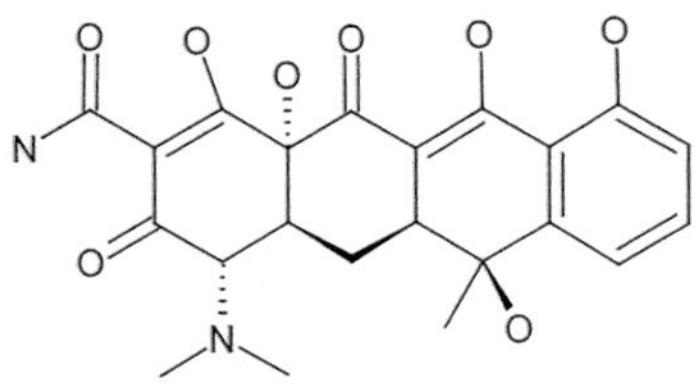

Tetracycline

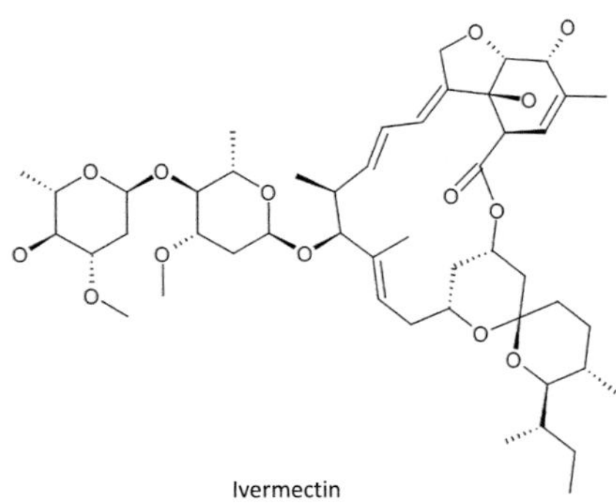

Ivermectin

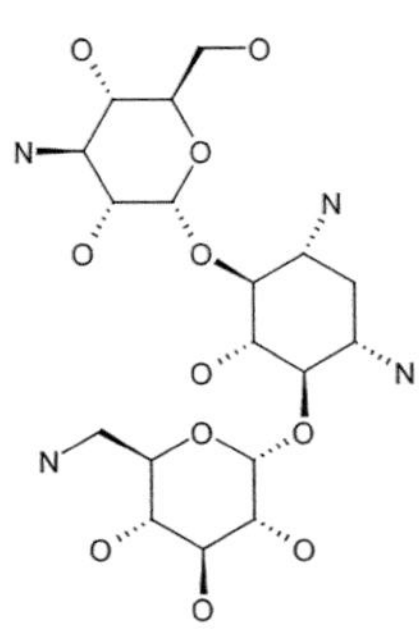

Kanamycin

Figure 7.2 (Continued)

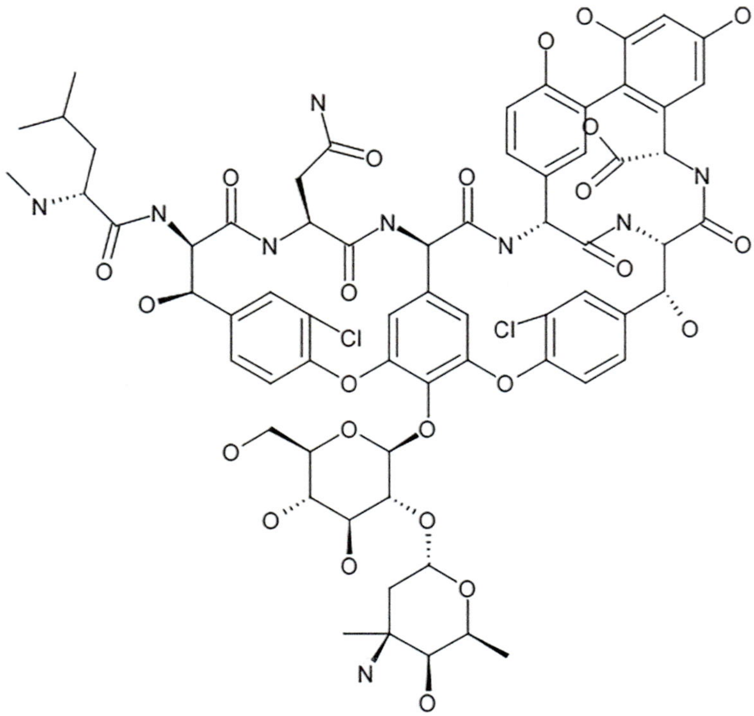

Figure 7.2 (Continued)

for the production of metabolites, such as BGCs (biosynthetic gene clusters) which encode different enzymes, transporter molecules and regulatory proteins that are responsible for the production and transportation of the metabolites (Medema and Fischbach 2015). According to WHO antibacterial resistance has reached an alarming level in different parts of the world (WHO 2014). In this context discovery of new bioactive compounds with new modes of action is a necessity to avoid unwanted circumstances.

7.3 APPLICATION OF ACTINOMYCETES IN MEDICINE

7.3.1 Antimicrobial activity

Streptomycetes are renowned for their antimicrobial activity, particularly their ability to produce antibiotics. Ever since streptomycin—the first antibiotic to treat tuberculosis—was discovered, these bacteria that inhabit the soil have played a crucial role in the field of antibiotic synthesis. They are producers of a diverse range of antibiotics that target various bacterial infections. These antimicrobial compounds inhibit essential biological processes

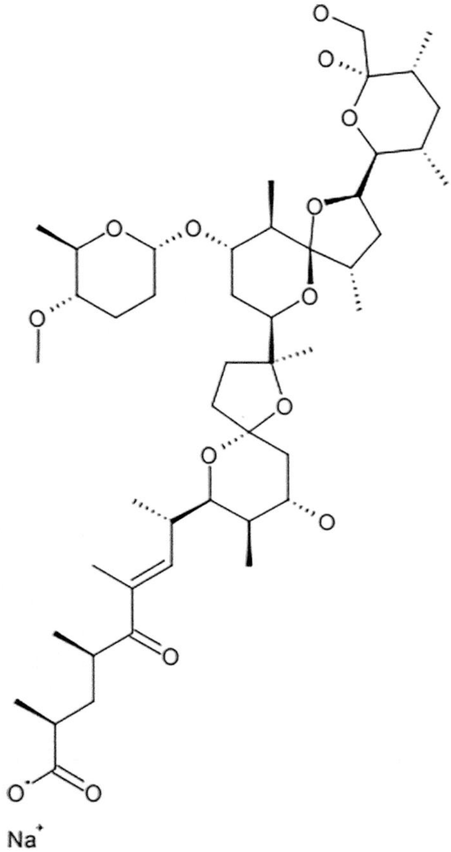

Nanchangmycin

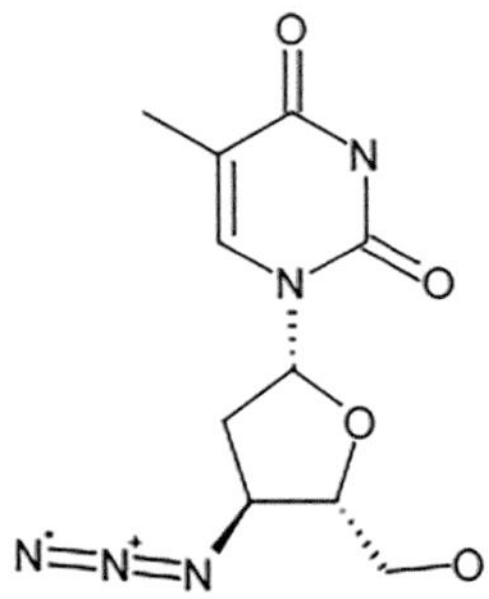

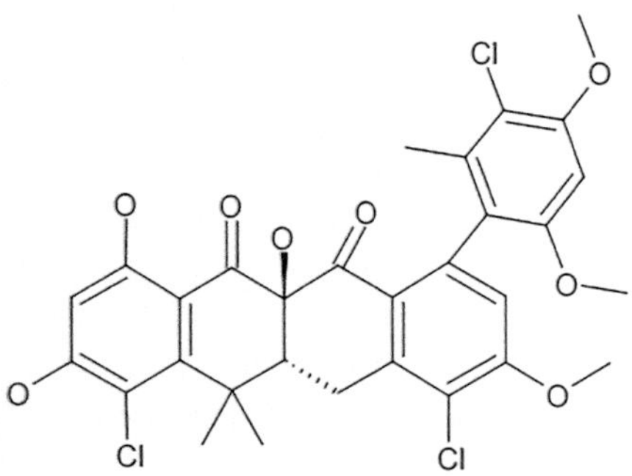

Formicamycin -E

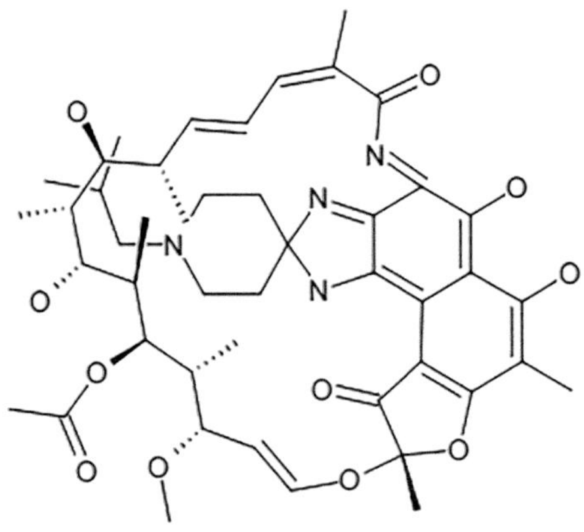

Ansamycin

Figure 7.2 (Continued)

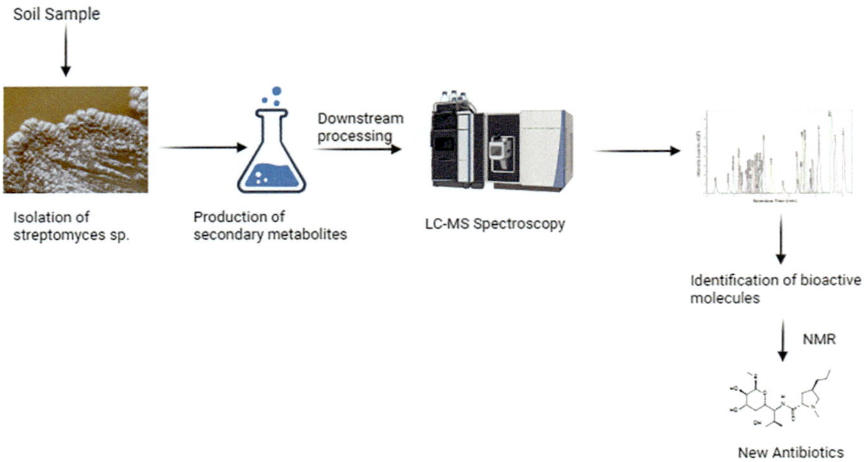

Figure 7.3 Identification of novel antibiotics from *Streptomycetes*.

in pathogenic microorganisms, such as protein synthesis, DNA replication, or cell wall formation. Hence, antimicrobial compounds produced by streptomycetes become invaluable in the treatment and prevention of infectious diseases. These bioactive compounds are analyzed and identified using LC–MS and NMR techniques (Figure 7.3).

Among novel bioactive compounds,'maklamicin', a polycyclic compound extracted from *Micromonospora* sp., showed antimicrobial activity against different Gram-positive bacteria including *Bacillus cereus*, *Staphylococcus aureus*, *Micrococcus luteus* and *Enterococcus faecalis* (Y. Igarashi et al. 2011). Another bioactive compound is 'nomimicin', isolated from the culture extract of *Actinomadura* sp., showed antimicrobial activity against *Kluyveromyces fragilis*, *C. albicans* and *M. luteus* (Y. Igarashi et al. 2012). Another three bioactive compounds are lobophorin A, B and G, which showed antimicrobial activity against *B.subtilis*, *Mycobacterium bovis* and *Mycobacterium tuberculosis*, from the lobophorin family (C. Chen et al. 2013). Ansamycin-type polyketides (chaxamycins A-D) showed the strongest activity against *Escherichia coli* and *S. aureus* (Rateb et al. 2011a). Asenjonamides A, B and C active compounds belonging to the β-diketone family of polyketides showed strong antibacterial activity (Abdelkader et al. 2018). New polyketides, for example formicamycins A-L, worked against the growth of the MRSA and vancomycin-resistant *Enterococcus faecium* (VRE) (Qin et al. 2017). Three new chaxalactins A–C belonging to macrolactone polyketides acted strongly against Gram-positive bacteria, such as *L. monocytogenes*, *S. aureus* and *B. subtilis* (Rateb et al. 2011b).

Another antibiotic, anthracimycin, displayed antibacterial activity against *H. influenzae, E. faecalis*, *Moraxella catarrhalis*, *S. pneumoniae* MRSA and vancomycin-resistant *S. aureus* (Jang et al. 2013) and also showed protection to mice from MRSA–induced mortality (Hensler et al. 2014). Compounds like amycolatopsins come under the family of polyketide macrolides. Only amycolatopsins A and C were active against *M. tuberculosis* and *M. bovis* (Khalil et al. 2017). Isoikarugamycin and 28-N-methylikarugamycin compounds coming under the class of polycyclic tetramic acid macrolactams showed antibacterial and antifungal activities (Lacret et al. 2015). Pseudonocardians A and B were active against bacterial strains as well as tumor cell lines (S. Li et al. 2011). Quinones, for example pyranonaphthoquinone compounds, displayed antibacterial activity (Song et al. 2013). Fradimycins A and B showed cyto-toxicity toward human cancer cell lines (Xin et al. 2012). Ageloline A belonging to class Quinolones showed anti-chlamydial and antioxidant effects (Cheng et al. 2016). A new pentacyclic xanthone, buanmycin, showed activity against both gGam-positive and Gram-negative bacteria as well as inhibited the *S. aureus* sortase A, an enzyme responsible for the adhesion and invasion of Gram-positive bacteria (Moon et al. 2015). Four novel xanthones (citreamicin θ A, citreamicin θ B, citreaglycon A and dehydrocitreaglycon) compounds displayed activity toward *S. haemolyticus*, *S. aureus* and *B. subtilis* and showed cytotoxicity against HeLa cells (Liu et al. 2012).

Some meroterpenoids showed antibacterial activity as well as cytotoxicity against human cancer cells (Wu et al. 2013). Actinomadurol, which is a secondary metabolite, showed major inhibitory activity against many bacterial strains such as *K. rhizophila* NBRC 12708, *S. aureus* ATCC 6538p, *B. subtilis* ATCC 6633, *P. hauseri* NBRC 3851 and *S. enterica* ATCC 14028 (Shin et al. 2016). A new thiazolyl peptide 'kocurin' inhibits the bacterial growth by blocking protein biosynthesis at the translation stage (Mahajan et al. 2013). New cyclic depsipeptides, ohmyungsamycins A and B showed antibacterial activites. On comparing these two components, ohmyungsamycins A was more potent than ohmyungsamycins B in antibacterial and cytotoxic activities (Um et al. 2013). The two lipopeptides from *Streptomyces spp.* potentially showed activity against *Staphylococcus epidermidis* HCCB20256 (Rao et al. 2013).

7.3.2 Antifungal activity

Humans can get infected by microorganisms such as filamentous fungus and yeasts. These diseases can be anything from skin infections to potentially fatal systemic illnesses. For instance, vaginal infections are frequently caused by the yeast *Candida albicans*. Those receiving chemotherapy or other compromised immune system patients may experience far more severe illnesses from the same pathogens.

Aspergillus fumigatus, a fungus that can cause aspergillosis, is another example of significant medical importance. The total annual death toll from fungal lung infections is comparable to that from tuberculosis, which is why developing novel antifungal medications is of great importance. Thankfully, some streptomycetes have the ability to create antifungal chemicals, which may help treat fungal infections. Nystatin is one instance of an antifungal substance derived from streptomycetes that works effectively against the yeast *Candida*. This substance has been used in medicine for a long time and is particularly significant since, when taken orally, it is not well absorbed from the gut. This allows it to function specifically within the digestive system to lessen disturbance to the regular flora of the body.

Interestingly, some streptomycetes that have been isolated from the environment have now been used in genetic engineering to produce large amounts of the antifungal compound griseofulvin. This is an example of the huge potential of genetic engineering techniques to allow increased yields of valuable compounds from streptomycetes.

Sceliphrolactam is a potent secondary metabolite isolated from *Streptomyces* spp. which was associated with the mud dauber from Wisconsin (United States). It showed potent antifungal activity against amphotericin B-resistant *C. albicans* (Oh et al. 2011). Another component, 15-glycidylfilipin III, demonstrated potent antifungal activity against *C. albicans* (Yang et al. 2016). Other compounds obtained from soil sample in Nanchang, Jiangxi, People's Republic of China were Enduspeptides A, B and C that showed strong antifungal activity against *C. glabrata* ATCC 90030 (Hu et al. 2017). At Nerima-ku, Tokyo, Japan, a strong metabolite, kribellosides A–D, isolated from soil samples showed antifungal activity against *S. cerevisiae* (M. Igarashi et al. 2017).

The identification of additional streptomycetes compounds will increase the quantity of these kinds of promising antifungal medications, which will lead to further therapeutic options. It is now feasible to fully analyze the genetic information found in streptomycetes to guide the creation of novel antifungal drugs through genetic engineering, thanks to DNA databases and advanced bioinformatics approaches. These collectively comprise the fascinating field of genomics, which is the study of organisms' genetic makeup as it relates to an organism's whole gene pool. It is exciting to note that the genes for a wide range of potentially helpful substances from streptomycetes are already being identified; it will not be long before some of these are turned into new drugs.

7.3.3 Potential applications in treating fungal infections

Although only a small portion of people worldwide suffer from fungal infections, the number of cases is rising quickly. Every year, over 1.5 million people worldwide pass away from severe systemic fungal infections, and there are 150 million cases of serious fungal infections and an estimated 1.7 billion cases of cutaneous fungal infections (Kainz et al. 2020). Antifungal drugs have demonstrated remarkable efficacy in treating many fungal infections, such as those resulting from *Candida* sp. (da Rosa Coelho Martins et al. 2023). The wide variety of antifungal chemicals that filamentous bacteria produce is astounding. Secondly, the majority of antifungal drugs utilized in modern medicine come from synthetic or fungal sources. However, the antifungal agents produced by *Streptomyces* show new promise in the development of potent and safer alternatives (Bobek et al. 2022). The identification of *Streptomyces* as a putative source of antifungals has opened up new avenues for medical research and a number of studies have been carried out to uncover the genes for antifungal compounds and investigate the mechanisms of the antifungal actions (Pacios-Michelena et al. 2021). Besides providing hope to the patients who are suffering from fungal diseases, the application of *Streptomyces* in producing antifungal agents offers significant benefits over current antifungal therapies as well. These agents exhibit varying degrees of antifungal activity and continue to be effective against various fungus species. *Streptomyces'* antifungal chemicals can specifically target the structural or metabolic pathways of fungi, reducing the likelihood of medication resistance developing. Conversely, fungal-produced antifungal drugs, such as Amphotericin B, prevent fungal growth by attaching themselves to the sterols found in fungal cell membranes, but at the same time, the same antifungal process inhibits the functioning of cholesterol found in human cell membranes causing serious side effects, thereby limiting the use of amphotericin B in therapeutic therapies. Therefore, the discovery and the future development of antifungal agents from *Streptomyces* can be considered as a breakthrough in this field of medical science (Oliveira et al. 2023).

7.3.4 Antiviral activity

The scientific community has conducted a great deal of study and analysis on the bacteria genus *Streptomyces*. Numerous thousand distinct compounds have been found and isolated thus far, demonstrating the genus' reputation for producing a wide variety of bioactive chemicals. In fact, more than two-thirds of all naturally occurring antibiotics used in clinical settings worldwide come from these microorganisms. But it's becoming more and more clear that *Streptomyces* species have additional positive bioactivity: they may be a source of antiviral chemicals. Compared to their antibacterial qualities, *Streptomyces'* antiviral activity has received less research attention; yet, there is evidence that they may have usefulness in this field. With the wide range of chemical diversity that *Streptomyces* produces in its secondary metabolites, it's possible that many more molecules with antiviral qualities are still to be found. The growth of viral diseases and the growing problem of viral resistance to current therapies call for the urgent development of novel antiviral medications, which research in this field may provide promising leads for. But it's crucial to remember that this is a new field, and more research is required to completely comprehend and utilize *Streptomyce's* antiviral potential.

The following *Streptomyces* strains have shown antiviral activity against various viruses; for instance, *Streptomyces jiujiangensis* NBERC-24992 extract exhibited significant antiviral activity against pseudorabies virus

(PRV) (Zulkifli et al. 2023). *Streptomyces* sp. zx10-19 (KH29) extract inhibited the replication of enterovirus 71 (EV71) (Liu et al. 2023). *Streptomyces* sp. strains IBSBF 2019 and IBSBF 2397 demonstrated antibacterial activity against *Staphylococcus aureus, Bacillus cereus, Pseudomonas aeruginosa*, and *Escherichia coli* (Jeong, Kim, and Lim 2022). Li et al. discusses the antiviral activity of a butanolide compound derived from *Streptomyces* sp. SMU03, which showed potent activity against various influenza viruses, including H1N1 and H3N2 subtypes (F. Li et al. 2018). Feces-derived *Streptomyces* sp. HTL16 produced bafilomycins with potent antiviral activity against influenza A and SARS-CoV-2 viruses with nanomolar range IC50 values by inhibiting endosomal ATP–driven proton pumps (Xie et al. 2021). Secondary metabolites antimycin A1a obtained from *Streptomyces* sp. isolated from coast of New Ireland showed powerful antiviral activity against Western equine encephalitis virus (Raveh et al. 2013). New compounds were named carbazole-bearing indolosesquiterpenoids from the topsoil of a Korean solar saltern. Among these compounds, xiamycin D showed the strongest antifungal activity (Kim et al. 2016). HIV-1 protease inhibition was showed by *N,N,N*-(trimethylated)-Tyr-L-Leu-L-Val-L-Leu-(dehydrated)-His. HIV-1 protease is essential for the life cycle of HIV (Liu et al. 2012). Ahmpatinin Bu, a linear peptide, also presents significant inhibitory activity against HIV-1 protease (M. H. Chen et al. 2018). Other compounds like daptomycin and nanchangmycin significantly displayed their antiviral activity against ZIKV (da Silva et al. 2018). The potential of secondary metabolites from *Streptomyces* have the ability to inhibit the multiplication of viruses.

A continuous and close collaboration between the virologist, microbiologist, bioinformatician, immunologist, structural biologist and medicinal chemist must be enhanced in order to develop new antiviral agents at the molecular level. In the genomics era, genomics and the latest bioinformatics technologies have been used to identify the gene cluster vs gene in the synthesis of antiviral agents, which brings factual evidence on the antiviral activity of those bioactive compounds mentioned in the present review. Moreover, the latest software and databases like Auto-Dock Vina, Swiss target prediction, and protein data bank also provide a platform for the molecular docking study, which is one of the parts of a rational drug design program. A better understanding of the interaction between the protein and bioactive compounds can be achieved. It is expected that more secondary metabolites with antiviral activity can be discovered using the aforementioned advanced technologies.

7.3.5 Anticancer activity

Resistance to programmed cell death is the main cause of cancer-related mortality. Antiapoptotic protein overexpression is the primary reason of cancer cells' frequent development of resistance to a variety of chemotherapeutic treatments. Cancer therapy remains one of the biggest medical challenges, despite breakthroughs in our understanding of the biology of the disease and the creation of more effective treatment alternatives. The possible anticancer activities of *Streptomycetes* have also attracted interest. Some of the secondary metabolites that these microbes are able to produce have been shown to have anticancer properties. This is caused by a number of different mechanisms of action, including the suppression of the growth of cancerous cells, the induction of programmed cell death, the disruption of the cell cycle, and the suppression of angiogenesis (the development of new blood vessels that provide oxygen to tumors). Well-known examples of anticancer medications made from *Streptomyces* include bleomycin, doxorubicin and daunorubicin. Similarly, *Streptomyces parvullus* provided the source of dactinomycin, a well-known anticancer drug that was first identified in 1950. Dactinomycin has been applied to treat numerous types of cancer. *Streptomyces peucetius* was the source of doxorubicin, sometimes referred to as Adriamycin, another anticancer drug, which was initially identified in the early 1960s. It is still often used in clinical settings today to treat a variety of cancer types, such as acute leukemia, lymphomas and other solid tumors.

From mangrove, *Streptomyces* strain 124092, five compounds were isolated out of which four compounds displayed cytotoxic activity against the human hepatoma SMMC-7721 cell line (Qiu et al. 2022). Streptocarbazole A showed potent cytotoxic effect against human cancer cells including leukaemia HL-60 cells and HeLa cells while Streptocarbazole B exhibited cytotoxic effects against HeLa cells only (Fu et al. 2012). Halichoblelide D from mangrove strain *Streptomyces* sp. exhibited significant cytotoxicity against HeLa cells (Han et al. 2016). Antimycins derived from streptomycetes of various origin are used in chemotherapeutic medicines. Their anticancer property was on lung cancer, breast cancer and glioblastoma cells (Hu et al. 2017). Streptomycetes species are a potential source of bioactive metabolites which contribute largely to the medical field. Studies have shown that there are number of streptomycetes and natural products that remain undiscovered. For isolation and production of a novel compound from *Streptomyces*, profound knowledge of microbial physiology and metabolism is necessary. Continued research into *Streptomyces* species and their metabolic products may provide a rich source of novel anticancer agents that could be effective against various types of cancer. Moreover, other species of *Streptomyces* should be isolated from the marine environment for their bioactivity and new drug discoveries. By harnessing the potential of novel technological approaches, such as genomics, transcriptomics, proteomics and metabolomics, and through the establishment of antibacterial and anticancer high-throughput screening (HTS) assays using various model systems, the discovery of bioactive compounds from *Streptomyces* will continue to flourish well into the foreseeable future. As with antibiotics and other pharmaceuticals derived from *Streptomyces*, there is significant interest in further investigating and harnessing these properties for medical therapeutics.

The study has made a substantial contribution to our understanding of the mechanisms underlying the bioactive compounds created by *Streptomyces* that demonstrate deadly effects on multiple human cancer cell lines by using a number of state-of-the-art biological and biochemical methodologies. The findings suggest that the secondary metabolites produced by the *Streptomyces* strains may be used to develop novel anticancer drugs. Additionally, the research has opened up new directions for understanding the molecular mechanisms underlying *Streptomyces'* bioactive compounds' ability to inhibit the growth of cancer cells. More innovative cancer treatments derived from natural sources, such as *Streptomyces*, are predicted to be discovered and developed as research techniques advance. Patients with cancer who are unable to tolerate the adverse side effects of the current chemotherapy medications might find hope in these treatments. In recent years, many chemo preventive agents have been isolated from microorganisms, marine organisms, higher plants and other natural sources. These agents inhibit the process of carcinogenesis at different stages and help to delay the development of invasive cancer. As shown in this study, many of the bioactive compounds that have been isolated from *Streptomyces* have already been employed successfully as chemo preventive agents for various types of cancer. This not only underpins the importance of the discovery of *Streptomyces*-derived novel anticancer drugs, but also reinforces the future prospects of using *Streptomyces* in cancer research and treatment. On the other hand, our current knowledge of the medicinal properties of *Streptomyces* is just the tip of the iceberg. With the continuous advancement in microbiology, molecular biology and biotechnology, many more novel bioactive compounds will be discovered and more genetic mechanisms that control the production of these compounds will be unraveled. It is envisaged that the genus *Streptomyces* will continue to contribute immensely to the field of medicinal biochemistry and to the pharmaceutical industry. One could then imagine an increasingly important role of the *Streptomyces*-derived natural products in drug discovery and development and perhaps a substantial improvement in the general wellness of human beings through less detrimental chemotherapy in the near future.

7.4 FUTURE PROSPECTS: EXPLORATION OF NOVEL COMPOUNDS

Scientists and researchers have shifted their attention to studying and analyzing the *Streptomyces* genome sequence in an effort to find previously unidentified genetic pathways that may facilitate the synthesis of novel bioactive chemicals. It has been demonstrated that the gene manipulation approach is useful for finding novel bioactive substances. With this technique, the parent strain's DNA is often randomly mutated using a variety of mutagenesis agents, such as UV radiation or specific chemicals. After that, the modified strains are screened to find the ones that are overproducing and may be able to create more novel or known antibiotics. Finally, the novel strains' secondary metabolites are isolated and put through bioassay, which enables the discovery of possible new medications. What is groundbreaking in this method is that most of the mutated sites will be genetically mapped and the specific gene cluster responsible for the production of the metabolites of interest can be identified by using modern molecular genetics and bioinformatics techniques. Following the identification of the gene clusters, scientists can focus on learning more about the roles played by individual genes and how they interact to produce the final products. By doing this, novel compounds can be created through transformation with changed DNA fragments carrying the desired genes, or by employing genetic engineering tools to change the precursor availability in the new strains and introduce new pathways. To guarantee its safety and efficacy, the newly discovered substance must pass a routine laboratory infection test, even if its pure form would not be utilized as a new medication. Due to the exponential rise in antibiotic resistance in pathogens, streptomycetes are able to stay at the forefront of antibiotic research and can be utilized to continuously identify new bioactive chemicals. As these techniques are used, there is increasing excitement about the possibilities for new discoveries. The potential for creating new bioactive substances is endless for scientists due to the combination of closely examining sequenced DNA through bioinformatics inspection and attempting gene manipulation. New researches and discoveries in the discovery of the interesting bioactive materials from the genome sequences of *Streptomyces* are helping to shape novel antibiotics that can potentially be used in human and animal medicine in the future.

7.5 CONCLUSION

Streptomyces are soil-dwelling bacteria that have been recognized for their significant medical contributions, particularly in the area of antibiotic synthesis. Its use for therapeutic purposes outside of antibiotic synthesis has been investigated more recently. There are now more therapy options available since it has been observed that its secondary metabolites can be utilized to create antiviral, antifungal and anticancer medications. Researchers are investigating the biotechnological innovations that may harness the potential of *Streptomyces* to facilitate drug development and pharmaceutical innovation. Advances in the understanding of protein-bioactive chemical interactions and the fields of genomics and bioinformatics have made it easier to target gene clusters that produce therapeutic compounds in the form of secondary metabolites. This presents opportunities for novel therapeutic approaches to address a range of illnesses and diseases. Further advancements in the use of *Streptomyces* for human health benefits are expected as our knowledge of these microbes deepens and biotechnology methods continue to progress. It is anticipated that the discovery of such novel bioactive compounds will include better human health and fewer chemotherapy side effects. There are exciting days ahead for the continued investigation of *Streptomyces'* potential in medicines and pharmaceuticals.

REFERENCES

Abdelaziz, Rewan, Hassnaa Mahmoud Elsheshtawy, Walaa El-Houseiny, Abeer S. Aloufi, Khairiah Mubarak Alwutayd, Abdallah Tageldein Mansour, Ghada Hadad, Ahmed H. Arisha, Abd Elhakeem El-Murr, and Amany M. Yassin. 2024. "A Novel Metabolite of Streptomyces Coeruleorubidus Exhibits Antibacterial Activity against Streptococcus Agalactiae through Modulation of Physiological Performance, Inflammatory Cytokines, Apoptosis, and Oxidative Stress-Correlated Gene Expressions in Nile Tilapia (Oreochromis Niloticus)." *Fish & Shellfish Immunology* 148 (May). Academic Press: 109496. doi:10.1016/J.FSI.2024.109496

Abdelkader, Mohamed S.A., Thomas Philippon, Juan A. Asenjo, Alan T. Bull, Michael Goodfellow, Rainer Ebel, Marcel Jaspars, and Mostafa E. Rateb. 2018. "Asenjonamides A-C, Antibacterial Metabolites Isolated from Streptomyces Asenjonii Strain KNN 42.f from an Extreme-Hyper Arid Atacama Desert Soil." *Journal of Antibiotics* 71 (4). Nature Publishing Group: 425–431. doi:10.1038/s41429-017-0012-0

Balasubramanian, Srikkanth, Joseph Skaf, Ulrike Holzgrabe, Richa Bharti, Konrad U. Förstner, Wilma Ziebuhr, Ute H. Humeida, Usama R. Abdelmohsen, and Tobias A. Oelschlaeger. 2018. "A New Bioactive Compound from the Marine Sponge-Derived Streptomyces Sp. SBT348 Inhibits Staphylococcal Growth and Biofilm Formation." *Frontiers in Microbiology* 9 (JUL). Frontiers Media S.A.: 376846. doi:10.3389/fmicb.2018.01473

Bobek, Jan, Eliška Filipová, Natalie Bergman, Matouš Čihák, Miroslav Petříček, Ana Catalina Lara, Vaclav Kristufek, et al. 2022. "Polyenic Antibiotics and Other Antifungal Compounds Produced by Hemolytic Streptomyces Species." *International Journal of Molecular Sciences* 23 (23). MDPI: 15045. doi:10.3390/ijms232315045

Chakravorty, Jharna, Sampat Ghosh, and Victor B. Meyer-Rochow. 2011. "Practices of Entomophagy and Entomotherapy by Members of the Nyishi and Galo Tribes, Two Ethnic Groups of the State of Arunachal Pradesh (North-East India)." *Journal of Ethnobiology and Ethnomedicine* 7 (January). J Ethnobiol Ethnomed. doi:10.1186/1746-4269-7-5

Chen, Caixia, Jian Wang, Hui Guo, Weiyuan Hou, Na Yang, Biao Ren, Mei Liu, et al. 2013. "Three Antimycobacterial Metabolites Identified from a Marine-Derived Streptomyces Sp. MS100061." *Applied Microbiology and Biotechnology* 97 (9). Appl Microbiol Biotechnol: 3885–92. doi:10.1007/s00253-012-4681-0

Chen, Ming Hua, Shan Shan Chang, Biao Dong, Li Yan Yu, Ye Xiang Wu, Ren Zhong Wang, Wei Jiang, Zeng Ping Gao, and Shu Yi Si. 2018. "Ahmpatinin IBu, a New

HIV-1 Protease Inhibitor, from Streptomyces Sp. CPCC 202950†." *RSC Advances* 8 (10). The Royal Society of Chemistry: 5138–44. doi:10.1039/c7ra13241g

Cheng, Cheng, Eman M. Othman, Anastasija Reimer, Matthias Grüne, Vera Kozjak-Pavlovic, Helga Stopper, Ute Hentschel, and Usama R. Abdelmohsen. 2016. "Ageloline A, New Antioxidant and Antichlamydial Quinolone from the Marine Sponge-Derived Bacterium Streptomyces Sp. SBT345." *Tetrahedron Letters* 57 (25). Pergamon: 2786–89. doi:10.1016/j.tetlet.2016.05.042

Conti, Raphael, Fernanda Oliveira Chagas, Andrés Mauricio Caraballo-Rodriguez, Weilan Gomes da Paixão Melo, Andréa Mendes do Nascimento, Bruno Coêlho Cavalcanti, Manoel Odorico de Moraes, et al. 2016. "Endophytic Actinobacteria from the Brazilian Medicinal Plant Lychnophora Ericoides Mart. and the Biological Potential of Their Secondary Metabolites." *Chemistry and Biodiversity* 13 (6). Chem Biodivers: 727–36. doi:10.1002/cbdv.201500225

Fu, Peng, Chunli Yang, Yi Wang, Peipei Liu, Yiming Ma, Lei Xu, Mingbo Su, Kui Hong, and Weiming Zhu. 2012. "Streptocarbazoles A and B, Two Novel Indolocarbazoles from the Marine-Derived Actinomycete Strain Streptomyces Sp. FMA." *Organic Letters* 14 (9). Org Lett: 2422–25. doi:10.1021/ol3008638

Gos, Francielly M.W.R., Daiani C. Savi, Khaled A. Shaaban, Jon S. Thorson, Rodrigo Aluizio, Yvelise M. Possiede, Jürgen Rohr, and Chirlei Glienke. 2017. "Antibacterial Activity of Endophytic Actinomycetes Isolated from the Medicinal Plant Vochysia Divergens (Pantanal, Brazil)." *Frontiers in Microbiology* 8 (SEP). Front Microbiol. doi:10.3389/fmicb.2017.01642

Han, Ying, Erli Tian, Dongbo Xu, Min Ma, Zixin Deng, and Kui Hong. 2016. "Halichoblelide D, a New Elaiophylin Derivative with Potent Cytotoxic Activity from Mangrove-Derived Streptomyces Sp. 219807." *Molecules* 21 (8). Molecules. doi:10.3390/molecules21080970

Hensler, Mary E., Kyoung Hwa Jang, Wdee Thienphrapa, Lisa Vuong, Dan N. Tran, Evaristus Soubih, Leo Lin, et al. 2014. "Anthracimycin Activity against Contemporary Methicillin-Resistant Staphylococcus Aureus." *Journal of Antibiotics* 67 (8). Nature Publishing Group: 549–53. doi:10.1038/ja.2014.36

Hu, Chen, Shi Wen Zhou, Fang Chen, Xin Heng Zheng, Hui Fang Shen, Bi Run Lin, and Guang Xiong Zhou. 2017. "Neoantimycins A and B, Two Unusual Benzamido Nine-Membered Dilactones from Marine-Derived Streptomyces Antibioticus H12-15." *Molecules* 22 (4). Molecules. doi:10.3390/molecules22040557

Igarashi, Masayuki, Ryuichi Sawa, Manabu Yamasaki, Chigusa Hayashi, Maya Umekita, Masaki Hatano, Toshinobu Fujiwara, Kiyohisa Mizumoto, and Akio Nomoto. 2017. "Kribellosides, Novel RNA 5′-Triphosphatase Inhibitors from the Rare Actinomycete Kribbella Sp. MI481-42F6." *Journal of Antibiotics* 70 (5). J Antibiot (Tokyo): 582–89. doi:10.1038/ja.2016.161

Igarashi, Yasuhiro, Takako Iida, Naoya Oku, Hiroyuki Watanabe, Kazuo Furihata, and Koji Miyanouchi. 2012. "Nomimicin, a New Spirotetronate-Class Polyketide from an Actinomycete of the Genus Actinomadura." *Journal of Antibiotics* 65 (7). Nature Publishing Group: 355–59. doi:10.1038/ja.2012.30

Igarashi, Yasuhiro, Hiromu Ogura, Kazuo Furihata, Naoya Oku, Chantra Indananda, and Arinthip Thamchaipenet. 2011. "Maklamicin, an Antibacterial Polyketide from an Endophytic Micromonospora Sp." *Journal of Natural Products* 74 (4). American Chemical Society and American Society of Pharmacognosy: 670–74. doi:10.1021/np100727h

Jang, Kyoung Hwa, Sang Jip Nam, Jeffrey B. Locke, Christopher A. Kauffman, Deanna S. Beatty, Lauren A. Paul, and William Fenical. 2013. "Anthracimycin, a Potent Anthrax Antibiotic from a Marine-Derived Actinomycete." *Angewandte Chemie - International Edition* 52 (30). Angew Chem Int Ed Engl: 7822–24. doi:10.1002/anie.201302749

Jeong, Bo Young, Hong Ki Kim, and Byung Kwan Lim. 2022. "Streptomyces Extract Inhibits Enterovirus 71 Replication by Activation of PKB/AKT Signaling." *Journal of Bacteriology and Virology* 52 (3). The Korean Society for Mocrobiology / The Korean Society of Virology: 94–102. doi:10.4167/JBV.2022.52.3.094

Kainz, Katharina, Maria A. Bauer, Frank Madeo, and Didac Carmona-Gutierrez. 2020. "Fungal Infections in Humans: The Silent Crisis." *Microbial Cell* 7 (6). Shared Science Publishers: 143. doi:10.15698/MIC2020.06.718

Khalil, Zeinab G., Angela A. Salim, Daniel Vuong, Andrew Crombie, Ernest Lacey, Antje Blumenthal, and Robert J. Capon. 2017. "Amycolatopsins A-C: Antimycobacterial Glycosylated Polyketide Macrolides from the Australian Soil Amycolatopsis Sp. MST-108494." *The Journal of Antibiotics* 70 (12). J Antibiot (Tokyo): 1097–1103. doi:10.1038/JA.2017.119

Kim, Seong Hwan, Thi Kim Quy Ha, Won Keun Oh, Jongheon Shin, and Dong Chan Oh. 2016. "Antiviral Indolosesquiterpenoid Xiamycins C-E from a Halophilic Actinomycete." *Journal of Natural Products* 79 (1). J Nat Prod: 51–58. doi:10.1021/ACS.JNATPROD.5B00634

Lacret, Rodney, Daniel Oves-Costales, Cristina Gómez, Caridad Díaz, Mercedes De La Cruz, Ignacio Pérez-Victoria, Francisca Vicente, Olga Genilloud, and Fernando Reyes. 2015. "New Ikarugamycin Derivatives with Antifungal and Antibacterial Properties from Streptomyces Zhaozhouensis." *Marine Drugs* 13 (1). Multidisciplinary Digital Publishing Institute (MDPI): 128. doi:10.3390/MD13010128

Latha, Selvanathan, Govindhan Sivaranjani, and Dharumadurai Dhanasekaran. 2017. "Response Surface Methodology: A Non-Conventional Statistical Tool to Maximize the Throughput of Streptomyces Species Biomass and Their Bioactive Metabolites." *Critical Reviews in Microbiology* 43 (5). Crit Rev Microbiol: 567–82. doi:10.1080/1040841X.2016.1271308

Li, Fangfang, Daiwei Chen, Shengsheng Lu, Guang Yang, Xiaoling Zhang, Zhao Chen, Sheng Fan, Shaohua Wu, and Jian He. 2018. "Anti-Influenza A Viral Butenolide from Streptomyces Sp. Smu03 Inhabiting the Intestine of Elephas Maximus." *Viruses* 10 (7). Viruses. doi:10.3390/V10070356

Li, Sumei, Xinpeng Tian, Siwen Niu, Wenjun Zhang, Yuchan Chen, Haibo Zhang, Xianwen Yang, et al. 2011. "Pseudonocardians A–C, New Diazaanthraquinone Derivatives from a Deap-Sea Actinomycete Pseudonocardia Sp. SCSIO 01299." *Marine Drugs 2011, Vol. 9, Pages 1428-1439* 9 (8). Molecular Diversity Preservation International: 1428–39. doi:10.3390/MD9081428

Liu, Xiao, Maoluo Gan, Biao Dong, Tian Zhang, Yongzhen Li, Yuqin Zhang, Xiuyong Fan, et al. 2012. "4862F, a New Inhibitor of HIV-1 Protease, from the Culture of Streptomyces I03A-04862." *Molecules 2013, Vol. 18, Pages 236-243* 18 (1). Multidisciplinary Digital Publishing Institute: 236–43. doi:10.3390/MOLECULES18010236

Maciejewska, Marta, Delphine Adam, Loïc Martinet, Aymeric Naômé, Magdalena Całusińska, Philippe Delfosse, Monique Carnol, et al. 2016. "A Phenotypic and Genotypic Analysis of the Antimicrobial Potential of Cultivable Streptomyces

Isolated from Cave Moonmilk Deposits." Frontiers in Microbiology 7 (SEP). Front Microbiol. doi:10.3389/FMICB.2016.01455

Mahajan, Girish, Becky Thomas, Rajashri Parab, Zarine E. Patel, Sandip Kuldharan, Vijayaphanikumar Yemparala, Prabhu Dutt Mishra, et al. 2013. "In Vitro and In Vivo Activities of Antibiotic PM181104." *Antimicrobial Agents and Chemotherapy* 57 (11). American Society for Microbiology1752 N St., N.W., Washington, DC: 5315–19. doi:10.1128/AAC.01059-13/ASSET/E0BBF2F3-201B-4EE7-82F1-005174F63958/ASSETS/GRAPHIC/ZAC0111322650005.JPEG

Medema, Marnix H., and Michael A. Fischbach. 2015. "Computational Approaches to Natural Product Discovery." *Nature Chemical Biology* 11 (9). Nat Chem Biol: 639–48. doi:10.1038/NCHEMBIO.1884

Moon, Kyuho, Beomkoo Chung, Yoonho Shin, Arnold L. Rheingold, Curtis E. Moore, Sung Jean Park, Sunghyouk Park, et al. 2015. "Pentacyclic Antibiotics from a Tidal Mud Flat-Derived Actinomycete." *Journal of Natural Products* 78 (3). J Nat Prod: 524–29. doi:10.1021/NP500736B

Nalini, M. S., and H. S. Prakash. 2017. "Diversity and Bioprospecting of Actinomycete Endophytes from the Medicinal Plants." *Letters in Applied Microbiology* 64 (4). Lett Appl Microbiol: 261–70. doi:10.1111/LAM.12718

Oh, Dong Chan, Michael Poulsen, Cameron R. Currie, and Jon Clardy. 2011. "Sceliphrolactam, a Polyene Macrocyclic Lactam from a Wasp-Associated Streptomyces Sp." *Organic Letters* 13 (4). American Chemical Society: 752–55. doi:10.1021/OL102991D/SUPPL_FILE/OL102991D_SI_001.PDF

Oliveira, Manuela, Diana Oliveira, Carmen Lisboa, José Laerte Boechat, and Luís Delgado. 2023. "Clinical Manifestations of Human Exposure to Fungi." *Journal of Fungi* 9 (3). MDPI: 381. doi:10.3390/JOF9030381/S1

Pacios-Michelena, Sandra, Cristobal N. Aguilar González, Olga B. Alvarez-Perez, Raul Rodriguez-Herrera, Mónica Chávez-González, Roberto Arredondo Valdés, Juan A. Ascacio Valdés, Mayela Govea Salas, and Anna Ilyina. 2021. "Application of Streptomyces Antimicrobial Compounds for the Control of Phytopathogens." *Frontiers in Sustainable Food Systems* 5 (September). Frontiers Media S.A.: 696518. doi:10.3389/FSUFS.2021.696518/BIBTEX

Qin, Zhiwei, John T. Munnoch, Rebecca Devine, Neil A. Holmes, Ryan F. Seipke, Karl A. Wilkinson, Barrie Wilkinson, and Matthew I. Hutchings. 2017. "Formicamycins, Antibacterial Polyketides Produced by Streptomyces Formicae Isolated from African Tetraponera Plant-Ants." *Chemical Science* 8 (4). The Royal Society of Chemistry: 3218–27. doi:10.1039/C6SC04265A

Qiu, Ziyan, Yinshuang Wu, Kunyan Lan, Shiyi Wang, Huilin Yu, Yufei Wang, Cong Wang, and Shugeng Cao. 2022. "Cytotoxic Compounds from Marine Actinomycetes: Sources, Structures and Bioactivity." *Acta Materia Medica* 1 (4). NIH Public Access: 445. doi:10.15212/AMM-2022-0028

Rangseekaew, Pharada, and Wasu Pathom-Aree. 2019. "Cave Actinobacteria as Producers of Bioactive Metabolites." *Frontiers in Microbiology* 10 (MAR). Front Microbiol. doi:10.3389/FMICB.2019.00387

Rao, Min, Wei Wei, Mei Ge, Daijie Chen, and Xiafang Sheng. 2013. "A New Antibacterial Lipopeptide Found by UPLC-MS from an Actinomycete Streptomyces Sp. HCCB10043." *Natural Product Research* 27 (23). Taylor & Francis: 2190–95. doi:10.1080/14786419.2013.811661

Rateb, Mostafa E., Wael E. Houssen, Markus Arnold, Mostafa H. Abdelrahman, Hai Deng, William T.A. Harrison, Chinyere K. Okoro, et al. 2011b. "Chaxamycins A

- D, Bioactive Ansamycins from a Hyper-Arid Desert Streptomyces Sp." *Journal of Natural Products* 74 (6). J Nat Prod: 1491–99. doi:10.1021/np200320u

Rateb, Mostafa E., Wael E. Houssen, William T.A. Harrison, Hai Deng, Chinyere K. Okoro, Juan A. Asenjo, Barbara A. Andrews, et al. 2011a. "Diverse Metabolic Profiles of a Streptomyces Strain Isolated from a Hyper-Arid Environment." *Journal of Natural Products* 74 (9). J Nat Prod: 1965–71. doi:10.1021/np200470u

Raveh, Avi, Phillip C. Delekta, Craig J. Dobry, Weiping Peng, Pamela J. Schultz, Pennelope K. Blakely, Andrew W. Tai, et al. 2013. "Discovery of Potent Broad Spectrum Antivirals Derived from Marine Actinobacteria." *PLoS ONE* 8 (12). PLoS One. doi:10.1371/journal.pone.0082318

Rosa Coelho Martins, Nathália Da, Aldo Rodrigues Da Silva, Norman Ratcliffe, Victor Gustavo Oliveira Evangelho, Helena Carla Castro, and Gerry A. Quinn. 2023. "Streptomyces: A Natural Source of Anti-Candida Agents." *Journal of Medical Microbiology* 72 (11). Microbiology Society: 001777. doi:10.1099/JMM.0.001777/CITE/REFWORKS

Sarmiento-Vizcaíno, Aida, Julia Espadas, Jesús Martín, Alfredo F. Braña, Fernando Reyes, Luis A. García, and Gloria Blanco. 2018. "Atmospheric Precipitations, Hailstone and Rainwater, as a Novel Source of Streptomyces Producing Bioactive Natural Products." *Frontiers in Microbiology* 9 (APR). Frontiers Media S.A. doi:10.3389/fmicb.2018.00773

Seabrooks, Lauren, and Longqin Hu. 2017. "Insects: An Underrepresented Resource for the Discovery of Biologically Active Natural Products." *Acta Pharmaceutica Sinica B*. doi:10.1016/j.apsb.2017.05.001

Shin, Bora, Byung Yong Kim, Eunji Cho, Ki Bong Oh, Jongheon Shin, Michael Goodfellow, and Dong Chan Oh. 2016. "Actinomadurol, an Antibacterial Norditerpenoid from a Rare Actinomycete, Actinomadura Sp. KC 191." *Journal of Natural Products* 79 (7). American Chemical Society: 1886–90. doi:10.1021/acs.jnatprod.6b00268

da Silva, Suely, Daniel Oliveira Silva Martins, and Ana Carolina Gomes Jardim. 2018. "A Review of the Ongoing Research on Zika Virus Treatment." *Viruses* 10 (5). Multidisciplinary Digital Publishing Institute (MDPI). doi:10.3390/V10050255

Song, Yongxiang, Hongbo Huang, Yuchan Chen, Jie Ding, Yun Zhang, Aijun Sun, Weimin Zhang, and Jianhua Ju. 2013. "Cytotoxic and Antibacterial Marfuraquinocins from the Deep South China Sea-Derived Streptomyces Niveus Scsio 3406." *Journal of Natural Products* 76 (12): 2263–68. doi:10.1021/np4006025

Terra, Luciana, Paul J. Dyson, Matthew D. Hitchings, Liam Thomas, Alyaa Abdelhameed, Ibrahim M. Banat, Salvatore A. Gazze, et al. 2018. "A Novel Alkaliphilic Streptomyces Inhibits ESKAPE Pathogens." *Frontiers in Microbiology* 9 (OCT). Frontiers Media S.A. doi:10.3389/fmicb.2018.02458

Um, Soohyun, Tae Joon Choi, Heegyu Kim, Byung Yong Kim, Seong Hwan Kim, Sang Kook Lee, Ki Bong Oh, Jongheon Shin, and Dong Chan Oh. 2013. "Ohmyungsamycins A and B: Cytotoxic and Antimicrobial Cyclic Peptides Produced by Streptomyces Sp. from a Volcanic Island." *Journal of Organic Chemistry* 78 (24): 12321–29. doi:10.1021/jo401974g

Wainwright, M. 1991. "Streptomycin: Discovery and Resultant Controversy." *History and Philosophy of the Life Sciences* 13 (1): 97–124.

Waksman, Selman A., and Hubert A. Lechevalier. 1949. "Neomycin, a New Antibiotic Active against Streptomycin-Resistant Bacteria, Including Tuberculosis Organisms." *Science* 109 (2830): 305–7. doi:10.1126/science.109.2830.305

WHO. 2014. "Antimicrobial Resistance. Global Report on Surveillance." *World Health Organization* 61 (3): 12–28. doi:10.1007/s13312-014-0374-3

Wu, Zhengchao, Sumei Li, Jie Li, Yuchan Chen, Kumar Saurav, Qingbo Zhang, Haibo Zhang, et al. 2013. "Antibacterial and Cytotoxic New Napyradiomycins from the Marine-Derived Streptomyces Sp. SCSIO 10428." *Marine Drugs* 11 (6): 2113–25. doi:10.3390/md11062113

Xie, Xi, Shengsheng Lu, Xiaoyan Pan, Min Zou, Fangfang Li, Haixing Lin, Jianan Hu, Sheng Fan, and Jian He. 2021. "Antiviral Bafilomycins from a Feces-Inhabiting Streptomyces Sp." *Journal of Natural Products* 84 (2). American Chemical Society: 537–43. doi:10.1021/ACS.JNATPROD.0C01243/SUPPL_FILE/NP0C01243_SI_001.PDF

Xin, Wenxiu, Xuewei Ye, Siran Yu, Xiao Yuan Lian, and Zhizhen Zhang. 2012. "New Capoamycin-Type Antibiotics and Polyene Acids from Marine Streptomyces Fradiae PTZ0025." *Marine Drugs* 10 (11): 2388–2402. doi:10.3390/md10112388

Yang, Jiayue, Zhijun Yang, Yu Yin, Min Rao, Yongheng Liang, and Mei Ge. 2016. "Three Novel Polyene Macrolides Isolated from Cultures of Streptomyces Lavenduligriseus." *The Journal of Antibiotics* 69 (1): 62–65. doi:10.1038/ja.2015.76

Zhang, Xiaoping, Lixin Zhang, Miaomiao Liu, Wael M. Abdel-Mageed, Biao Ren, Wenni He, Pei Huang, et al. 2014. "Endophytic Streptomyces Sp Y3111 from Traditional Chinese Medicine Produced Antitubercular Pluramycins." *Applied Microbiology and Biotechnology* 98 (3). Springer Verlag: 1077–85. doi:10.1007/S00253-013-5335-6/METRICS

Zhao, Huimin, Aiping Yang, Nan Zhang, Shiyang Li, Tianjie Yuan, Ning Ding, Siwang Zhang, et al. 2020. "Insecticidal Endostemonines A-J Produced by Endophytic Streptomyces from Stemona Sessilifolia." *Journal of Agricultural and Food Chemistry* 68 (6). American Chemical Society: 1588–95. doi:10.1021/acs.jafc.9b06755

Zulkifli, Nurfatihah, Jasmine Elanie Khairat, Adzzie Shazleen Azman, Nur Faralyza Mohd Baharudin, Nurul Adila Malek, Syafiq Asnawi Zainal Abidin, Sazaly AbuBakar, and Pouya Hassandarvish. 2023. "Antiviral Activities of Streptomyces KSF 103 Methanolic Extracts against Dengue Virus Type-2." *Viruses* 15 (8). doi:10.3390/v15081773

Chapter 8

Streptomycetes

A promising biological tool for bioremediation of environmental contaminants

Kehinde Caleb Omidoyin
Department of Agricultural Chemistry, Chonnam National University, Gwangju, South Korea

Oluwabusola Omolola Fatade-Ogunseye
Department of Microbiology, University of Ibadan, Ibadan, Nigeria

8.1 INTRODUCTION

The ever-increasing global anthropogenic activities have resulted in serious environmental pollution. Reports have shown an increasing occurrence and prevalence of environmental contaminants at hazardous concentrations (Arihilam & Arihilam, 2019). Microplastics, heavy metals, pesticides, dyes, and hydrocarbons are common examples of environmental contaminants of global concern (Khan et al., 2022). Although these substances are beneficial to humans, they also constitute a significant hazard to the environment and human health. For instance, many organic and inorganic contaminants possess carcinogenic, mutagenic, and teratogenic effects on aquatic and terrestrial organisms (Parida et al., 2021). Given these issues, several treatment methods have evolved and are adopted to eradicate contaminants from the environment. Treatment methods such as adsorption, membrane separation, and advanced oxidation processes are referred to as physicochemical techniques of environmental remediation as they rely on the physicochemical properties of contaminants. Contaminant removal methods using biological systems such as plants, microorganisms, and enzymes are generally called bioremediation. Figure 8.1 shows the common techniques for environmental contaminant removal.

Generally, the use of physicochemical techniques for contaminant removal is faster when compared with the use of biological techniques. For instance, lime softening (chemical precipitation method) used to remove heavy metals and phosphates from wastewater, occurs relatively faster than treatment with microorganisms. Also, the treatment of wastewater with activated carbon (adsorption method) and coagulants (coagulation method) to remove organic compounds and heavy metals is a relatively rapid process (Sathya et al., 2022). However, major setbacks with these physicochemical techniques include phase transfer and incomplete removal of

DOI: 10.1201/9781032694634-8

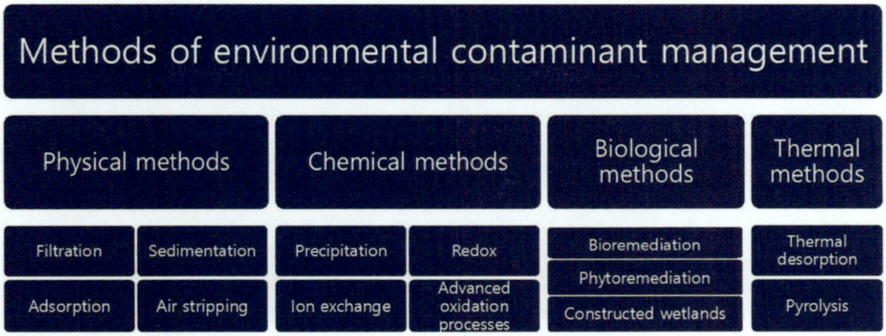

Figure 8.1 Methods of environmental contaminant removal.

the contaminants, cost ineffectiveness, and generation of toxic residues or by-products (Nazari et al., 2022). Despite being a slower process, bioremediation is considered to be a more acceptable contaminant removal strategy because it is less disruptive and offers greater benefits to the environment. Furthermore, the use of biological agents for contaminant removal is more adaptable since they employ living organisms and/or their metabolites, allowing them to undertake complex processes such as decomposing organic contaminants and transforming inorganic substances into nontoxic compounds (Nazari et al., 2022).

Microorganisms are promising agents for bioremediation as they are metabolically inclined to degrade many organic and inorganic compounds in the environment, using them as a source of energy (Lin et al., 2023). Actinomycetes, a group of gram-positive bacteria, have been identified with great bioremediation potential owing to their ubiquitous nature and complex metabolic frameworks. Of this group, streptomycetes stands out with high degradation efficiency of organic and inorganic compounds such as microplastics, heavy metals, pesticides, dyes, phenolic compounds, and hydrocarbons. *Streptomyces*, the most abundant and possibly most significant actinomycetes genus, is an excellent source of bioactive compounds, extracellular enzymes, and antibiotics (Omidoyin & Femi-Ola, 2020). This group of bacteria has demonstrated great potential to advance the prospects of environmental bioremediation over time. In addition to their ubiquity, several streptomycetes strains outperform other commonly utilized microorganisms such as *Rhodococcus* (Nazari et al., 2022), *Bacillus*, *Pseudomonas* (Liu et al., 2023), *Penicillium*, *Aspergillus* (Omidoyin and Jho, 2023), *Chlorella*, and *Scenedesmus* (Mustafa et al., 2021) in bioremediation of persistent contaminants, due to their capacity to adjust to harsh environmental conditions and tolerance to hazardous chemicals.

8.2 SOURCES OF ENVIRONMENTAL POLLUTION

Pollution is constantly increasing through anthropogenic sources while pollution from natural sources such as volcanic eruptions, wildfires, dust storms, and natural decay is seasonal. Anthropogenic sources of environmental pollution will be mainly considered in this chapter due to their larger scale occurrence, greater intensity, and more direct impacts on human health and ecosystems. Figure 8.2 presents a summary of the sources of environmental pollution.

8.2.1 Industrialization and urbanization

Unquestionably, the fast pace of industrialization and urbanization over the years has accelerated human progress and ushered in a period of economic expansion and technological improvement. However, because industrialization and urbanization are two of the main causes of environmental degradation, this remarkable development has come at a great cost to the environment. Urbanization, which generally involves the movement of people from villages to more developed urban centers, has contributed immensely to air pollution. Gaseous pollutants like nitrogen oxides, volatile organic compounds, and sulfur dioxide, as well as particulate matter, are consistently emitted from industrial activities, energy production and consumption, and emissions from transportation networks. These sources have contributed to the rapid deterioration of air quality and overall environmental pollution.

Industrialization has equally played a significant role in environmental pollution. Pollutants such as heavy metals, toxic dyes, and other chemicals contained in industrial effluents that are discharged into water bodies have led to the widespread contamination of aquatic environments. For instance, around 100,000 commercial dyes are produced every year by the textile industry with over 1 million tons produced in 2023 (Lin et al., 2023). During manufacturing and processing activities, about 10–15% of these dyes are released as effluents into the environment, particularly the water bodies, where they are potentially hazardous to the aquatic biota. Inappropriate waste disposal methods and runoff from industrial facilities worsen the issue by affecting aquatic ecosystems, endangering biodiversity, and lowering the supply of clean water for agricultural and human use (Abu-Hussien et al., 2022).

Beyond air and water pollution, industrialization and urbanization are leading the way in the severe destruction of habitats by felling of trees for wood, building of roads, and construction of houses. These actions all sum up to the devastation of ecosystems, degradation of land, and extinction of certain plant and animal species. Deforestation, ecological fragmentation, and the transformation of agricultural land into concrete jungles are all frequent outcomes of urban growth. Industrial operations lead to soil contamination, which lowers fertility and presents long-term risks to agricultural output.

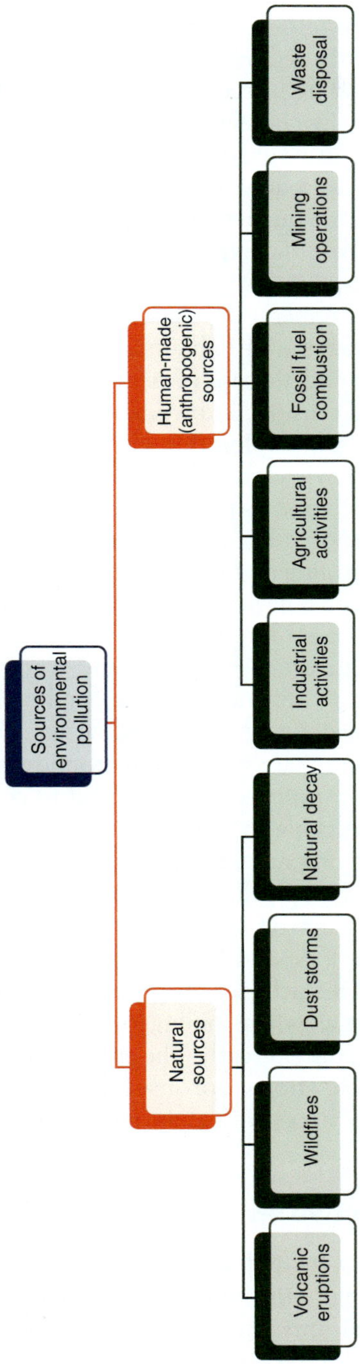

Figure 8.2 Sources of environmental pollution.

8.2.2 Mining and exploration

The continuous growth of resource extraction and industrialization-related production has significantly accelerated economic development in this era. The fast depletion of resources, ecological degradation, and environmental pollution brought on by the ever-expanding consuming culture, however, have substantially outweighed the economic benefits. The degree of environmental pollution caused by mining and exploration is dependent on the phase and the extent of the activities. Mining and exploration can be categorized into two phases, the exploration phase and the extraction and processing phase. The exploration phase, which often involves a detailed search and assessment of mineral deposits, frequently results in the disturbance of natural ecosystems and habitats. Exploration activities such as drilling, seismic surveys, and the use of heavy exploration machinery can lead to deforestation, soil erosion, and wildlife disturbance. In addition, the discharge of wastewater and exploration chemicals may introduce pollutants into water and soil environments, hence causing long-term hazardous effects. The extraction and processing phase, whether done by underground mining, open-pit mining, or other means, requires an enormous resource and presents serious environmental problems.

During mining operations, hazardous chemicals, heavy metals, and particulate matter can be released into the air, water, and soil, thereby causing pollution. Aquatic habitats are also altered, and water shortage is exacerbated by the use of water in mineral processing and tailings pond construction. Sulfide minerals reacting with air and water can cause acid mine drainage, which can contaminate surface and groundwater bodies with runoff that is generally acidic. Tailings dams are built to contain mining waste, but they are usually vulnerable to failure. If they fail, contaminants could leak into streams, destroying aquatic ecosystems and endangering human lives. The vandalization of oil pipelines in oil-producing countries, especially in developing ones, results in oil spills that contaminate the environment. Soil erosion, biodiversity loss, and landscape alteration are common outcomes of large-scale land disturbance for mining reasons. Effective land reclamation becomes essential when mining operations end. The results of these initiatives, however, differ greatly. The environmental effects of mining can persist if reclamation efforts are insufficient or incomplete, leaving behind degraded, desolate areas with enduring ecological scars.

8.2.3 Agricultural activities

Agriculture is conceivably the oldest activity that man has done to ensure the survival and well-being of the human race. From simply gathering food, agriculture has developed into a massive, technologically advanced industry. The industrialization of agriculture in recent decades has had negative effects on the environment and ecology. Without a doubt, the development of modern agricultural techniques has been crucial to the global increase in food production. Agricultural practices have also been a significant contributor to waste production and environmental contamination, much like other aspects of development. These two closely linked agricultural outcomes arise from a variety of efforts and resources utilized to boost productivity and raise agronomic output worldwide.

Prevention of weeds, pests, and plant diseases as well as improvement of food product quality can be achieved with pesticides. A vast variety of chemicals are included under the umbrella word "pesticide," such as herbicides, insecticides, fungicides, rodenticides, nematicides, molluscicides, and plant growth regulators. A pesticide should ideally have lethal effects on the target pests, but not harm non-target organisms. Unfortunately, this is inaccurate, which is why the controversy over the appropriate and inappropriate use of pesticides has surfaced. Owing to the misconception that "if little is good, a lot more will be better," these chemicals are extensively misused and harm both humans and other living things in the environment. For instance, the total amount of pesticides used worldwide in 2019 was estimated to be 4.19 million tons, with China consuming the most (1.76 million tons), followed by the United States (408 thousand tons), Brazil (377 thousand tons), and Argentina (204 thousand tons) (Sharma et al., 2019). In 2022, the Food and Agriculture Organization (FAO) of the United States reported the total pesticides used globally at 4.2 million tons (Shattuck et al., 2023). Both the aquatic and terrestrial environments can be contaminated with pesticides, and they can be detrimental to a wide range of living organisms, including beneficial insects, fish, birds, and non-target plants, in addition to affecting the microbiota.

Overuse of chemical fertilizers in agriculture leads to an array of environmental issues because some fertilizers have high concentrations of radionuclides and heavy metals (like chromium and cadmium) and accumulate inorganic pollutants. High amounts of nitrogen and phosphorus are introduced into the soil by the overuse of synthetic fertilizers, leading to nutritional imbalances and soil degradation. Pesticides and fertilizers are carried into neighboring water bodies via surface runoff from agricultural fields, thereby causing pollution. This runoff has a negative impact on marine life by aiding the development of dead zones in aquatic environments. Other agricultural activities such as the burning of agricultural wastes, use of farm machinery, and livestock farming can lead to air pollution. Excessive land plowing and monoculture also lead to soil erosion, and the expansion of land for agricultural purposes leads to deforestation and destruction of natural habitats.

8.2.4 Fossil fuel combustion

Even before they burn, fossil fuels have the potential to release toxic air pollutants. Several air pollutants are released during the burning of fossil fuels, contributing to environmental pollution and the subsequent devastation of the ecosystem. The burning of gas, coal, and oil to suit human energy needs contributes significantly to the current issue of global warming. Burning fossil fuels releases a wide range of primary and secondary pollutants, such as nitrous oxides, hydrocarbons, organic compounds, fluorinated gases, sulfur dioxide, carbon dioxide, methane, and airborne particulates. The main greenhouse gases, such as methane, carbon dioxide, and nitrous oxide, trap atmospheric heat which contributes to global warming and climate change. During the combustion of fossil fuels, the release of sulfur dioxide and nitrogen oxides can lead to acid rain formation after their combination with atmospheric moisture. This has a damaging effect on aquatic and terrestrial ecosystems.

8.2.5 Plastic wastes

Within the category of anthropogenic compounds that are environmentally hazardous, plastic materials have a particular place. Massive pollution has been documented in both terrestrial and marine environments (Walker and Fequet, 2023). This is due to a combination of factors, including mass production, improper waste disposal, and recalcitrance of plastics. A recent study found that 60–80% of marine debris is made up of plastic garbage, with over 10 million tons entering the ocean each year (Walker and Fequet, 2023). Regretfully, plastic materials end up in the environment at every stage of their life cycle. When released into the environment, a series of physicochemical and biological activities cause the weathering of plastic polymers which leads to the fragmentation of larger plastic items to smaller particles called microplastics (i.e., smaller than 5 mm).

In the past years, the number of reports regarding the effects of plastic pollution on the environment has increased dramatically. In addition to aesthetic issues, plastic waste can be consumed by marine animals or entangled in them, causing harm, suffocation, or even death. When ingested by aquatic and terrestrial organisms, microplastics can cause significant physical, chemical, and biological damage to organisms. Plant growth can also be negatively affected by microplastics. Concerns associated with microplastics have come to light more recently, and they could pose a significant threat to aquatic and terrestrial ecosystems, as well as human health. Additionally, plastics frequently contain chemical additives, including plasticizers, stabilizers, and flame retardants that can leak into the environment and cause harm to aquatic and terrestrial organisms, reduce soil and groundwater quality, and result in numerous health problems. In addition to microplastics, other pollutants commonly found in the environment are shown in Figure 8.3.

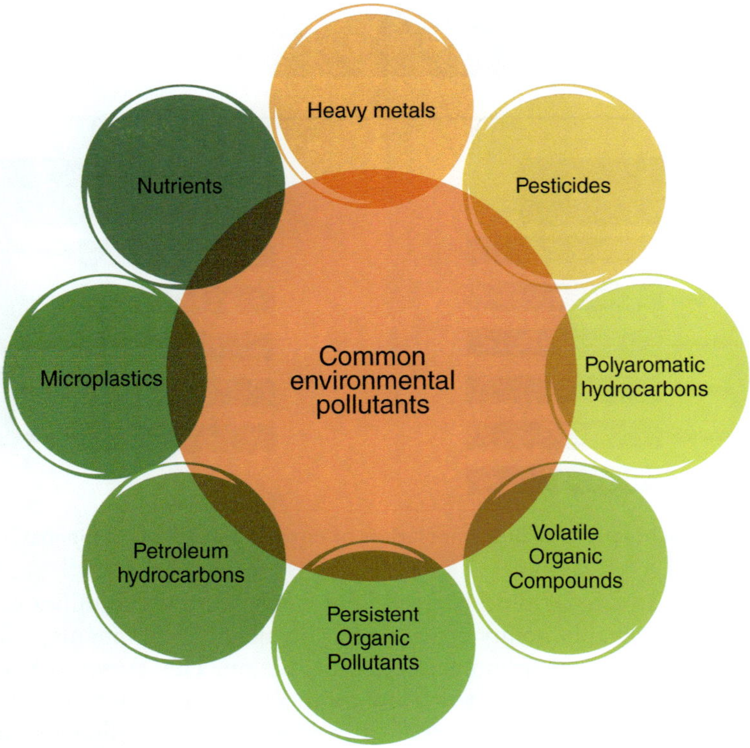

Figure 8.3 Pollutants commonly found in the environment.

8.3 BIOREMEDIATION POTENTIAL OF *STREPTOMYCES* AND THEIR MECHANISMS OF CONTAMINANT DEGRADATION

Streptomyces, one of the most common genera of actinobacteria, are well known for their production of natural compounds (Subramani & Detmer, 2019). *Streptomyces* possess the ability to adapt to various environmental conditions and produce secondary metabolites that are useful in biotechnology. These features make them good agents for bioremediation. Generally, bioremediation techniques are in two categories: in situ, and ex-situ. The in-situ bioremediation methods take place within the contaminated zone. The primary techniques used include bioaugmentation, biostimulation, natural attenuation, bioventing, and biosparging. For ex-situ bioremediation, treatment is done away from polluted sites. Bioreactors, land-farming, biopiles, and composting are common examples of ex-situ bioremediation methods. The techniques for bioremediation of environmental contaminants are represented in Figure 8.4.

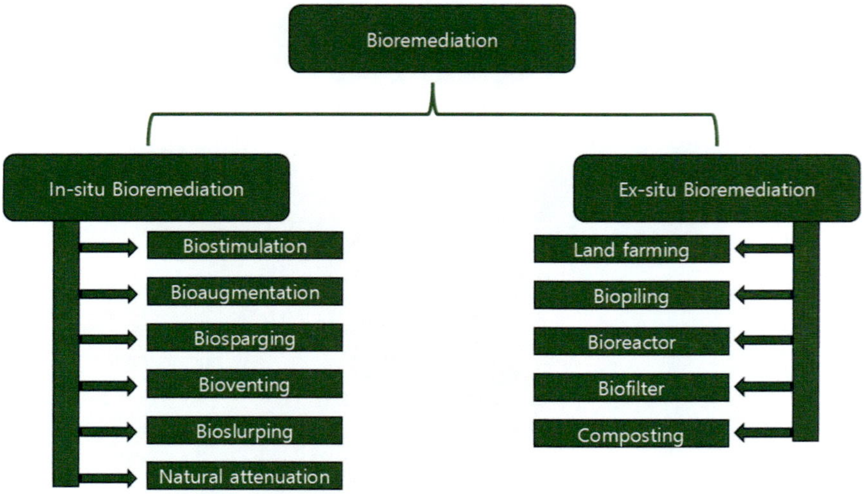

Figure 8.4 Techniques for bioremediation of environmental contaminants.

Streptomyces are aerobic bacteria and under aerobic conditions, they exhibit a remarkable ability to degrade polymeric contaminants, making them invaluable for environmental bioremediation. This group of filamentous bacteria thrive in oxygen-rich environments, where they secrete a wide array of extracellular enzymes specifically designed to break down complex polymeric substances. Data presented in Table 8.1 show that different strains of *Streptomyces* are able to degrade various environmental contaminants, including microplastics, dyes, hydrocarbons, heavy metals, and pesticides. Enzymes such as polyesterases, PET hydrolases, peroxidases, and laccases enable *Streptomyces* to efficiently hydrolyze pollutants such as plastics, dyes, and pesticides (Lokesh et al., 2023). Figure 8.5 depicts the general mechanism of degradation of polymers by *Streptomyces* under aerobic conditions. Enzymes produced by *Streptomyces* are able to cleave polymer chains into smaller units (monomers and oligomers), which can be further metabolized or assimilated by *Streptomyces* strains or other microorganisms in the ecosystem (Cuebas-Irizarry and Grunden 2023; Lokesh et al., 2023).

LDPE: Low-density polyethylene; PE: Polyethylene; PP: Polypropylene; PET: Polyethylene terephthalate; HDPE: High-density polyethylene; PAH: Polycyclic aromatic hydrocarbons.

8.3.1 *Streptomyces* in bioremediation of dyes

Pollutants, mainly from chemicals and untreated wastes from anthropogenic activities and industrial discharges have been a source of pollution in different environmental compartments. Dyes are common chemicals

Table 8.1 Some *Streptomyces* strains involved in contaminant degradation

Category of Contaminant	*Compound*	*Degrading Strain*	*References*
Microplastics	LDPE	*Streptomyces labedae*	Rajapandi and Rajamanickam, 2024
	PE	*Streptomyces* sp.	Han et al. 2020
	PP	*Streptomyces ardesiacus* NBI0111	Wichatham et al., 2024
	PET	*Streptomyces* spp	Farzi et al., 2019
	HDPE	*Streptomyces* sp.	Farzi et al., 2017
Dyes	Reactive Blue 59	*Streptomyces krainskii*	El-Sersy et al., 2011
	Acid fast red	*Streptomyces alanaosinicus* *Streptomyces ruber*	El-Sersy et al., 2011
	Amido Black 10B	*Streptomyces cyaneus* 3335	Popovic et al., 2021
	Congo Red	*Streptomyces sviceus* KN3	Chakravarthi et al., 2021
Heavy metals	Lead	*Streptomyces rimosus*	Sahmoune, 2018.
	Copper	*Streptomyces* sp. WR1L1S8	Djinni and Djoudi, 2022
	Cadmium	*Streptomyces* sp. R25.	Zulfiqar et al., 2023
Hydrocarbons	n-alkanes	*Streptomyces* sp.	Baoune et al., 2018
	PAH	*Streptomyces rochei*	Chaudhary et al., 2011
	Used motor oil	*Streptomyces ginkgonis*	Soumeya et al., 2022
Pesticides	Cypermethrin	*Streptomyces diastaticus*	Janarthanan et al., 2018
	Deltamethrin	*Streptomyces rimosus*	Khajezadeh et al., 2020
	Atrazine	*Streptomyces nigra LM01*	Zhao and Wang, 2024
	Dichlorodiphenyltrichloroethane DDT	*Streptomyces* sp.	Ito et al., 2021

that are contained in industrial effluents discharged into the environment (Khalilzadeh Shirazi et al., 2020). Their unwanted presence in waterbodies results in the pollution of the aquatic environment as well as health problems (such as cancer, skin irritation, and respiratory) for humans (Al-Tohamy et al., 2020). Physicochemical processes such as adsorption, coagulation, oxidation, and membrane filtration have been used to remove dyes. However, these treatment methods face limitations such as high cost, toxic sludge generation, incomplete degradation, and high energy consumption (Yahiaoui et al., 2021), and these highlight the need for more sustainable and efficient

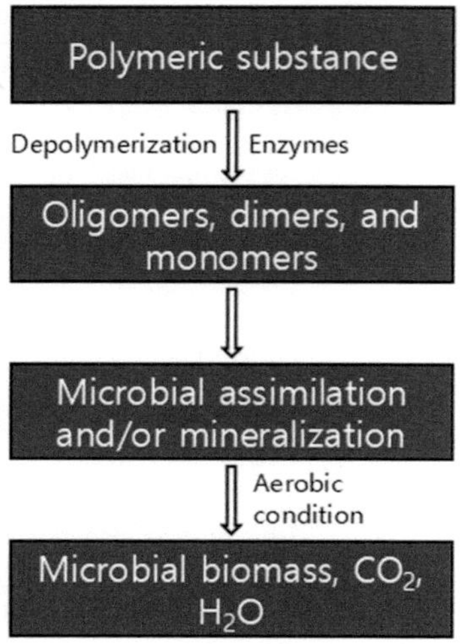

Figure 8.5 General mechanism of degradation of polymers by *Streptomyces* under aerobic conditions.

dye removal strategies. The new focus of researchers is the use of biological agents that are environmentally friendly, less expensive, and easy to apply.

Different microorganisms including *Streptomyces* have been reported to produce enzymes capable of hydrolyzing dyes. Such enzymes include azo reductase, manganese-lignin peroxidase, and ascorbate oxidase. *Streptomyces* produce extracellular peroxidases and play a part in lignin biodegradation resulting in the production of water-soluble polymeric compounds. Peroxidase produced by *Streptomyces* oxidizes lignin. The presence of glucose in the growth medium supports the enhancement of extracellular peroxidase production by *Streptomyces* sp. (Ngo and Tischler, 2022). Detergent-stable dye-decolorizers and peroxidases produced by *Streptomyces* have the potential to replace sodium perborate in the destaining of synthetic textile dyes. Peroxidases, which are heme-containing proteins, are specifically utilized for dye decolorization. These enzymes facilitate the oxidation of substrates by reacting with hydrogen peroxide or organic hydroperoxides (R-OOH) (Ngo and Tischler, 2022). Their application for multiple biological processes such as dye decolorization is due to their ability to oxidize various substrates.

Mechanisms involved in dye removal by *Streptomyces* are biosorption and biodegradation. The removal of dyes by *Streptomyces* is attributed to the adsorption of dyes onto live or dead cell biomass, as well as biodegradation by living cells. In the biodegradation of malachite green, *Streptomyces exfoliates* reduced malachite green to leucomalachite green, which was further oxidized to form heterocyclic ring structures (Abu-Hussien et al., 2022). The dyes with two methyl substitution groups in the ortho position relative to the hydroxyl group and a hydroxyl group in the para positions related to the azo linkage are the most susceptible to *Streptomyces chromofuscus*.

Streptomyces chromofuscus removes acid yellow 9 through biodegradation. *Streptomyces krainskii* uses biodegradation to remove reactive blue 59 while *Streptomyces alanosinicus* and *Streptomyces ruber* remove acid-fast red and Congo red through biosorption (Mane et al., 2008; El-Sersy et al., 2011). During dye biodegradation, strains of *Streptomyces* use the dyes as a nutrient source and for metabolic activities. Two species of *Streptomyces*, *Streptomyces violaceoruber* and *Streptomyces spiroverticillatus*, are able to utilize the dye Poly R-478 as sole carbon and nitrogen sources. Increasing the inoculum size and reducing the starch content of the medium enhances the decolorization potential of *Streptomyces*. By using potato peel as an inexpensive source of nutrients, *Streptomyces globosus* can produce the maximum amount of cellular biomass needed for dye adsorption (El-Sersy et al., 2011). In maximizing decolorization efficiency, a consortium of *Streptomyces* strains can also be used. *Streptomyces* consortium degraded reactive blue 222 (reactive sulfonated di azo dye) and the biodegradation rate was enhanced by 89% (Pillai, 2017).

Aside from biosorption and biodegradation processes, enzymes from *Streptomyces* have also been used for dye decolorization and detoxification. Purified laccase-like multicopper oxidase produced by *Streptomyces* sp. C1 showed potential in azo dye decolorization (Lu et al., 2014). *Streptomyces cyaneus* 3335 was used to create a laccase-mediator system that allowed for the decolorization and detoxification of azo dyes (Moya et al., 2009). Azoreductases are another type of enzyme produced by *Streptomyces*. They are oxidoreductases mostly known for the degradation of azo dyes by reducing the azo linkage. The immobilization of dye-decolorizing enzymes by *Streptomyces* is also an effective method for dye decolorization and detoxification. Immobilized laccase produced by *Streptomyces cyaneus* 3335 was able to decolorize Remazol brilliant blue (100%), reactive black 5 (100%), Evans blue (90%), and amido black 10B (61%). Congo red was similarly mineralized and detoxified using immobilized laccase from *Streptomyces sviceus* KN3. (Chakravarthi et al., 2021).

8.3.2 *Streptomyces* in bioremediation of hydrocarbons

Hydrocarbons are used all over the world as a crucial source of energy. However, their extensive usage has led to serious environmental pollution, causing significant harm to the ecosystem. Various hazardous chemicals and wastes are released into the environment through the combustion of hydrocarbons, drilling activities, oil spills from tankers and damaged pipelines, and other anthropogenic activities. Moreover, these hydrocarbons linger in the environment for long periods where they impact the environment negatively. The common clean-up methods used for hydrocarbon pollution including in-situ burning, physical/mechanical remediation, and chemical remediation do not result in the complete removal of the pollutants and might cause further pollution (Sathya et al., 2022).

Actinobacteria are among the microorganisms that have exhibited promising bioremediation potential of hydrocarbons. Two main methods are generally used to enhance the efficiency of bioremediation. The bio stimulation method entails the alteration of environmental conditions to stimulate the activity of existing contaminant-degrading indigenous microorganisms. This can be achieved by adding various rate-limiting nutrients such as carbon, nitrogen, and oxygen to the environment and ensuring that the environmental conditions are suitable for microbial growth and natural biodiversity. The other method, bioaugmentation, involves the introduction of cultured microorganisms that can degrade contaminants of interest in the pollution area. Specific strains are employed that may be isolated or genetically modified to increase the efficiency of contaminant degradation.

Streptomyces have been reported to be responsible for the breakdown of alkanes and aromatic hydrocarbons. A surfactant-producing strain, *Streptomyces rochei*, was reported to degrade polycyclic aromatic hydrocarbons containing three to four rings (Chaudhary et al., 2011). Up to 98% of the aromatic and polycyclic aromatic hydrocarbons, as well as n-alkanes (C6–C30), were degraded by *Streptomyces* sp. isolated from plants in polluted soil (Baoune et al., 2018). Also, biosurfactants from *Streptomyces* have been used to increase the bioavailability of hydrocarbon contaminants thereby facilitating their degradation. A polysaccharide MOE6-EPS that is produced extracellularly by *Streptomyces* sp. MOE6 can emulsify motor oil, making it a valuable tool for bioremediation of polluted areas (Elnahas et al., 2021).

Despite being a promising method for hydrocarbon and oil spill cleanup, bioremediation faces several limitations. Environmental factors such as temperature, pH, and nutrient availability significantly impact its effectiveness, with extreme conditions hindering microbial activity. The process is slow, often taking months to years for substantial degradation, particularly in large spills. Additionally, certain oil components are toxic to microbes, and some hydrocarbons are resistant to degradation, prolonging cleanup efforts. Unpredictable outcomes due to complex microbial communities, regulatory constraints, and the high costs associated with scaling up for large spills further impede widespread use (Radhakrishnan et al., 2023).

8.3.3 *Streptomyces* in bioremediation of pesticides

Pesticides are transformed during the biodegradation process with enzymes such as hydrolases, peroxidases, and oxygenases catalyzing the biochemical processes. Due to their extensive use in agricultural activities, pesticides are widespread pollutants in the environment. Pesticides such as carbamates, triazinones, organochlorines, organophosphates, and others have had an increase in use over the years. The chemical properties of these chemicals result in the negative impact they exhibit on the ecosystem. Organochlorine pesticides (OCPs), which are the most widely used pesticides, contain chlorine atoms that replace one or more hydrogen atoms in their molecular structure. The degree of chlorination of OCPs plays a crucial role in their degradability in the environment (Derbalah et al., 2014). Several strains of *Streptomyces* have been examined for their ability to degrade OCPs.

In liquid medium, *Streptomyces* are known to degrade pesticides in a matter of days or hours. Pesticide breakdown by *Streptomyces* is evaluated only when the compound is used as a source of nutrients (carbon, nitrogen, phosphorus, etc.). Cypermethrin was totally broken down by *Streptomyces* sp. HU-S-01 in 24–30 hours at a concentration of less than 50 mgL–1. Even after 48 hours, decomposition was not complete at higher concentrations (Lin et al., 2011). To degrade pesticides, the culture medium of *Streptomyces* containing mineral salts such as ammonium chloride, potassium phosphate, magnesium sulphate, and trace elements can also be supplemented with additional carbon sources such as glucose (Lu et al., 2013). Numerous hydrolytic enzymes and oxidoreductases, such as laccase, xylanase, esterase, dehydrogenase, protease, and amylase, are known to be present in *Streptomyces* and may be implicated in the breakdown of pesticides (Khalilzadeh Shirazi et al., 2020). Chitin has been proposed as a semi-selective substrate that, when mixed with contaminated soils, can stimulate the metabolism for pesticide degradation.

The use of microbial consortium has been reported to enhance pesticide degradation. A hybrid *Streptomyces* sp. culture comprising the strains AC5, AC9, GA11, and ISP13 was successfully used in a continuous stirred tank bioreactor to remove a pesticide mixture that contained organophosphorus, chlorpyrifos, and diazinon (Briceño et al., 2020). Similarly, a combination of bioaugmentation and bio-stimulation strategies enhanced the biodegradation of the organochlorine pesticide, lindane (Raimondo et al., 2020).

Other approaches in pesticide bioremediation include the use of enzymes, bioreactors with immobilized cells, and the expression of recombinant enzymes (Santillan et al., 2020). The creation of ester bonds in the organic phase and their breaking in the aqueous phase are catalyzed in large part by carboxylesterases (Wang et al., 2020). These enzymes are widely distributed in nature among microbes, plants, and animals, and are considered as detoxifying enzymes due to their ability to degrade organophosphorus pesticides, pyrethroids, and organochlorides (Kumar et al., 2019). *Streptomyces* are known for producing numerous hydrolases in the process of secondary metabolism. *Streptomyces lividans* TK24 expressed *EstC*, a novel carboxylesterase that showed potential for removal of chlorpyrifos residues (Wang et al., 2020).

8.3.4 *Streptomyces* in bioremediation of heavy metals

Heavy metals in soil, air, water, and living organisms originate from both natural and anthropogenic sources, posing significant risks to the environment and human health. Metals such as cobalt, lead, mercury, nickel, chromium, and cadmium are particularly hazardous. Within contaminated sites, microbial communities develop tolerance to these contaminants, resulting in genomic and genetic material modifications that facilitate bioremediation. Microorganisms employ various survival mechanisms in polluted environments, including bioaccumulation, biomineralization, bioleaching, and biotransformation.

A nickel enzyme, superoxide dismutase produced by *Streptomyces seoulensis* acts as a regulator of specific gene expression (Shin et al., 2020). Heavy metal–polluted soil can be remedied by certain microbes by producing chelating agents such as glomalin, siderophores, and biosurfactants. In order to bind certain heavy metals and shield microorganisms from heavy metal toxicity, microbial strains produce siderophores, which include hydroxamate. *Streptomyces* possess the ability to produce a wide range of siderophores which is important for growth.

The biotechnological process of extracting heavy metals from sediment ores using organic acids generated by acidophilic bacteria is known as "bioleaching." This process can be achieved through contact and noncontact mechanisms. In contact bioleaching, physical contact occurs between the mineral sulfide and bacterial cells. The oxidation to sulfate is catalyzed by enzymes, resulting in electron transfer from the mineral surface. *Streptomyces albidoflavus* TN10 has been found to effectively leach out various metal ions from printed circuit boards (Kaliyaraj et al., 2019). These methods are employed to remove heavy metals from solid particles or soils.

Bioaccumulation, on the other hand, is a natural metabolic process in which heavy metals accumulate within intracellular compartments of living bacterial cells. This process, which relies on an import-storage system, occurs when the rate of metal absorption exceeds the rate of metal loss. Adsorption of chemicals from a solution onto biological materials via physiochemical processes like ion/proton displacement and electrostatic forces is known as biosorption. The achievement of this process relies on the variety of structures found in the cell walls. Within the cell wall of bacteria, the functional groups assume responsibility for the execution of binding functions. Heavy metals such as lead, cadmium, and copper have been removed using *Streptomyces* sp. (Hamdan et al., 2021).

8.3.5 *Streptomyces* in bioremediation of microplastics

Plastics have become indispensable in human lives as they have found applications in various sectors, replacing materials like glass, wood, and metals. Common types of plastics used in different applications include polyethylene terephthalate (PET), high-density polyethylene (HDPE), polystyrene (PS), polyurethane (PU), polyvinyl chloride (PVC), low-density polyethylene (LDPE) and polypropylene (PP), which are resistant to environmental degradation. Currently, one of the most serious environmental threats is plastic pollution in all its manifestations. The detrimental impacts of small plastic particles on the environment and public health—such as microplastics and nanoplastics originating from primary and secondary sources—are a global issue (Li et al., 2021).

Different physical and chemical methods are commonly used for plastic waste disposal and treatment, for example landfilling, recycling, and incineration. Landfilling, a method of waste disposal that entails the dumping of plastic waste in open grounds, is a common practice, especially in developing countries. This method of plastic waste treatment results in the release of plastic leachates such as phthalates which are known to cause harm to aquatic and terrestrial organisms, reduce soil and groundwater quality, and result in numerous health problems. Incineration increases energy demand because it requires high temperatures. It also accounts for the release of toxic gases and compounds into the environment. The challenges of cost-ineffectiveness, low product quality, and environmental pollution have equally made plastic recycling undesirable. Figure 8.6 provides a summary of the common plastic waste treatment methods and their limitations. Due to these, attempts are being made by scientists and researchers to develop effective, eco-friendly, and sustainable plastic waste management strategies (Omidoyin & Jho, 2023).

Even though microplastics are recalcitrant and highly resistant to degradation, some microbial strains, including *Streptomyces*, have been found to have the ability to degrade different types of microplastics. These microorganisms have been isolated from natural environments such as soil, landfills, marine water, sewage sludge, and the guts of insects that consume plastic (Omidoyin & Jho, 2023). Microorganisms can withstand a wide range of environmental conditions, including harsh ones, and are ubiquitous in nature. Their potential to degrade a wide range of environmental contaminants is attributed to their complex enzyme system, ability to grow on a variety of organic and inorganic substrates, and ability to quickly adapt to harsh environmental circumstances. As such, they offer an environmentally friendly and sustainable substitute for the management of plastic waste.

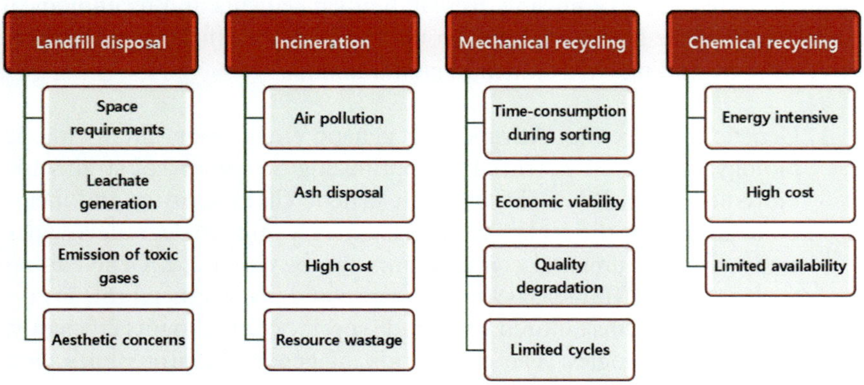

Figure 8.6 Common plastic waste treatment methods and their limitations.

Extracellular polymeric compounds secreted by microorganisms colonizing microplastic surfaces function as bio-adhesives, enhancing their attachment (Ogonowski et al., 2018). The ability of microorganisms to colonize and form biofilms on microplastic surfaces enhances their degradation activities. Biofilm formation and biofilm-mediated degradation of microplastics occur in four stages (Yuan et al., 2020). These are 1) adhesion of microorganisms to the surface of microplastics and modification of their surface characteristics; 2) microplastic deterioration occurring by the release of microbial enzymes that accelerate degradation and release monomers and chemical additives from the microplastic; 3) microplastic fragmentation as a result of microbial attack that causes the microplastics to lose their mechanical stability and strength and become brittle and fragile; and 4) assimilation of microplastic polymer, involving the penetration of microbial filaments and water, followed by the microbial decomposition and utilization of MPs as a nutrient source.

The breakdown of microplastics is mostly carried out by microbial extracellular enzymes such as laccases, lignin peroxidase, manganese peroxidase, lipases, and esterases. The activities of these enzymes increase the hydrophilicity of microplastics and subsequent degradation (Taniguchi et al., 2019). Hydrolase enzymes (lipase, cutinase, esterase) degrade microplastics by acting on the plastic surface and enhancing the chain cleavage reactions. Microplastic degradation by *Streptomyces* is influenced by the polymer properties and environmental conditions. *Streptomyces* strains that have been reported to be associated with microplastic degradation include *Streptomyces scabies*, *Streptomyces badius*, *Streptomyces viridosporus*, and *Streptomyces* sp. (Omidoyin & Jho, 2023).

8.4 FUTURE PERSPECTIVE

The potential advancements and applications of streptomycetes in bioremediation are highly promising. Research advancements and technological innovations in the following areas would enhance the efficiency of streptomycetes in degrading environmental contaminants.

1. Genetic Engineering: Advances in genetic engineering and synthetic biology hold great potential for optimizing *Streptomyces* strains for bioremediation. By modifying their genetic characteristics and metabolic pathways, the ability of *Streptomyces* strains to degrade a wider array of environmental contaminants can be enhanced. Gene-editing tools such as CRISPR-Cas9 technology could play a crucial role in creating super-strains tailored to degrade specific contaminants efficiently.
2. Omics Technologies: The application of genomics, proteomics, and metabolomics will provide deeper insights into the metabolic capabilities of streptomycetes. Understanding the genetic and molecular

mechanisms underlying their biodegradative processes will enable the design of more effective bioremediation strategies. These technologies can help identify key genes and regulatory networks involved in the breakdown of contaminants.
3. Consortia and Synergistic Approaches: Future bioremediation efforts may benefit from the use of microbial consortia, where *Streptomyces* strains work in synergy with other microorganisms. Such consortia can provide a more comprehensive degradation of complex contaminants. Research into microbial interactions and community dynamics will be crucial for developing effective consortia-based bioremediation systems.
4. Environmental Applications and Field Trials: Scaling up laboratory successes to real-world applications is a critical step forward. Field trials and pilot studies will provide valuable data on the effectiveness of streptomycetes in diverse environmental conditions. These trials will help refine application methods, optimize conditions for microbial activity, and address practical challenges in deploying *Streptomyces* strains for bioremediation on a large scale.
5. Bioreactor Design and Optimization: The development of advanced bioreactors tailored for streptomycetes-based bioremediation can revolutionize the treatment of contaminated water and soil. Innovations in reactor design, including immobilization techniques and continuous flow systems, can enhance the degradation rates and scalability of bioremediation processes.

8.5 CONCLUSION

The relentless increase in global anthropogenic activities has led to the continuous spread of contaminants, resulting in significant negative impacts on the environment and human health. This pressing issue has prompted extensive research into effective waste management strategies and the development of methods to ameliorate polluted environments. Among the various approaches, bioremediation using *Streptomyces* species stands out as a particularly advantageous and environmentally friendly method for restoring contaminated sites.

Streptomycetes, a group of microorganisms known for their remarkable ability to degrade or remove a wide range of organic and inorganic contaminants, owe their efficiency to their unique enzymatic capabilities and metabolic versatility. Along with the critical experimental variables involved in the bioremediation process (pH, temperature, dosage of the contaminant, concentration of the inoculum, speed of agitation, extra sources of carbon and nitrogen, among others), the bioremediation efficiency varies depending on the species of *Streptomyces*, the remediation technique used, and the contaminants evaluated.

Even though the *Streptomyces* genus has been thoroughly tested to treat a wide range of pollutants, including pesticides, heavy metals, microplastics, polycyclic aromatic hydrocarbons, and dyes, there are still some newly discovered contaminants that need to be investigated further. These include novel synthetic chemicals and complex waste products that have recently been recognized for their potential ecological and health impacts. Continued research into the bioremediation potential of streptomycetes, with a focus on optimization and expanding the range of treatable contaminants, is essential for advancing this field and enhancing the effectiveness of bioremediation strategies in addressing contemporary environmental challenges.

CONFLICT OF INTEREST

The authors declare that they have no conflict of interest.

REFERENCES

Abu-Hussien, S. H., Hemdan, B. A., Alzahrani, O. M., Alswat, A. S., Alatawi, F. A., Alenezi, M. A., Darwish, D. B., Bafhaid, H. S., Mahmoud, S. F., Ibrahim, M. F. M., and El-Sayed, S. M. 2022. Microbial degradation, spectral analysis and toxicological assessment of malachite green dye by *Streptomyces exfoliatus*. *Molecules* 27(19): 6456.

Al-Tohamy, R., Kenawy, E.-R., Sun, J., and Ali, S. S. 2020. Performance of a newly isolated salt-tolerant yeast strain *Sterigmatomyces halophilus* SSA-1575 for Azo dye decolorization and detoxification. *Frontiers in Microbiology* 11: 1163.

Arihilam, N. H., and Arihilam, E. 2019. Impact and control of anthropogenic pollution on the ecosystem–A review. *Journal of Bioscience and Biotechnology Discovery* 4(3): 54–59.

Baoune, H., Ould El Hadj-Khelil, A., Pucci, G., Sineli, P., Loucif, L., and Polti, M. A. 2018. Petroleum degradation by endophytic *Streptomyces* spp. isolated from plants grown in contaminated soil of southern Algeria. *Ecotoxicology and Environmental Safety* 147: 602–609.

Briceño, G., Levio-Raiman, M., González, M., Saez, J. M., Schalchli, H., and Diez, M. 2020. Performance of a continuous stirred tank bioreactor employing an immobilized actinobacteria mixed culture for the removal of organophosphorus pesticides. *3 Biotech* 10: 252.

Chakravarthi, B., Mathkala, V., and Palempalli, U. M. D. 2021. Degradation and detoxification of congo red azo dye by immobilized laccase of *Streptomyces sviceus*. *Journal of Pure and Applied Microbiology* 15(2): 864–876.

Chaudhary, P., Sharma, R., Singh, S., and Nain, L. 2011. Bioremediation of PAH by *Streptomyces* sp. *Bulletin of Environmental Contamination and Toxicology* 86: 268–271.

Cuebas-Irizarry, M.F., and Grunden, A.M. 2023. *Streptomyces* spp. as biocatalyst sources in pulp and paper and textile industries: Biodegradation, bioconversion and valorization of waste. *Microbial Biotechnology* 17(1): e14258.

Derbalah, A., Ismail, A., Hamza, A., and Shaheen, S. 2014. Monitoring and remediation of organochlorine residues in water. *Water Environment Research* 86(7): 584–593.

Djinni, I., and Djoudi, W. 2022. *Streptomyces* sp. WR1L1S8 a potent endophytic marine strain for heavy metal resistance and copper removal enhanced by RSM modeling. *Acta Ecologica Sinica* 42(2): 80–89.

El-Sersy, N., Abou-Elela, G., Hassan, S., and Abd-Elnaby, H. 2011. Bioremediation of acid fast red dye by *Streptomyces globosus* under static and shake conditions. *African Journal of Biotechnology*, 10: 3467–3474.

Elnahas, M., Hou, L., Wall, J., and Majumder, E. 2021. Bioremediation potential of *Streptomyces* sp. MOE6 for toxic metals and oil. *Polysaccharides* 2:47–68.

Farzi, A., Dehnad, A., and Fotouhi, A. F. 2019. Biodegradation of polyethylene terephthalate waste using Streptomyces species and kinetic modeling of the process. *Biocatalysis and Agricultural Biotechnology* 17: 25–31.

Farzi, A., Dehnad, A., Shirzad, N., and Norouzifard, F. 2017. Biodegradation of high density polyethylene using *Streptomyces* species. *Journal of Coastal Life Medicine* 5:474–479.

Hamdan, A. M., Abd-El-Mageed, H., and Ghanem, N. 2021. Biological treatment of hazardous heavy metals by *Streptomyces rochei* ANH for sustainable water management in agriculture. *Scientific Reports* 11(1):9314.

Han, Y., Wei, M., Han, F., Fang, C., Wang, D., Zhong, Y., Guo, C., Shi, X., Xie, Z., and Li, F. 2020. Greater biofilm formation and increased biodegradation of polyethylene film by a microbial consortium of *Arthrobacter* sp. and *Streptomyces* sp. *Microorganisms* 8(12): 1979.

Ito, K., Mahmood, A., Kataoka, R., and Takagi, K. 2021. Dichlorodiphenyltrichloroethane (DDT) degradation by Streptomyces sp. isolated from DDT contaminated soil. *Bioremediation Journal* 25(2): 148–158.

Janarthanan, R., Gopikrishnan, V., Kavitha, K., Murugan, A. and Balagurunathan, R. 2018. Biodegradation of cypermethrin metabolites using terrestrial actino bacterium, Streptomyces diastaticus (PA2) and its GC-MS analysis. *International Journal of ChemTech Research* 11(5): 509–520.

Kaliyaraj, D., Rajendran, M., Angamuthu, V., Antony, A. R., Kaari, M., Thangavel, T., Venugopal, G., Joseph, J., and Manikkam, R. 2019. Bioleaching of heavy metals from printed circuit board (PCB) by Streptomyces albidoflavus TN10 isolated from insect nest. *Bioresources and Bioprocessing* 6: 47.

Khajezadeh, M., Abbaszadeh-Goudarzi, K., Pourghadamyari, H., and Kafilzadeh, F. 2020. A newly isolated Streptomyces rimosus strain capable of degrading deltamethrin as a pesticide in agricultural soil. *Journal of Basic Microbiology* 60(5): 435–443.

Khalilzadeh Shirazi, E., Metzger, J., Fischer, K., and Hassani, A. 2020. Removal of textile dyes from single and binary component systems by Persian bentonite and a mixed adsorbent of bentonite/charred dolomite. *Colloids and Surfaces A: Physicochemical and Engineering Aspects* 598: 124807.

Khan, S., Naushad, M., Govarthanan, M., Iqbal, J., and Alfadul, S. M. 2022. Emerging contaminants of high concern for the environment: Current trends and future research. *Environmental Research* 207: 112609.

Kumar, V., Singh, S., and Upadhyay, N. 2019. Effects of organophosphate pesticides on siderophore producing soils microorganisms. *Biocatalysis and Agricultural Biotechnology* 21: 101359.

Li, P., Wang, X., Su, M., Zou, X., Duan, L., and Zhang, H. 2021. Characteristics of plastic pollution in the environment: A review. *Bulletin of Environmental Contamination and Toxicology* 107(4): 577–584.

Lokesh, P., Shobika, R., Omer, S., Reddy, M., Saravanan, P., Rajeshkannan, R., Saravanan, V. and Venkatkumar, S. 2023. Bioremediation of plastics by the help of microbial tool: A way for control of plastic pollution. *Sustainable Chemistry for the Environment* 3: 100027.

Lin, J., Ye, W., Xie, M., Seo, D. H., Luo, J., Wan, Y., and Bruggen, B. V. 2023. Environmental impacts and remediation of dye-containing wastewater. *Nature Reviews Earth and Environment* 4: 785–803.

Lin, Q.S., Chen, S.H., Hu, M.Y., Rizwan-ul-Haq, M., Yang, L., Li, H. 2011. Biodegradation of Cypermethrin by a newly isolated actinomycetes HU-S-01 from wastewater sludge. *International Journal of Environmental Science and Technology* 8: 45–56.

Liu, Y., Zhang, J., Yang, S., and Yang, H. 2023. Biodegradation and removal of heavy oil using Pseudomonas sp. and Bacillus spp. isolated from oily sludge and wastewater in Xinjiang Oilfield, China. *Journal of Environmental Chemical Engineering* 11(3): 110123.

Lu, L., Zeng, G., Fan, C., Chen, A., Chen, M., Jiang, M., Yuan, Y., Wu, H., Lai, M., and He, Y. 2014. Diversity of two-domain laccase-like multicopper oxidase genes in *Streptomyces* spp.: Identification of genes potentially involved in extracellular activities and lignocellulose degradation during composting of agricultural waste. *Applied and Environmental Microbiology* 80(11): 3305–3314.

Lu, L., Zeng, G., Fan, C., Ren, X., Wang, C., Zhao, Q., Zhang, J., Chen, M., Chen, A., and Jiang, M. 2013. Characterization of a laccase-like multicopper oxidase from newly isolated *Streptomyces* sp. C1 in agricultural waste compost and enzymatic decolorization of azo dyes. *Biochemical Engineering Journal* 72: 70–76.

Mane, U. V., Gurav, P. N., Deshmukh, A. M., and Govindwar, S. P. 2008. Degradation of textile dye reactive navy-blue Rx (Reactive blue-59) by an isolated Actinomycete Streptomyces krainskii SUK-5. *Malaysian Journal of Microbiology* 2008: 4:1–5.

Moya, R., Hernandez, M., Garcia Martin, A., Ball, A., and Arias, M. 2009. Contributions to a better comprehension of redox-mediated decolouration and detoxification of azo dyes by a laccase produced by *Streptomyces cyaneus* CECT 3335. *Bioresource Technology* 101: 2224–2229.

Mustafa, S., Bhatti, H. N., Maqbool, M., and Iqbal, M. 2021. Microalgae biosorption, bioaccumulation and biodegradation efficiency for the remediation of wastewater and carbon dioxide mitigation: Prospects, challenges and opportunities. *Journal of Water Process Engineering* 41: 102009.

Nazari, M. T., Simon, V., Machado, B. S., Crestani, L., Marchezi, G., Concolato, G., Ferrari, V., Colla, L. M., and Piccin, J. S. 2022. *Rhodococcus*: A promising genus of actinomycetes for the bioremediation of organic and inorganic contaminants. *Journal of Environmental Management* 323: 116220.

Ngo, A., and Tischler, D. 2022. Microbial degradation of azo dyes: Approaches and prospects for a hazard-free conversion by microorganisms. *International Journal of Environmental Research and Public Health* 19: 4740.

Ogonowski, M., Motiei, A., Ininbergs, K., Hell, É., Gerdes, Z., Udekwu, K. I., Bacsik, Z., and Gorokhova, E. 2018. Evidence for selective bacterial community structuring on microplastics. *Environmental Microbiology* 20: 2796–2808.

Omidoyin, K. C., and Femi-Ola, T. O. 2020. Isolation and screening of *Streptomyces* spp from soil samples of Ekiti State University Nigeria for antibacterial activity. *International Journal of Research Innovation and Entrepreneurship* 1(2): 48–54.

Omidoyin, K. C., and Jho, E. H. 2023. Effect of microplastics on soil microbial community and microbial degradation of microplastics in soil: A review. *Environmental Engineering Research* 28(6): 220706–220710.

Parida, V. K., Saidulu, D., Majumder, A., Srivastava, A., Gupta, B., and Gupta, A. K. 2021. Emerging contaminants in wastewater: A critical review on occurrence, existing legislations, risk assessment, and sustainable treatment alternatives. *Journal of Environmental Chemical Engineering* 9(5): 105966.

Pillai, H. P. J. 2017. Optimization of process conditions for effective degradation of azo blue dye by *Streptomyces* DJP15. *Journal of Pure and Applied Microbiology* 11:1757–1765.

Popovic, N., Pržulj, D., Mladenović, M., Prodanović, O., Ece, S., Ilic Djurdjic, K., Ostafe, R., Fischer, R., and Prodanovic, R. 2021. Immobilization of yeast cell walls with surface displayed laccase from *Streptomyces cyaneus* within dopamine-alginate beads for dye decolorization. *International Journal of Biological Macromolecules* 181: 1072–1080.

Radhakrishnan, A., Balaganesh, P., Vasudevan, M., Natarajan, N., Chauhan, A., Arora, J., Ranjan, A., Rajput, V., Sushkova, S., Minkina, T., Basniwal, R., Kapardar, R., and Srivastav, R. 2023. Bioremediation of hydrocarbon pollutants: Recent promising sustainable approaches, Scope and Challenges. *Sustainability* 15 (7): 5847.

Raimondo, E. E., Saez, J. M., Aparicio, J. D., Fuentes, M. S., and Benimeli, C. S. 2020. Bioremediation of lindane-contaminated soils by combining of bioaugmentation and biostimulation: Effective scaling-up from microcosms to mesocosms. *Journal of Environmental Management* 276: 111309.

Rajapandi, J.D., and Rajamanickam, U. 2024 Low–density polyethylene management by using selective bacterial strains from garbage soil. *Biologia* 79; 985–1001.

Sahmoune, N.S. 2018. Performnce of *Streptomyces rimosus* biomass in biosorption of heavy metals from aqueous solutions. *Microchemical Journal* 141: 87–95.

Santillan, J., Muzlera, A., Molina, M., Lewkowicz, E., and Iribarren, A. 2020. Microbial degradation of organophosphorus pesticides using whole cells and enzyme extracts. *Biodegradation* 31:423–433.

Sathya, K., Nagarajan, K., Carlin Geor, Malar, G., Rajalakshmi, S., and Raja Lakshmi, P. 2022. A comprehensive review on comparison among effluent treatment methods and modern methods of treatment of industrial wastewater effluent from different sources. *Applied Water Science* 12: 70.

Sharma, A., Kumar, V., Shahzad, B., Tanveer, M., Sidhu, G. P. S., Handa, N., Kohli, S. K., Yadav, P., Bali, A. S., and Parihar, R. D. 2019. Worldwide pesticide usage and its impacts on ecosystem. *SN Applied Sciences* 1:1–16.

Shattuck, A., Werner, M., Mempei, F., Dunivin, Z., and Galt, R. 2023. Global pesticide use and trade database (GloPUT): New estimates show pesticide use trends in low-income countries substantially underestimated. *Global Environmental Change* 81: 102693.

Shin, J., Park, S., Lee, J.-S., Lee, E.-J., and Youn, H.-D. 2020. Complete genome sequence and comparative analysis of *Streptomyces seoulensis*, a pioneer strain of nickel superoxide dismutase. *Genes & Genomics* 42(3):273–281.

Soumeya, S., Allaoueddine, B., and Hocine, A. 2022. Biodegradation of used motor oil by *Streptomyces ginkgonis* KM-1-2, isolated from soil polluted by waste oils in the region of Azzaba (Skikda-Algeria). *Journal of Biotechnology* 349: 1–11.

Subramani, R., and Detmer, S. 2019. Marine rare actinomycetes: A promising source of structurally diverse and unique novel natural products. *Marine Drugs* 17(5):249.

Taniguchi, I., Yoshida, S., Hiraga, K., Miyamoto, K., Kimura, Y., and Oda, K. 2019. Biodegradation of PET: Current status and application aspects. *ACS Catalysis* 9(5): 4089–4105.

Walker, T.R., and Fequet, L. 2023. Current trends of unsustainable plastic production and micro(nano)plastic pollution. *TrAC Trends in Analytical Chemistry* 160: 116984.

Wang, B., Wu, S., Chang, X., Chen, J., Ma, J., Wang, P., and Zhu, G. 2020. Characterization of a novel hyper-thermostable and chlorpyrifos-hydrolyzing carboxylesterase EstC: A representative of the new esterase family XIX. *Pesticide Biochemistry and Physiology* 170, 104704.

Wichatham, K., Piyaviriyakul, P., Boontanon, N., Surinkul, N., Visvanathan, C., Fujii, S., and Boontanon S.K. 2024. Biodegradation of polypropylene plastics *in vitro* and natural condition by *Streptomyces* sp. isolated from plastic-contaminated sites. *Environmental Technology and Innovation* 35:103681.

Yahiaoui, C., Kameche, M., Innocent, C., and Khenifi, A. 2021. Conception of yeast microbial desalination cell: Applications to dye wastewater treatment and lead removal. *Chemical Engineering Communications* 208: 364–375.

Yuan, J., Ma, J., Sun, Y., Zhou, T., Zhao, Y., and Yu, F. 2020. Microbial degradation and other environmental aspects of microplastics/plastics. *Science of the Total Environment* 715: 136968.

Zhao, S., and Wang, J. 2024. Biodegradation of atrazine and nicosulfuron by Streptomyces nigra LM01: Performance, degradation pathway and possible genes involved. *Journal of Hazardous Materials* 471:134336.

Zulfiqar, U., Haider, F.U., Maqsood, M.F., Mohy-Ud-Din, W., Shabaan, M., Ahmad, M., Kaleem, M., Ishfaq, M., Aslam, Z., and Shahzad, B. 2023. Recent advances in microbial-assisted remediation of cadmium-contaminated soil. *Plants* 12(17): 3147.

Chapter 9

Biological features of streptomycetes biofilms and their role in the biotechnology industry

Antonina Nikolayeva
Medical University of Karaganda, Kazakhstan

9.1 INTRODUCTION

In our environment, there are many dangers that microbes face at every stage of their existence. Physical and chemical environmental factors influence the metabolism of bacteria and their ability to adapt to external conditions (Rothschild and Mancinelli, 2001). Microorganisms can survive under the harsh conditions of environments with extreme radiation, vacuum pressure or salinity (Horneck et al., 2010; Yamagishi et al., 2018). Other dangerous factors for microbes are non-optimal temperature or non-optimal pH level, nutrient limitation in the environment and antibacterial drugs used in medicine to treat infectious diseases. Microorganisms are afraid of these conditions; they form consortia that allow them to be resistant to extreme exposures (Blanco et al., 2019).

These consortia or biofilms consist of a large number of monospecies or microbes belonging to different species and this community allows them to survive in many stressful situations. Biofilms serve as a natural form of existence of microbes, while planktonic forms is only one of the stages of development of the microbial community. As a result of many studies on biofilms, it has been proven that biofilm formation is suitable for many pathogenic and opportunistic microorganisms (Stoodley et al., 2002) and the mechanism of biofilm formation is assessed as a pathogenicity factor (Costerton et al., 1999).

Biofilm is a stable association of microorganisms surrounded by an extracellular matrix, the basis of which is polysaccharides produced by bacterial cells. Biofilm is the preferred form of existence of bacteria: more than 95% of all microorganisms in our world live as such community. The ability to form biofilms is an integral part of the life cycle of most microorganisms and a successful strategy for protecting bacteria from adverse environmental factors. The formation of biofilms provides a suitable habitat for microbial populations and creates effective metabolic pathways between microbes for their productive activities (Chen et al., 2013). The three-dimensional structure of a biofilm is a very convenient strategy for survival and the formation

DOI: 10.1201/9781032694634-9

of a biofilm is an advantageous position for life compared to planktonic cells (Van Houdt and Michiels, 2010).

The current review presents modern views on the molecular mechanism of biofilm formation in streptomycetes, the role of biofilm in extreme conditions and the ability to use biofilm communities as active producers of substances necessary for humans.

9.2 STRUCTURE OF BIOFILMS

Microbial biofilms are communities formed by related and unrelated bacteria isolated from the external environment. At the same time, cells within communities have specialization and contact with each other. The key features of biofilms are isolation from the environment, formation of extracellular matrix, the presence of bacterial interactions, cooperation of cells forming a biofilm and the presence of different characteristics. All members of normal microflora in the human body are parts of biofilms. Microbes within the biofilm maintain their composition and spread due to cells that are released periodically and migrate, contributing to the spread of infection. Biofilms of different microbes have similar structural compositions. The main components of biofilms are the bacteria, intercellular matrix and surface membrane that isolates the community from the environment (Tets and Tets, 2013; Costerton et al., 2003; Sutherland, 2001).

The composition of the surface membrane and matrix of biofilms presented in Table 9.1. The percentage of polysaccharides is 40–95% (dextran, hyaluronic acid, cellulose, etc.). The concentration of other chemical components varies greatly. The proportion of proteins can go up to 60%, lipids up to 40% and nucleic acids (extracellular DNA and RNA) 1–20%. These compounds are in a hydrated state because 80–90% of the biofilm volume is occupied by water (Tets and Tets, 2013).

It is known that extracellular DNA and proteins of bacterial biofilms play an important role in morphogenesis and functional changes in biofilms (Chebotar et al., 2016). The DNA double helix is a rigid, extended molecule that can act as a support for the jelly-like matrix of the biofilm. Bacterial

Table 9.1 Biofilm chemical composition

Components	*Percentage of matrix*
Polysaccharides	40–95%
Proteins, enzymes	Up to 60%
Lipids	Up to 40%
DNA and RNA	1–20%
Water	80–90%

cells within a biofilm exchange signaling molecules with each other (this phenomenon is known as quorum sensing). Under signals released by neighboring cells, some of the cells are destroyed (autolysis). When cells are lysed, their DNA is released and remains in the biofilm matrix (L. Montanaro et al., 2011). In biofilms, DNA is bound to the DNABII family proteins IHF and HU, which maintain its double helix. Experiments have shown that in the absence of DNABII, the biofilm disintegrates and the bacteria encounter a threatening environment and die (Buzzo et al., 2021).

The matrix, extracellular polymeric substances and the internal environment of biofilm communities play important roles in the life of the microbial community. The matrix also ensures the circulation of nutrients, transport of toxins, metabolites, gases, signaling molecules, and more (Chebotar' et al., 2016; Tets et al., 2017; McDougald et al., 2012; Tetz et al., 1990).

The biofilm matrix can interact with antibiotics, and as a result prevents the penetration of antibacterial drugs into microorganisms of the microbial biofilm. Microbial cells at the border of the biofilm can become victims by absorbing the antibiotic. As a result, microbial cells in the inner layer of the biofilm will be protected from the effects of these antibiotics. In addition, in biofilms, many cells have increased activity of pumps (membrane proteins), which push antibiotics out and make them inactive against biofilm microbes (Venkatesan et al., 2015).

The surface membrane consists of the cell membranes of the members of the community and has similar properties. Bacteria within isolated biofilms gain additional advantages compared to isolated cells. For practical medicine, it is especially important that microbes in biofilms are characterized by an increased survival rate in the presence of antimicrobial substances, immune cells, antibiotics and antiseptics. Bacteria in biofilms survive in the presence of antibiotics in doses up to 500–1000 times higher than their minimum inhibitory concentration (Knorring et al., 2008; Gristina, 1994).

The surface membrane provides protection from external influences, is a barrier to many substances that can negatively affect the formation of a biofilm, is involved in the active removal of antibacterial substances from the cell, and is the site of the presence of the receptors. Extracellular DNA promotes the transportation of genetic information between cells of biofilm. The matrix creates an environment for the microbial community, promotes the accumulation of metabolites and toxins and participates in the transport of signaling molecules, interaction and binding with substances (antibiotics) that can penetrate the microbial community (Khryanin, 2020).

Microbes in biofilms have very high resistance to antibiotics. The efficiency of penetration of antibiotics is related to their ability to penetrate the surface membrane and intercellular matrix of biofilms, which contain a significant amount of various lipids (Tetz et al., 2004; Chambless et al., 2006).

It has been established that ampicillin cannot penetrate effectively into *Klebsiella pneumoniae* communities, and ampicillin, co-trimaxazole and vancomycin cannot penetrate into *Enterococcus faecalis* biofilms (Anderl et al., 2000; Sandoe et al., 2006).

However, there are also special forms of resistance in biofilms that may be associated with certain mechanisms. The first of them is the formation of persister cells. Persisters are cells with inhibited metabolism, therefore, persisters are resistant to almost all drugs (Lewis, 2010; L'jyuis, 2005; Plakunov et al., 2010).

According to various data, the proportion of persisters is different, ranging from 15 to 85%, while it is noted that in unfavorable conditions it increases significantly (Plakunov et al., 2010).

Community resistance may be related to the filtering capacity of the matrix. The matrix forms the internal environment of the community, connecting cells into a single structure, filling all intercellular spaces, forming a three-dimensional complex system that matures along with the community. As a result, the membrane and matrix are a "molecular filter" that is responsible for the functioning of the biofilm (Chebotar' et al., 2016).

Many articles describe the delayed diffusion of antibiotics into biofilms. Superficial cells of the biofilm are more susceptible to the effects of the antibiotic than deep-lying cells (Amorena et al., 1999).

The mechanism of these effects may be either the binding of antibacterial agents or their inactivation. In addition, the antibiotic cannot act on persisters that are inactive and resistant to antibacterial drugs (BalcaZar et al., 2015). The second level of resistance is related to cell wall properties and the presence of efflux pumps that are responsible for antimicrobial resistance. The third level is cytoplasmic, associated with changes in the availability of the antibiotic target and gene regulation.

It has also been shown that spore-forming bacteria associated with the resistance of bacterial spores to the action of antibiotics and antiseptics/disinfectants (Tetz and Tetz, 2017).

The "cheating" phenomenon is a very important stage, which is associated with microbes that are not capable of synthesizing the biofilm matrix and forming biofilms, but can participate in biofilm formation using biofilm components produced by other microbes (Chebotar' et al., 2016; Boyle et al., 2013).

Another key feature of biofilms is Quorum sensing (QS). QS is a process of collective regulation of the expression of certain genes, mediating specific cell behavior and the synthesis of substances.

The formation, growth and differentiation of biofilm cells in biofilms are regulated through mechanisms of intercellular communication (Mehta et al., 2009).

It has been established that intercellular interactions influence the expression of virulence genes, regulate the growth of community members, apoptosis and toxin formation. Microbes use different signaling systems: peptides and cyclopeptides are characteristic of gram-positive microorganisms, and acyl homoserine lactones (AHL) are characteristic of gram-negative microorganisms (Tets and Tets, 2013).

QS mediate relationships in a microbial population, and such interactions are also possible among unrelated microorganisms. Biofilm is also the basis for the exchange of genetic information between bacteria. Antibiotic-resistant bacteria are capable of not only secreting protective enzymes or proteins that can protect other antibiotic-sensitive bacteria in the biofilm (McDougald et al., 2012) but also transmitting genes responsible for antibiotic resistance to other, even unrelated bacteria (Tetz et al., 2004).

9.3 MECHANISM OF BIOFILM DEVELOPMENT

Figure 9.1 shows the diversity of the internal environment of biofilms. In addition to microbial cells with signs of differentiation, the matrix contains proteins and polysaccharides, fatty acids and other products, fragments of DNA and RNA. Development of the biofilm consists of a certain number of stages (Figure 9.2). At the first stage, adhesion of microorganisms occurs on the surface of the biotic or abiotic substrate. This stage is reversible, since the attached cells can return to planktonic form of existence.

The second phase is the final attachment of cells to the surface and it is called fixation or colonization. At this stage, microbes secrete extracellular polymers responsible for strong adhesion.

At the third stage, they have formed microcolonies or individual clusters of adherent cells. At this stage cells are actively dividing, and the released matrix holds all microbial cells together as an adult colony.

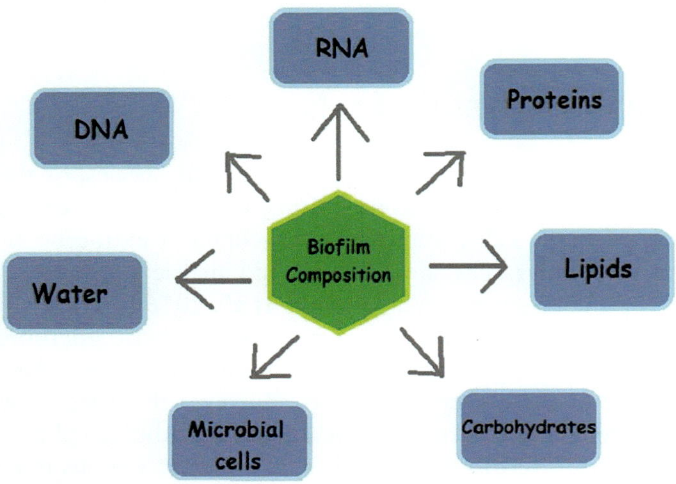

Figure 9.1 Composition of biofilm.

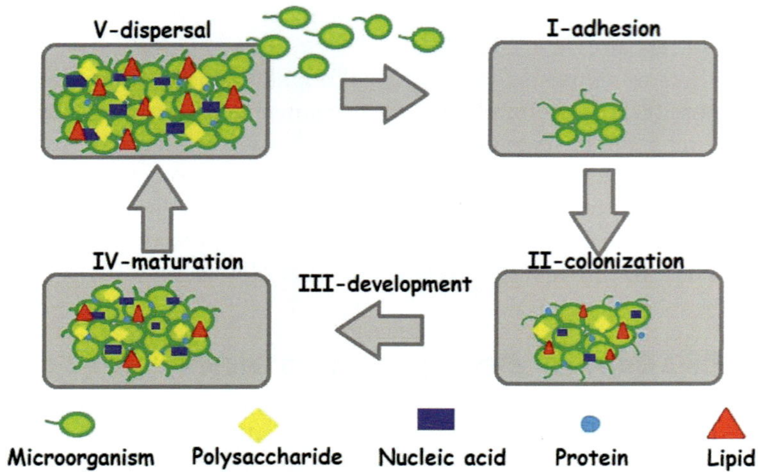

Figure 9.2 Model of microbial biofilm formation.

The microcolonies merge and a mature biofilm is produced that has complex organized structure. This microbial composition can change its own size and shape. The extracellular matrix protects it from external threats.

Microbial cells are able to spread along the substrate using various cell surface elements (pili, fimbriae). Under uncomfortable situations (for example, low content of nutrients), cells are able to leave the biofilm and turn into planktonic form. It is called dispersion (release of bacteria). As a result of dispersion, individual cells periodically go away from the biofilm, capable of attaching to another surface and forming a new colony (Wen Yin et al., 2019).

9.4 STREPTOMYCETES BIOFILMS

Streptomycetes biofilm consists of members of one species, or the biofilm is a community with other microorganisms, which exists as an aggregated population of many cells, immersed in a common extracellular matrix, which consist of polysaccharides, proteins and nucleic acids and is attached to a biotic or abiotic surface (Vinogradova et al., 2016).

Members of the genus *Streptomyces* are saprophytic microorganisms. Their main importance for medicine is the accumulation of natural antibiotics produced by these microbial agents. However, there are a few reports of streptomycetes, which are causative agents of pathological processes in humans. Infections resulting from the installation of various implants are an actual problem. It has been established that streptomycetes found on implants can form biofilms in the human body (Shanmughapriya et al., 2012).

When examining microbial biofilms on buildings that include different materials such as cement, concrete and brick, streptomycetes are very actively involved in the formation of biofilms. Biofilms cause mechanical and chemical damage to objects and destroy different types of buildings (Scheerer et al., 2009; Lupan and Popescu, 2012).

Streptomycetes are found in microbial communities on the surface of seaweed, being the main and permanent components of these biofilms. Marine streptomycetes are adapted to extreme living conditions, and to isolate them it is necessary to select special conditions for cultivation. Some of them are capable of synthesizing antibiotic substances, such as actinomycin D produced by *S. parvullus* (Sarkar et al., 2008).

Streptomycetes are also found in biofilms developing in wastewater facilities. It is known that the treatment of industrial wastewaters effectively occurs with the participation of biofilms. A wide variety of streptomycetes have been identified in biofilms used to treat industrial contaminated waters (El-Shatoury et al., 2004).

Members of the genus *Streptomyces* such as *S. parvullus* (Sarkar et al., 2008) and *S. griseus* (Winn et al., 2014) are capable of forming biofilms. The ability to form biofilms has also been shown for other streptomycetes that have not been identified to species (Winn et al., 2014).

9.5 STRUCTURE OF STREPTOMYCETES BIOFILMS

An important feature of bacterial biofilms is the coexistence of many cells, their formation of an extracellular matrix and the ability to adhere. In addition, biofilms are forms of bacterial multicellular community that are formed during the process of aggregation and exist as a substance attached to the substrate, immersed in the extracellular matrix (Claessen et al., 2014).

Many scientific experiments about streptomycetes biofilms focus on the ability of these microorganisms to adhere to the substrate (El Othmany et al., 2021).

One of the studied streptomycetes that is capable of forming biofilms is *S. coelicolor*, which forms biofilms in liquid media in the presence of calcium salts and other insoluble particles. The authors showed that pellets or mycelium play a role in the formation of biofilms, which are formed when the *S. coelicolor* grows in a liquid medium with spores introduced into the medium. The extracellular matrix plays a large role in this process, which is involved in the adhesion of hyphae in the pellets. When pellets are treated with DNase and hyaluronidase, these structures are destroyed and different fragments of hyphae appear. When kaolin and calcium phosphate are added to the medium, these substances are included by the streptomycetes in the forming of pellets, which is the basis for biofilm formation. Thus, pellets of *S. coelicolor* have characteristics of biofilms: a combination of many cells that contain an extracellular matrix with DNA and hyaluronic acid. Biofilms formed by one species of streptomycetes are called monofilms. Streptomycetes form biofilms on the surface of soil particles, since the aggregation of clay particles together with microorganisms living in the soil contributes to the formation of soil (Kim and Kim, 2004).

It has also been shown that adhesion of streptomycetes occurs with the participation of amyloid proteins chaplins, which are involved in the formation of aerial mycelium. Chaplins proteins are extracellular hydrophobic proteins and adhesion of streptomycetes, as *S.coelicolor* occurs with the participation of these proteins (Vinogradova et al., 2016).

Cellulose plays an important role in the adhesion of *Streptomyces*. Cellulose is involved in the attachment of amyloid fimbriae to the surface of hyphae and takes part in the formation of these fimbriae. Fimbriae are attached to hyphae in the area where these special formations are present. In these formations, cellulose is synthesized and cellulose fibrils are formed. When exposed to the enzyme cellulase, the fimbriae are detached from the surface of the hyphae and the streptomycetes are detached from the substrate. As a result, cellulose plays a significant role in the adhesion of streptomycetes to the substrate (de Jong et al., 2009).

Winn M. and coauthors describe the stages of biofilm formation in *S. griseus*. The aim of this study was to investigate the cultivation of the bacterium *Streptomyces griseus* using a tubular biofilm reactor. It has been established that the development of the *Streptomyces griseus* biofilm occurs continuously and begins with primary adhesion, followed by the development of the biofilm community and the division of the biofilm into separate fragments. When biofilm was co-cultured with *Bacillus amyloliquefaciens*, no biofilm detachment was observed after 96 hours of cultivation, although film detachment from the substrate did occur after exposure to flow. This biofilm revealed a network of fimbriae-like structures that developed poorly in a single-species biofilm and developed significantly better when the microorganisms were co-cultured. The formation of *S. griseus* biofilm did not depend on the speed of liquid medium movement and the chemical composition of this medium. The structure of mono biofilm is a very dense network, which consists of a large number of hyphae. The specific different formations were found on the surface of the hyphae, which are necessary for linking the hyphae into a network (Winn et al., 2014).

9.6 CHARACTERIZATION OF STREPTOMYCETES BIOFILMS

The metabolic activity of microorganisms of biofilms is very different from the metabolism of planktonic cells, and this phenomenon is applied in the biotechnology industry (Vinogradova et al., 2015).

When a microbial biofilm is cultured to produce a desired substance, the amount of that substance in microbial biofilms is greater than in planktonic cells. For example, acetic acid is produced in large proportions using biofilm producers (Maksimova, 2013).

Streptomycetes are active producers of the antibiotic actinomycin D. A marine microorganism *S. parvallus*, when cultivated as a biofilm, exhibits a much higher level of actinomycin D synthesis than a *S. parvallus* isolated from soil. Under experimental conditions, a bioreactor was created for cultivating a microbe, which provides conditions for the growth of this microbe in the form of a biofilm. These conditions for the development of the microbe are similar to those conditions of the ecological niche in which the streptomycetes form biofilms. The marine streptomycetes in the form of a biofilm begins to synthesize actinomycin D in 20 hours of its development, and the maximum accumulation of the antibiotic was detected in 45 hours. In turn, the beginning of antibiotic synthesis by microbes taken from a soil sample occurred in 55 hours of development, and maximum accumulation of antibacterial drug was observed in 144 hours (Sarkar et al., 2008). Biofilm is a coordinated community of microorganisms that, in the shortest time, are able to produce important substances for growth, development and the formation of resistance to external factors.

Now laboratories are copying the conditions for growth and reproduction under which microorganisms live in the environment in order to create comfortable conditions for the formation of substances useful to the human population (Vinogradova et al., 2016).

For example, laboratory staff have developed conditions for cultivating a microorganism in the form of a biofilm capable of oxidizing petroleum. In this case, a positive result was achieved, and it showed an increase in the efficiency of utilization of n-hexadecane (Zhurina et al., 2008).

S. mirabilis, capable of existing in the form of biofilms, has a high ability to detoxify the atmosphere, removing a toxic compound K2Cr2O7 from the environment by enzymatically converting this substance into trivalent chromium. The detoxification activity of biofilms of these microorganisms is better than that of planktonic cells. Removal of toxic compounds from the environment is better achieved by microbial biofilms *S. mirabilis*. Microbial biofilm has tolerance to toxic compound K2Cr2O7, but microbial culture without biofilm formation is sensitive to this compound (Morales et al., 2007). Microbial planktonic cells are more sensitive to external influences than microorganisms in the form of a biofilm. This is due to the presence of factors characteristic of a biofilm: the presence of an extracellular matrix (Flemming and Wingender, 2010), nature of the substrate on which the biofilm is formed (Flemming et al., 2016) and the production of resistance genes characteristic of a biofilm (Ghigo, 2001).

Streptomyces strains were tested for biofilm formation and activation of plant growth ability. Biofilms were induced on the perlite carrier and colonize the rhizosphere and roots of *Triticum aestivum*, resulting in increase of total biomass of these plants. Successful formation of microbial consortia was observed when mixed strains of *Streptomyces* were applied, which suggests long-term interaction with plant cells with subsequent activation of their growth (Domínguez-González et al., 2022). Many unfavorable environmental factors cause the formation of microbial biofilms, and microorganisms are always trying to find the optimal way to exist in the environment and biofilms provide them this opportunity.

9.7 CONCLUSION

The existence of *Streptomyces* as biofilm communities gives them many advantages. In biofilms, microorganisms are able to survive in difficult and extreme conditions. Microorganisms living in the form of biofilms have a higher resistance to different environmental factors, such as resistance to a wide variety of stress factors, exposure to antibiotics, the ability to actively consume nutritional resources and to protect themselves from other microorganisms, colonize new habitats and acquire beneficial metabolic activity.

Studying the distribution of streptomycetes biofilms, identifying environmental factors that contribute to the formation of biofilms and studying the processes of their development is the basis for the study of microbial biofilms. An actual problem is the prevention of microbial biofilms formation, control of the formation of unwanted biofilms, since they cause damage to industrial processes and cause many chronic diseases. Treatment of chronic infectious diseases associated with the formation of biofilm communities represents a significant problem for the human population, as there is a need to find new methods of effective treatment. On the other hand, the ability of streptomycetes to form biofilms is applied for the effective synthesis of a number of useful substances, such as antibiotics or the disposal of toxic compounds that can cause significant harm to the human population.

REFERENCES

Amorena, B., Gracia, E., Monzon, M. et al. 1999. Antibiotic susceptibility assay for Staphylococcus aureus in biofilms developed in vitro. *J Antimicrob Chemother*; 44: 43–55. doi: 10.1093/jac/44.1.43

Anderl, J.N., Franklin, M.J., Stewart P.S. 2000. Role of antibiotic penetration limitation in Klebsiella pneumoniae biofilm resistance to ampicillin and ciprofloxacin. *Antimicrob Agents Chemother*; 44: 1818–1824.

BalcaZar, J. L., Subirats, J., Borrego C.M. 2015. The role of biofilms as environmental reservoirs of antibiotic resistance. *Front Microbiol*; 6: 1216. doi: 10.3389/fmicb.2015.01216

Blanco, Y., Rivas, L.A., Gonzalez-Toril, E., et al. 2019. Environmental parameters, and not phylogeny, determine the composition of extracellular polymeric substances in microbial mats from extreme environments. *Sci Total Environ* 650: 384–393.

Boyle, K.E., Heilmann, S., van Ditmarsch, D., Xavier, J.B. 2013. Exploiting social evolution in biofilms. *Curr Opin Microbiol*; 16 (2): 207–212. doi: 10.1016/j.mib.2013.01.003

Buzzo, John R., Devaraj, Aishwarya, Gloag, Erin S., et al. 2021. Z-form extracellular DNA is a structural component of the bacterial biofilm matrix. *Cell*; 184(23): 5740–5758.e17. DOI: 10.1016/j.cell.2021.10.010

Chambless, J.D., Hunt, S.M., Philip, S.S. 2006. A three-dimensional computer model of four hypothetical mechanisms protecting biofilms from antimicrobials. *Appl Environ Microbiol*; 72: 2005–2013. doi: 10.1128/AEM.72.3.2005-2013.2006

Chebotar', I.V., Mayanskij, A.N., Mayanskij, N.A. 2016. Matriks mikrobnykh bioplenok. *Klinicheskaya mikrobiologiya i antimikrobnaya khimioterapiya*; 18 (1): 9–19. [in Russian]

Chen, Y.; Yan, F.; Chai, Y., et al. 2013. Biocontrol of tomato wilt disease by *Bacillus subtilis* isolates from natural environments depends on conserved genes mediating biofilm formation. *Environ Microbiol*, 15, 848–864.

Claessen, D., Rozen, D.E., Kuipers, O.P. et al. 2014. Bacterial solutions to multicellularity: a tale of biofilms, filaments and fruiting bodies. *Nat Rev Microbiol*; 12: 2: 115–124.

Costerton, J.W., Stewart, P.S., Greenberg, E.P. 1999. Bacterial biofilms: a common cause of persistent infections. *Science*; 284: 1318–1322. doi: 10.1126/science.284.5418.1318

Costerton, W., Veeh, R., Shirtliff, M. et al. 2003. The application of biofilm science to the study and control of chronic bacterial infections. *Clin Investig*, 112: 1466–1477. doi: 10.1172/JCI20365

de Jong, W., Wosten, H.A.B., Dijkhuizen, L., Claessen, D. 2009. Attachment of *Streptomyces coelicolor* is mediated by amyloidal fimbriae that are anchored to the cell surface via cellulose. *Mol Microbiol*; 73: 6: 1128–1140.

Domínguez-González, K.G.; Robledo-Medrano, J.J.; Valdez-Alarcón, J.J. et al. 2022. *Streptomyces spp.* biofilmed solid inoculant improves microbial survival and plant-growth efficiency of *Triticum aestivum*. *Appl Sci 12*, 11425. https://doi.org/10.3390/app122211425

El Othmany, R., Zahir, H., Ellouali, M., Latrache, H. 2021. Current understanding on adhesion and biofilm development in *Actinobacteria*. *Int J Microbiol*: 6637438. doi: 10.1155/2021/6637438. PMID: 34122552; PMCID: PMC8166509.

El-Shatoury, S., Mitchell, J., Bahgat, M., Dewedar, A. 2004. Biodiversity of *Actinomycetes* in constructed wetland for industrial effluent treatment. *Actinomycetologica*; 18: 1: 1–7.

Flemming, H. C., Wingender, J., Szewzyk, U., Steinberg, P., Rice, S. A., Kjelleberg, S. 2016. Biofilms: an emergent form of bacterial life. *Nat Rev Microbiol.* 14: 563–575.

Flemming, H.C., Wingender, J. 2010. The biofilm matrix. *Nat Rev Microbiol*; 8:632–633.

Ghigo, J.M. 2001. Natural conjugative plasmids induce bacterial biofilm development. *Nature* 412 (6845): 442–445.

Gristina, A.G. 1994. Biofilms and chronic bacterial infections. *Clin Microbiol Newsl*; 16 (22): 171–176. doi: //doi.org/10.1016/0196-4399(94)90037-X

Horneck, G., Klaus, D. M., Mancinelli, R. L. 2010. Space microbiology. *Microbiol Mol Biol Rev*; 74: 121–156. doi://doi.org10.1128/MMBR.00016-09

Khryanin, A. A. 2020. Microbial Biofilms: Modern Concepts. *Novosibirsk State Medical University, Novosibirsk* doi: 10.37489/0235-2990-2020-65-5-6-70-77

Kim, Y.M., Kim, J.H. 2004. Formation and dispersion of mycelial pellets of *S. coelicolor* A3 (2). *J Microbiol*; 42: 1: 64–67.

Knorring, G.Jyu., Sternin, Jyu.I., Minaev, S.V. et al. 2008. Intensifikatsiya antibakterial'noj terapii pri gnojnovospalitel'nykh zabolevaniyakh. *Voenno-Meditsinskij Zhurnal*; 329 (10): 35–41. [in Russian]

Lewis, K. 2010. Persister cells. *Annu Rev Microbiol*; 64: 357–372. doi: 10.1146/annurev.micro.112408.134306

L'jyuis K. 2005. Persistirujyushchie kletki i zagadka vyzhivaniya bioplenok. *Biokhimiya*; 70 (2): 327–336. [in Russian]

Lupan, I., Popescu, O. 2012. Metagenomics and future perspectives for biodeterioration and biodegradation studies. *Annals of RSCB*; 17: 2: 37–42.

Maksimova, Ju.G. 2013. Mikrobnye bioplenki v biotehnologicheskih processah. *Biotehnologija*; 4: 9–23. [in Russian]

McDougald, D., Rice, S., Barraud, N. et al. 2012. Should we stay or should we go: mechanisms and ecological consequences for biofilm dispersal. *Nat Rev Microbiol*; 10: 39–50. doi: https://doi.org/10.1038/nrmicro2695

Mehta, P. et al. 2009. Information processing and signal integration in bacterial quorum sensing. *Mol Syst Biol*; 5: 325. doi: 10.1038/msb.2009.79

Montanaro, L, Poggi, A, Visai, L, et al. 2011. Extracellular DNA in Biofilms. *Inter J Artif Org*; 34(9): 824–831. doi:10.5301/ijao.5000051

Morales, D.K., Ocampo, W., Zambrano, M.M. 2007. Efficient removal of hexavalent chromium by the toxic effect of metal exposure. *Appl Microbiol*; 193: 6: 2704–2712.

Plakunov, V.K., Strelkova, E.A., Zhurina, M.V. 2010. Persistentsiya i adaptivnyj mutagenez v bioplenkakh. *Mikrobiologiya*; 79(4): 447–458. [in Russian]

Rothschild, L.J., Mancinelli, R.L. 2001. Life in extreme environments. *Nature*; 409: 1092–1101.

Sandoe, J.A.T., Wysome, J., West, A.P. et al. 2006. Measurement of ampicillin, vancomycin, linezolid and gentamicin activity against enterococcal biofilms. *J Antimicrob Chemother*; 57: 767–770.

Sarkar, S., Saha, M., Roy, D. et al. 2008. Enhanced production of antimicrobial compounds by three salt-tolerant actinobacterial strains isolated from the Sundarbans in a niche-mimic bioreactor. *Mar Biotechnol (NY)*; 10: 5: 518–526.

Scheerer, S., Ortega-Morales, O., Gaylarde, C. 2009. Microbial deterioration of stone monuments — an updated overview. *Adv Appl Microbiol*; 66: 97–139.

Shanmughapriya, S., Francis, A.L., Kavitha, S., Natarajaseenivasan, K. 2012. In vitro actinomycete biofilm development and inhibition by polyene antibiotic, nystatin, on IUD copper surfaces. *Biofouling*; 28: 9: 929–935.

Stoodley, P. Sauer, K., Davies, D. G. et al. 2002. Biofilms as complex differentiated communities. *Annu Rev Microbiol*; 56: 187–209. doi: 10.1146/annurev.micro.56.012302.160705

Sutherland, I.W. 2001. Biofilm exopolysaccharides: a strong and sticky framework. *Microbiology*; 147: 3–9. doi: 10.1099/00221287-147-1-3

Tets, G.V., Artemenko, N.K., Vecherkovskaya, M.F., et al. 2017. Rol' matrichnykh komponentov vo vzaimodejstvii bakteriofagov s bioplenochnymi bakteriyami. *Prakticheskaya pul'monologiya*; 3: 55–57. [in Russian]

Tets, V.V., Tets, G.V. 2013. Mikrobnye bioplenki i problemy antibiotikoterapii. *Atmosfera Pul'monologiya i Allergologiya*; 4: 60–64. [in Russian]

Tetz, G.V., Tetz, V.V. 2017. Introducing the sporobiota and sporobiome. *Gut Pathogens*; 9: 38. doi: 10.1186/s13099-017-0187-8

Tetz, V.V., Korobov, V.P., Artemenko, N.K. et al. 2004. Extracellular phospholipids of isolated bacterial communities. *Biofilms*; 1: 149–155. doi: 10.1017/S147905050400136X

Tetz, V.V., Rybalchenko, O.V., Savkova, G.A. 1990. Ultrastructural features of microbial colony organization. *J Basic Microbiol*; 30 (8): 597–607. doi: 10.1002/jobm.3620300819

Van Houdt, R., Michiels, C.W. 2010. Biofilm formation and the food industry, a focus on the bacterial outer surface. *J Appl Microbiol*; 109, 1117–1131.

Venkatesan Nandakumar, Perumal Govindaraj, Doble Mukesh. 2015. Bacterial resistance in biofilm-associated bacteria. *Future Microbiol* 10: 1743–1750.

Vinogradova K.A., Bulgakova V.G., Polin A.N., Kozhevin P.A. 2015. O bioplenkah streptomicetov. II. *Biotehnologicheskoe ispol'zovanie Antibiotiki i himioter*, 60: 5–6: 27–33. [in Russian]

Vinogradova, K.A., Bulgakova, V.G., Polin, A.N. 2016. Streptomycetes in the light of the concept of multicellularity of bacteria. *Antibiot Khimioter*, 61(7–8): 33–47. [in Russian] PMID: 29533559.

Wen Yin, Yiting Wang, Lu Liu et al. 2019. Biofilms: The Microbial "Protective Clothing" in Extreme Environments. *Int J Mol Sci*; 20(14), 3423; https://doi.org/10.3390/ijms20143423

Winn, M., Casey, E., Habimana, O., Murphy, C.D. 2014. Characteristic of *Streptomyces griseus* biofilms in continuous flow tubular reactors. *FEMS Microbiol Lett*; 352: 2: 157–164.

Yamagishi. A., Kawaguchi. Y., Hashimoto. H., et al. 2018. Environmental data and survival data of *Deinococcus aetherius* from the exposure facility of the Japan experimental module of the international space station obtained by the tanpopo mission. *Astrobiology*; 18: 1369–1374. 10.1089/ast.2017.1751

Zhurina, M.V., Strelkova, E.A., Plakunov, V.K., Beljaev, S.S. 2008. Vlijanie sostava rekonstruirovannyh bioplenok na aktivnost' parafinokisljajushhih bakterij. *Mikrobiologija*; 77: 5: 701–703. [in Russian]

Chapter 10

Streptomyces as source of secondary metabolites and bioprocessing strategies for enhanced production of secondary metabolites

Vivek Keshri
Kashi Institute of Pharmacy, Varanasi, India

Archana
Kashi Institute of Technology, Varanasi, India
Motilal Nehru National Institute of Technology Allahabad, U.P., India

Deepanshi Rajput
Jawaharlal Nehru University, New Delhi, India

Punit Kumar
Karaganda Medical University, Karaganda, Kazakhstan

10.1 INTRODUCTION

Streptomyces species are identified as one of the most prolific sources of large number of secondary metabolites including antibiotics. Due to their complex secondary metabolism, *Streptomyces* are considered a biofactory for a diverse range of natural products (Figure 10.1). It is assumed that about 17% of biologically active compounds are produced from *Streptomyces*. Moreover, *Streptomyces hygroscopicus* is credited to produce about 180 bioactive metabolites with diverse activities (Harir et al. 2018). Many of the world's antibiotics come from the bacterial genus *Streptomyces*, which makes them a highly significant group of bacteria (de Lima Procópio et al. 2012). The aerobic, Gram-positive bacterium *Streptomyces* has a filamentous form that resembles fungi and may grow in a variety of settings. They can be found practically anywhere, from high mountains to the deep sea (Ribeiro da Cunha et al. 2019; Sarmiento-Vizcaíno et al. 2018; Waksman et al. 2010). *Streptomyces*' capacity to generate bioactive secondary metabolites, such as antimicrobials (antibiotics, antifungals, antivirus), immunosuppressive agents, antitumor, antihypertensives, anti-obesity, and so on, is perhaps its most intriguing characteristic (Omura et al. 2001; Khan 2011; Patzer & Volkmar 2010; Harir et al. 2018). *Streptomyces* develop secondary

DOI: 10.1201/9781032694634-10

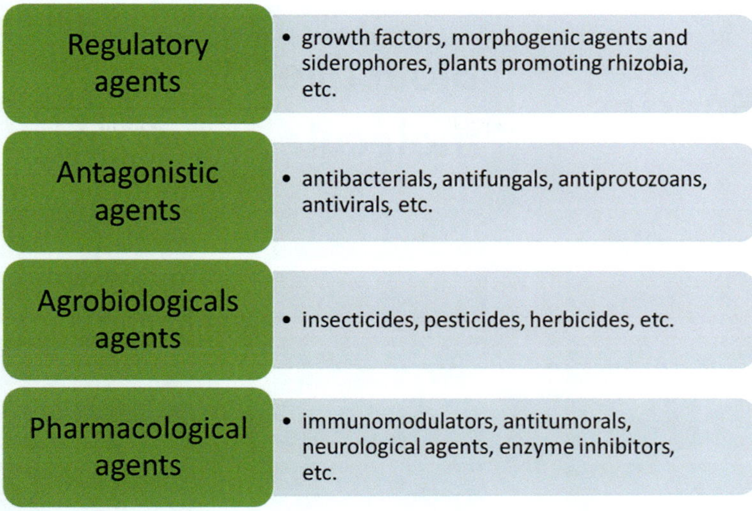

Figure 10.1 Metabolites from *Streptomyces* may be divided into four broad categories; regulatory, antagonistic, agrobiologicals, and pharmacological agents (Harir et al. 2018).

metabolites during this growth phase, which are substances that can provide the organism a competitive advantage and are not strictly required for growth or reproduction (Chater 2016).

Not mobile, *Streptomyces* spread by forming thread-like projections called hyphae that pierce materials in search of nourishment. *Streptomyces* produces dividing aerial hyphae that tolerate adverse conditions and spread rapidly to new locations or nutrient sources when resources are scarce. Because of *Streptomyces*' relatively big genome—which can be up to four times larger than certain other bacterial genomes—they are able to exhibit this molecular diversity. In this chapter we will discuss the different types of bioactive compounds produced from *Streptomyces* spp. and different source of *Streptomyces*. Approaches for optimization of culture conditions for the enhanced production of bioactive compounds are also discussed. Furthermore, advanced optimization approaches and characterization techniques for bioactive compounds are also mentioned.

10.2 *STREPTOMYCES* AS SOURCE OF ANTIBIOTICS

Streptomyces bacteria are a well-recognized source for the production of different types of antibiotics (Figure 10.2). Even currently, *Streptomyces* genus is providing many clinically used antibiotics. *Streptomyces* is also the source of the important antibiotic streptomycin which was discovered in 1943. Albert Schatz discovered streptomycin with the help of Doris

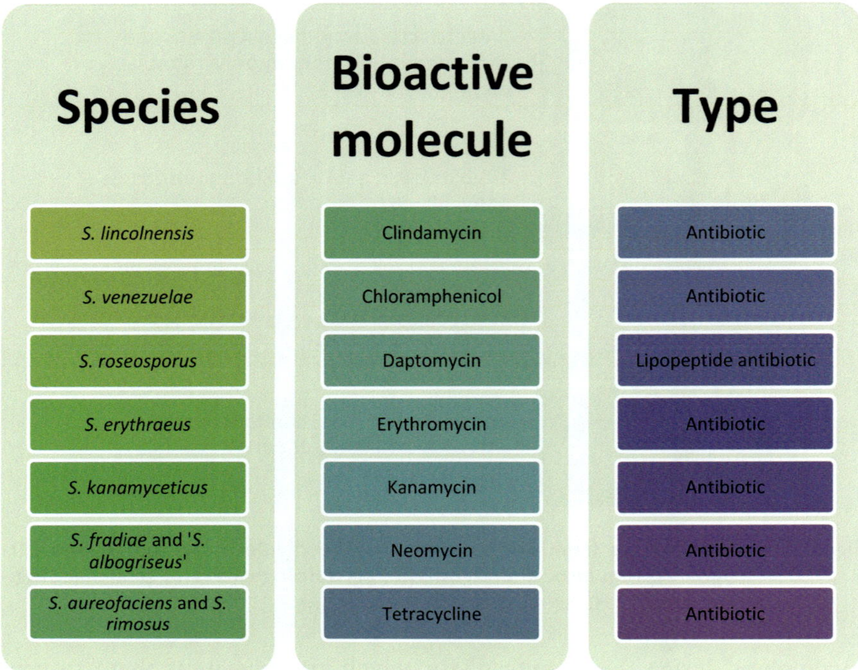

Figure 10.2 Production of different types of antibiotics from *Streptomyces* sp.

Ralston, Elizabeth Bougie, Christine Riley, and others (Wainwright 1991; Quinn et al. 2020). Streptomycin was discovered in two samples—one from *Streptomyces griseus* obtained from highly fertilized compost soil and the other from chicken stomach (Wainwright 1991). Streptomycin was further analyzed for its effectiveness against tuberculosis, bubonic plague, typhus, and cholera (Wainwright 1991; Waksman et al. 2010; Schatz et al. 1944; Waksman & Schatz 1946).

Although it was relatively simple to isolate new biochemical agents from *Streptomyces* in the laboratory, only three new classes of antibiotics have been identified since 1985. A new class of antibiotics, platensimycin produced from *Streptomyces platesonsis*, is one such compound. This was discovered by the Merck group (Martens & Demain 2011, Wang et al. 2006, Davies 2006). A *Streptomyces* strain with thermotolerant antibiotics was isolated. *Streptomyces* sp. TM32 was isolated from the rhizosphere of a medicinal plant *Curcuma longa* L. (Nakaew et al. 2019). A recently isolated strain of *Streptomyces sioyaensis* has been identified to possess strong antibacterial capabilities against plant and human pathogens, including the antibiotic-resistant strain of *Staphylococcus haemolyticus* MR-CoNS. It is believed that *Streptomyces sioyaensis* may also provide a new platform for the discovery of novel and advantageous bioactive compounds.

10.3 VARIOUS SOURCES OF *STREPTOMYCES* AND RELATED ANTIBIOTICS

10.3.1 *Streptomyces* from soils

Soils comprises a biologically and nutritionally complex system for the growth of microorganisms. Soil is the natural habitat of *Streptomyces* and thus different species of *Streptomyces* genus are found widely distributed in soil (Oskay 2009). Even streptomycin-producing *Streptomyces* were isolated from the soil. *Streptomyces* from soil are a major source of bioactive metabolites. Different types of soil habitats have been explored for the isolation of *Streptomyces* sp. such as forest soil, garden soil, grasslands, dumpsites, riverside, and more (Antido & Climacosa 2022). Researchers are also exploring non-agriculture soils, unique and unexplored habitats for the isolation of novel strains. Large numbers of scientific publications have reported the isolation of novel *Streptomyces* sp. from soil and have been analyzed for production of novel bioactive compounds. Furthermore, isolation approaches are also modified. Here we are mentioning some recent examples; Antido & Climacosa (2022) reported the isolation of novel *Streptomyces* using a modified integrated approach from different soil habitats in Calamba City, Laguna, Philippines. Al-Joubori et al. (2023) isolated a novel strain of *Streptomyces* from terrestrial habitats. These strains were also reported for antimicrobial and cytotoxic activities. Saadoun et al. (2023) reported isolation of *Streptomyces* sp. from fuel-oil polluted soils (soils of gas stations). Some of the isolates also demonstrated diesel degradation properties. Maiti et al. (2020) reported isolation of *Streptomyces* sp. SM01 from India. This bacterial strain produced an antibiotic compound, picolinamycin, which exhibited antimicrobial activity against multidrug resistance strains.

10.3.2 *Streptomyces* from marine sources

Marine ecosystems are also important sources for isolation of actinomycetes. Tenebro et al. (2021) isolated 2212 actinomycete strains from marine sediment from 11 geographical sites, and out of them 92 strains demonstrated antimicrobial activities against multidrug-resistant *Escherichia coli*, *Pseudomonas aeruginosa*, and *Staphylococcus aureus*. The molecular analysis revealed strains from *Streptomyces* genus, with the most dominant species as *Streptomyces parvulus*. Shrestha et al. (2021) reported the isolation of antibiotic-producing *Streptomyces* from water and soil samples. In this study, it was also revealed that bacteria isolated from water sources demonstrated higher antimicrobial activity than soil sources. Baskaran and Muthukumarasamy (2017) reported the isolation of *Streptomyces* sp. from Kodiyakkarai marine segment of east coastal of India. These strains produced the amylase enzyme and demonstrated amylolytic enzyme activity.

10.3.3 *Streptomyces* from traditional medicinal plants

Traditional medicine research may have been influenced by Geoffrey Cordell's development of systematic searches for plant compounds to develop anticancer drugs. A method that utilized ethnomedical principles was employed to evaluate documentary or historical evidence from traditional medical practices (Cordell et al. 1994). Traditional plant medicines have been used to isolate many species of *Streptomyces* in areas where nutrient levels are low or under extreme physiological conditions (Wardecki et al. 2015; Oberhofer et al. 2019). The life cycles of *Streptomyces* can be either spent in plant tissue as endoparasites or epiphytes that live outside the plant, without causing any harm (Nalini & Prakash 2017; Zhao et al. 2020). Proteobacteria and Actinobacteria occupy the position of the most frequently occurring endophytic species (Liu et al. 2014). Despite the fact that many studies have focused on traditional medicine's most populous areas such as India and China, the discoveries have been made in South America and Africa (Colombo et al. 2019; Qin et al. 2009). Many strains of *Streptomyces* have been isolated from medicinal plants (Figure 10.3).

Endophytes are also considered to provide protection to host plants from abiotic and biotic stresses and produce bioactive secondary metabolites. These endophyte-derived secondary metabolites are assumed to be source for drugs with various applications. It is also suggested that approaches like metabolic engineering and genome mining may be used for identification of novel compounds (Zotchev 2024).

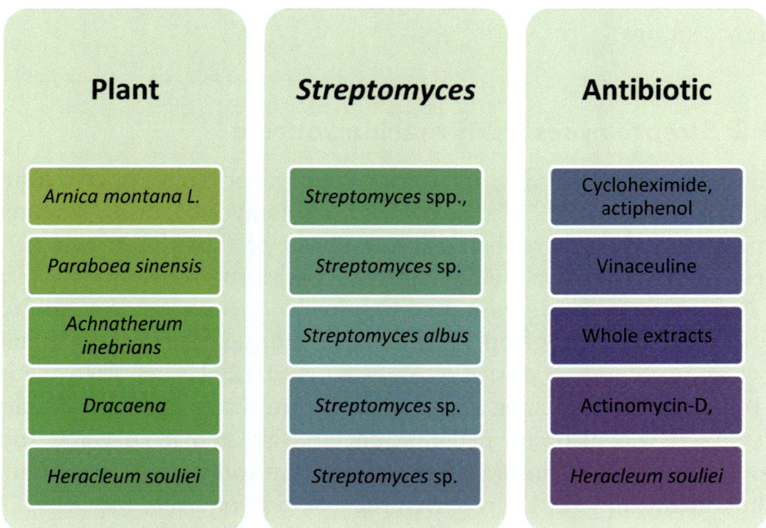

Figure 10.3 *Streptomyces* sp. obtained from medical plants.

10.3.4 Invertebrates as source of *Streptomyces*-linked traditional medicine

Streptomyces and traditional medicine are frequently linked to invertebrates (Chakravorty et al. 2011; Liu et al. 2018; Chevrette et al. 2019). A paste containing crushed black ants (*Bothroponera rufipes*) is reported to be effective in treating boils, scabies, and wounds in northern India (Seabrooks & Hu 2017). Additionally, ground ants mixed with water can be utilized to ease toothache (Chakravorty et al. 2011). These factors are intriguing in that *Streptomyces* is found associated with ant exoskeleton. At times, ants may rub their feet and later on apply it to the soil where the fungi can thrive (Qin et al. 2017; Currie et al. 1999). The leaf-cutter ant (Acromyrmex) is an excellent illustration of this as it employs *Pseudonocardia*, a type of actinomycete, to safeguard its fungal garden against invasive parasites (Poulsen et al. 2010).

African *Tetraponela pendigiari*, which lives in burrows in acacia trees called domatia, was also found to harbor *Streptomyces formicae*, which produces a powerful antibiotic known as formicamycin. This antibiotic has been found effective against methicillin-resistant *Staphylococcus aureus* (MRSA) and vancomycin-resistant enterococci (Qin et al. 2017). The interaction between ants and *Streptomyces* led researchers to investigate other symbiotic relationships in invertebrates. For example, mud wasps were recently discovered to be associated with *Streptomyces*, which produces a polyunsaturated polyoxygenated macrocyclic lactam (sicelifloractam), an antifungal agent (Poulsen et al. 2011).

10.3.5 *Streptomyces* from sponges

Although sponge fungi have contributed to traditional medicine for a long time (Müller et al. 2004), they have been discovered to have a symbiotic relationship with the antibiotic-producing *Streptomyces* genus (Pronzato & Manconi 2008). *Streptomyces* from the marine environment are considered to be a potential source of new antimicrobial compounds (Shaik et al. 2017). A group of Chinese researchers reported the isolation of an actinobacterium strain HNM0039^{T} from a marine sponge sample. Its name was proposed as *Streptomyces tirandamycinicus* sp. nov. The crude extract of this isolated strain demonstrated antibacterial activity against *Streptococcus agalactiae*. Including this, active components were identified as tirandamycins A and B (Huang et al. 2019). Other groups of researchers reported the isolation of active compounds from the genus *Streptomyces*, associated with sponges in the Vietnam Sea. Some of these novel compounds exhibited activity against both Gram-negative bacteria and Gram-positive bacteria, as well as *Mycobacterium tuberculosis* (Cao et al. 2019). These new bioactive compounds also inhibited bacterial formation of biofilm (Balasubramanian et al. 2018).

10.3.6 *Streptomyces* from caves

In recent decades, karst and cave environments have emerged as new areas for antimicrobial *Streptomyces* bioprospecting (Gosse et al. 2019; Rangseekaew and Pathom-Aree 2019). Many of such areas are generally associated with traditional medicine. In the ancient texts, it was mentioned that moon milk (milky exudate that coats the surface of some caves) is able to treat some diseases (Maciejewska et al. 2016). Upon detailed examination, studies have shown that moon milk is rich in *Streptomyces* sp., which has antibacterial effects against various bacteria and fungi (Rangseekaew and Pathom-Aree, 2019; Maciejewska et al. 2016), and is associated with invasive mycoses. It exhibits strong growth inhibition against the multidrug-resistant *Rasamsonia argiracea*, which is a pathogen, causing cystic fibrosis and chronic granulomatous disease (Maciejewska et al. 2016). A similar study in Iran's Hampoeir Cave (dolomite with limestone), associated with Paleolithic settlement, revealed many streptomycetes that produce antimicrobial substances as well as other species (Hamedi et al. 2019).

10.3.7 *Streptomyces* from traditional soil-based medicine

Since most modern antibiotics are obtained from the soil-based *Streptomyces* genus, it is not surprising that this medium also plays an important role in traditional medicine (Falkinham et al. 2009; Behroozian et al. 2016). Unlike *Streptomyces*, which was discovered during the mid-20th century (also termed as golden age of antibiotic discovery), traditional medicinal soils are site-specific, such as bohemian clays, and type-specific, such as Canada's glacial clays (Behroozian et al. 2016; Terra et al. 2018). Although clay itself has long been considered to have therapeutic properties, clay also hosts a variety of *Streptomyces* (Falkinham et al. 2009; Behroozian et al. 2016; Terra et al. 2018; Svensson et al. 2017). Traditional glacial clay from Kisameet Bay, Canada, has been used by the Heiltsuk people for thousands of years to treat skin infections (Behroozian et al. 2016). When it was analyzed in laboratory, this soil exhibited the inhibition of the growth of six ESKAPE pathogens: *Acinetobacter baumannii, Enterobacter* sp., *Enterococcus faecium*, *Klebsiella pneumoniae*, *Pseudomonas aeruginosa*, and *Staphylococcus aureus* (Behroozian et al. 2016). Across the Atlantic, researchers in Northern Ireland isolated a new species of *Streptomyces* from ancient soil remedies in an area called Boho in West Fermanagh (Terra et al. 2018). This alkaline soil overlying carbonaceous limestone contained a new species, *Streptomyces* sp. Mirophorea inhibited multiple strains of MRSA, vancomycin-resistant enterococci, carbapenem-resistant *Pseudomonas aeruginosa*, and *Acinetobacter baumannii* (Terra et al. 2018).

10.4 PRODUCTION OF BIOACTIVE COMPOUNDS FROM *STREPTOMYCES*

The bioactive compounds from *Streptomyces* are produced as secondary metabolites during the stationary phase of bacterial growth. These metabolites are considered not necessary requirements for growth and development of the *Streptomyces*.

10.4.1 Antibiotic synthesizing gene clusters

After sequencing the entire genome of *Streptomyces coelicolor* A3(2), the investigators found that its genome synthesized 10 times as many antibiotics instead of the two or three that are typically identified in the laboratory (Bentley et al. 2002).

This clearly indicates that there is presence of other antibiotic synthesizing clusters which are not always involved in antibiotic synthesis; such clusters are also referred to as "silent clusters" (Chiang et al. 2011). Recently, it has been reported that stress and extreme environmental conditions may influence expression of these silent clusters to produce antibiotics (Reen et al. 2015; Okada & Seyedsayamdost 2017; Baral et al. 2018). It is sometimes referred to as the "one strain, many compounds" phenomenon (Romano et al. 2018). It is observed that many traditional medicines containing *Streptomyces* are found associated with extreme and typically nutrient-poor environments such as deserts, highlands, salt flats, and cold regions, where these bacteria interact with native flora, forming a symbiotic relationship (Zhao et al. 2020; Sivalingam et al. 2019). Such extreme conditions may also be explored in the laboratory to stimulate silent gene clusters. Researchers are trying to analyze these extreme environments in the hope that actinomycetes in these regions express a different repertoire of antibiotics. In Russia, one group of researchers examined the intestines of amphipods from the bottom of Lake Baikal, where temperature generally does not rise higher than 4°C. This group isolated many *Streptomyces* strains that produced antibiotics that were effective against Gram-positive and Gram-negative bacteria (Shishlyannikova et al. 2017; Terkina et al. 2006).

10.4.2 Streptomycetes producing secondary metabolites in a biotechnological manner

Since the *Streptomyces* genus produces secondary metabolites with biomedical uses, including immune-suppressants, antimicrobials, anthelmintic agents, antitumor medications, antihypertensive medications, and intriguing pigments, it plays a significant role in biotechnology (Sanchez et al. 2012). These days, *Streptomyces* strains produce two-thirds of all antibiotics available for use in medicine, veterinary care, and agriculture. These strains produce over 5000 different molecules, with a global market valued at over 46 billion dollars in 2018. Among them are aminoglycosides, polyenes, macrolides, lactams, and tetracyclines. Fermentation in sizable bioreactors is the commonly used approach to produce secondary metabolites from streptomycetes, but fermentation methods also have some limitations. Due to the fact that these compounds are mostly produced in low concentrations during the late growth or stationary phase, industrial methods are typically time-consuming and expensive with low yields, low productivities, and so on (Olmos et al. 2013).

A successful industrial process always starts with carefully considering the many important factors such as; strain selection, growth medium composition, temperature and pH, oxygen transfer rate within the reactor vessel, and aggregation of mycelium. More creative methods such as mycelium immobilization-based fermentation, metabolic pathway control and precursor supplementation, genetic modification of strain, and metabolic engineering approaches have been combined with traditional methods to regulate these parameters in recent years. From this angle, the examination of the morphology using techniques such as scanning electron microscopy, transmission electron microscopy, or confocal-laser scanning fluorescence microscopy has proven essential in both submerged and immobilized fermentation techniques. It has allowed researchers to correlate changes in the mycelium with the various techniques used and the production of secondary metabolites (Pereira et al. 2008; López-García et al. 2014; Chakravarty & Kundu 2016).

10.5 ISOLATION OF NOVEL *STREPTOMYCES* SP.

As it is well elucidated, *Streptomyces* species are one of the most prolific sources of many types of novel secondary metabolites (Sivalingam et al. 2019; Kumar & Dubey 2016). These compounds exhibit different medical and pharmaceutical benefits. Interestingly, most of the *Streptomyces* species are isolated from diverse soil habitats. Due to the health and economic benefits of secondary metabolites produced from *Streptomyces* species, the genus *Streptomyces* is extensively studied in search of new bioactive compounds. Subsequently new *Streptomyces* species are isolated and characterized.

Scientifically, isolation, identification and characterization of *Streptomyces* from their source are performed by following different methods. Initially soil samples are collected using random soil sampling methods. The collected samples are air-dried and pre-treated with calcium carbonate. The processed samples are dissolved and diluted in distilled water or 0.9% NaCl solution and diluted soil samples. The diluted samples are inoculated by pour-plate technique on Petri plates that contain sterilized culture media (such as starch casein agar medium, Actinomycetes isolation agar medium, etc.) for the growth of *Streptomyces* species. The inoculated culture plates are incubated at 28–30°C for about 7 days or till the visible bacterial colonies appear on the culture plate (Vijayakumar & Dhanasekaran 2022). Moreover, the culture medium may be supplemented with antibiotics (Ampicillin, cycloheximide, nystatin, etc.) for selective isolation of *Streptomyces* and inhibition of growth of unwanted fungal colonies (Antido & Climacosa 2022). The grown bacterial colonies may be further quantified as total bacterial count (CFU/g of soil) and total *Streptomyces* count (CFU/g of soil). For further purification, the isolated bacterial colonies are reinoculated on new culture plates ISP-2 ((International *Streptomyces* Project) plate. Including this, the isolated pure bacterial cultures of *Streptomyces* species are inoculated in culture slants and stored at 4°C for further study.

10.5.1 Identification and characterization of *Streptomyces* Isolates

After the purification, the *Streptomyces* isolates are identified by morphological (such as margin, texture, size and shape, optical property of colony, etc.), biochemical and molecular means. The most commonly used approach for isolation and characterization of bacteria is based on the guidelines recommended by Bergey's Manual of Systematic Bacteriology (Antido & Climacosa 2022).

Morphological features of *Streptomyces*, such as color of mycelia (Kelly & Judd 1965), and melanoid pigment production (Barka et al. 2016) from bacterial colonies are analyzed by visualizing the grown bacterial colonies on the culture plates. The isolated colonies were further categorized into six color series, that is, gray, white, red, yellow, green, and variable (brown, pink, orange, violet). These colors were determined by comparison with the Inter-Society Color Council-National Bureau of Standards (ISCC-NBS) color charts (Kelly & Judd 1965). Including this, Gram staining is performed to confirm the Gram-positive nature of *Streptomyces* and spore chain morphology. The isolated *Streptomyces* species are also grouped on the basis of microscopic analysis of morphological features of spore chains, for example retinaculiaperti (open loops, hooks, or spirals having one or two turns), rectiflexibiles (straight to flexuous), and spirales (tight spirals) (Goodfellow et al. 2012; Shirling & Gottlieb 1996). Moreover, electron microscopy methods (SEM, TEM) may be performed to analyze detailed morphology. The isolated strains may be grown in submerged culture to analyze pellet morphology.

The biochemical analysis of isolated *Streptomyces* species is based on biochemical tests such as catalase, citrate utilization, indole production, hydrogen sulfide (H2S) production, methyl red, nitrate reduction, urea hydrolysis, Voges-Proskauer, utilization of carbon sources, and gas production (Antido & Climacosa 2022).

The advanced method to identify the isolated strain is phylogenetic analysis method. The phylogenetic analysis of isolated *Streptomyces* species is done by 16S rRNA sequencing (Khushboo et al. 2023; Rathod et al. 2018).

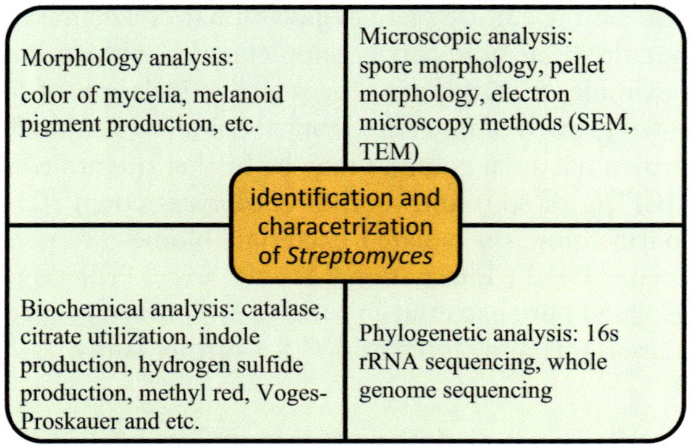

Figure 10.4 Different approaches used for identification and characterization of *Streptomyces*.

The sequence analysis is done by using the online available databases such as BLASTn program and EZbiocloud server (Maiti et al. 2020; Yoon et al. 2017). For phylogenetic analysis, bacterial DNA is isolated and 16rRNA gene is amplified by using appropriate primers. The amplified gene is sequenced and phylogenetic analysis is performed (Bo et al. 2019). This approach is widely used in identification of novel *Streptomyces* strains. In a recent study, Khushboo et al. (2023) isolated the *Streptomyces* strains and analyzed them morphologically. Further, sequencing of isolated *Streptomyces* sp. genome was performed by Illumina Nextseq500 platform. The analysis of sequenced genome revealed the presence of 25 biosynthetic gene clusters including genes associated with lipstatin biosynthesis. The identification of extracted metabolite was performed using the LC–MS analysis (Figure 10.4).

10.5.2 Optimization of culture conditions

Optimization of culture conditions is an important aspect of industrial application of *Streptomyces* species. Optimization of culture medium is performed to get a high amount of biomass and increased production of a desired metabolite. Optimization of culture conditions comprises two aspects: one is optimization of compositions of culture medium (growth medium, seed medium, fermentation medium, etc.), and optimization of growth conditions of *Streptomyces* (pH, incubation temperature, duration of incubation, rate of agitation in incubator shaker, etc.). The two approaches are employed for optimization, nonstatistical one-factor-at-a-time (OFAT) optimization method, and statistical experimental design method. OFAT approach uses the principle of optimizing an independent factor while keeping other factors constant. This approach is simple, but major drawbacks of OFAT approach are time consuming and require a large number of experiments and associated costs. Thus, major drawbacks of OFAT methods led to the utilization of statistical optimization approaches (Latha et al. 2017). The statistical experimental design approaches such as statistical design of experiments (DOE), central composite design (CCD), Plackett-Burman design (PBD) and response surface methodology (RSM) are commonly used for optimization of culture conditions. It is observed that statistical methods provide significant variables and produce high product yield, reduce time, and reduce operational cost (Arul Jose et al. 2013). Moreover, researchers are also exploring algorithm-based methods for optimization (Zhou et al. 2023). Advanced approaches such as artificial neural network (ANN), genetic algorithm (GA), and others are also being explored for culture condition optimization (Singh et al. 2017). It is also found that a combination of optimization techniques may be followed to get desirable results. Many research groups have explored different types of bioprocess optimization methods for the production of bioactive substances from microorganism including *Streptomyces* species. Significant numbers of scientific reports are available demonstrating use of OFAT and statistical approaches to produce high yield of desired metabolite and high amount of biomass.

Culture conditions were optimized using single-factor experiment for the production of Chrysomycin A from *Streptomyces* sp. 891. In this study, fermentation conditions (for example pH, seed age, inoculum, duration of fermentation period, shaking speed, and liquid loading) and culture medium composition (inorganic salts, nitrogen sources, and carbon sources) were optimized (Ni et al. 2021). In another study, actinomycetes were isolated and identified as *Streptomyces parvus* through 16S rRNA sequence analysis method. Furthermore, production of bioactive compound from *S. parvus* was enhanced by optimizing the culture conditions using Plackett–Burman design. Additionally, crude extract of *S. parvus* was determined by gas–liquid chromatography mass spectrometry (Abd-Elnaby et al. 2016). In another study, Ju et al. (2018) optimized the culture conditions for the growth of *Streptomyces rimosus* AG-P1441. OFAT approach identified that carbon and nitrogen sources (glucose, corn starch, and soybean) influence the antimicrobial productivity. Including this, RSM was used to optimize the amount of corn starch, glutamate, MgCl2, and soybean meal. Arul Jose et al. (2013) used Plackett-Burman design experiment and identified three culture media constituents ($CaCO_3$, KBr, and starch) that influence the antibiotic production from *Streptomyces* JAJ06. Response surface methodology with Box-Behnken design was used to get the optimal level of these factors. The combination of response surface methodology and Plackett–Burman design were also used for optimization of growth medium components. Additionally, OFAT approach was used for culture condition optimization to produce valinomycin from *Streptomyces* sp. ZJUT-IFE-354 (Zhang et al. 2023). The growth at optimized factors caused 26.8% increase in antibiotic production. Recently Yu et al. (2024) optimized the culture conditions for production of 4′-N-demethyl-vicenistatin by *Streptomyces parvus* SCSIO Mla-L010/ΔvicG using combination of approaches. In this study, OFAT and Plackett–Burman designs approaches were used and identified glycerol, Cassava starch, and seawater salt as key components of culture medium. Further, three-factor, five-level central composite design approach was used. Finally, RSM and artificial-neural-network-genetic-algorithm (ANN-GA) approaches optimized the culture medium components. This study revealed that ANN-GA model considerably increased the production of 4′-N-demethyl-vicenistatin. In another optimization study, culture conditions were optimized for the production of cholesterol oxidase from *Streptomyces rimosus* MTCC 10792 where initial optimization of culture medium components (yeast extract, dextrose, starch, and ammonium carbonate) was performed by Placket-Burman design. Further optimization was done using RSM followed by genetic algorithm (GA). RSM-GA–based optimization significantly increased the COD production (Srivastava et al. 2018).

10.5.3 Isolation and characterization of bioactive compounds

The selected microbial strain of *Streptomyces* species is grown in optimized culture medium under optimized culture conditions. The isolation of bioactive compounds from *Streptomyces* is performed on the basis of type of its intracellular and extracellular nature. Intracellular compounds are isolated after cell lysis and followed by extraction using different solvents. Isolated compounds are further purified by removing impurities and vacuum dried and analyzed for its properties.

The characterization of bioactive compounds is performed by biological, physical, and chemical methods (Figure 10.5). Physiochemical characterization of bioactive molecules is done by analyzing color, solubility, and pH. Further characterization of bioactive compounds is done by chromatography, Fourier transformation infrared (FTIR), nuclear magnetic resonance

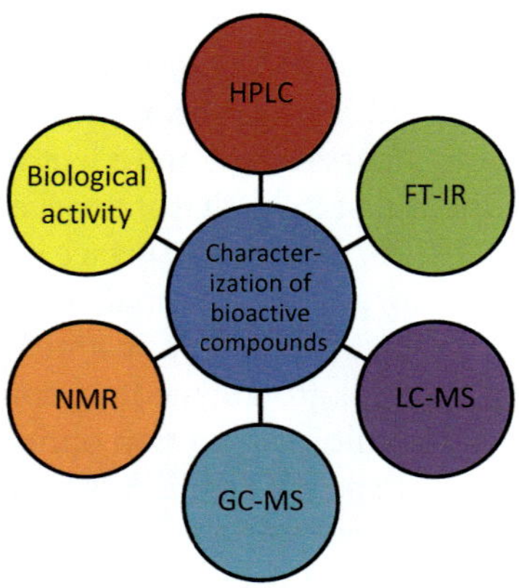

Figure 10.5 Some important approaches for characterization of bioactive compounds from *Streptomyces*.

(NMR), and UV–Vis spectroscopic methods. These molecules are categorized by different chromatography approaches like affinity chromatography, ion exchange chromatography, thin-layer chromatography (TLC), high performance liquid chromatography (HPLC), and gas chromatography. Investigators are analyzing bioactive molecules using advanced chromatographic approaches. Moreover, these molecules are also analyzed by mass spectroscopy approaches such as gas chromatography–mass spectrometry (GC-MS), and liquid chromatography–mass spectrometry (LC–MS) for further details. For example, lipstatin isolated from *Streptomyces* toxytricini has been analyzed by RP HPLC (Kumar & Dubey 2016), bioactive compounds isolated from *S. monomycini* RVE129 is characterized via NMR, UV-vis spectrophotometer, FTIR, and TLC (Elias et al. 2022).

The biochemical analysis is performed to find out different properties such as antioxidant properties, cytotoxicity, antimicrobial properties (antibacterial, anti-fungal), anticancer properties, and so on. Different research groups have reported antimicrobial activities of bioactive active compounds isolated from *Streptomyces* species. Elias et al. (2022) reported the isolation of antibiotic compound (RVE-02) from *S. monomycini* RVE129 and this RVE-02 demonstrated bioactivity against Gram-positive and Gram-negative bacteria. In-vitro cytotoxicity analysis revealed that RVE-2 also showed antiproliferative and morphological change in HeLa cells. Similarly bioactive compound (PJ85) isolated from *Streptomyces* sp. PJ85 demonstrated antimicrobial property against methicillin-resistant *Staphylococcus aureus* (Chanthasena et al. 2022). In another study, novel actinomycin group compounds (RSP 01 and RSP 02) have been isolated from *Streptomyces* sp. RAB12. These compounds were characterized by NMR and their antimicrobial analysis revealed that these compounds possess significant antimicrobial activity against variety of microorganisms (bacteria and fungi) such as *Bacillus subtilis, Micrococcus luteus, Pseudomonas aeruginosa, Staphylococcus aureus, Salmonella typhi*, and *Candida albicans*.

10.5.4 Role of artificial intelligence in product characterization

These days, artificial intelligence (AI) is being used in a variety of sectors such as engineering, business, medicine, healthcare, and research. It is increasing in automation, data acquisition and analysis, data modelling, and experimental planning. AI comprises different components such as machine learning, Deep learning, natural language processing, computer vision, robotics, and computational intelligence. Machine learning and robotics are widely used in automation in bioprocess development.

The bioprocess control and optimization of culture conditions are crucial components important for exploring the metabolic potential of microorganisms and production of commercially important biomolecules from microorganisms. Due to integration of digital technologies and vast applicability,

AI is also being used in bioprocessing to develop cost effective, ecofriendly and sustainable technologies. Sustainable development is the core essence of the bioeconomy. Though many advance models such as design of experiments, multivariate data analysis, principal component analysis, and partial least square regression have considerably optimized the bioprocesses but complexity, the nonlinear relationship of biological systems create challenges for optimization (Singh and Singhal, 2022). Thus, there is always chance of further improvements. Various machine learning models like artificial neural network, support vector machine, and principal coordinate analysis, etc. are being used in bioprocess engineering to scale up the production of bioproducts. These AI–guided modelling approaches are being used in optimization of bioprocess parameters (Cheng et al. 2023) and development of smart bioprocesses. Moreover, tools like machine learning and robotics are considered important in transforming biomanufacturing industries into Bioindustry 4.0 (Singh & Singhal, 2022). The digital transformation of bioprocesses, process automation, online monitoring, and real-time data acquisition are transforming bioprocessing industries into Industry 4.0.

10.6 CONCLUSION

Streptomyces are well-recognized sources for the production of a huge number of novel metabolites with a broad spectrum of bioactivities such as antimicrobial, anticancer, antioxidant, and antifungal activities as well as pigments. Novel strains of *Streptomyces* species are isolated from a wide spectrum of sources, mainly from soil. One widely used approach for identification of a novel strain is 16s rRNA sequencing and phylogenetic analysis. The production of metabolites from the selected strain is increased by optimization of culture conditions using non-statistical and statistical approaches. Now researchers are also exploring algorithm based advanced optimization methods. The bioactive compounds are isolated and characterized by physical and biochemical methods including spectroscopy and chromatography. Isolated compounds are further analyzed by advanced spectroscopy methods such as mass spectroscopy, nuclear magnetic resonance, and FTIR.

REFERENCES

Abd-Elnaby, H., Abo-Elala, G., Abdel-Raouf, U., Abd-elwahab, A. and Hamed, M. 2016. Antibacterial and anticancer activity of marine *Streptomyces* parvus: optimization and application. *Biotechnology & Biotechnological Equipment* 30(1):180–191.

Al-Joubori, B., Saadoun, I., Hotchin, N., Cunningham, D., and Alderwick, L. 2023. The isolation of novel terrestrial *Streptomyces* strains with antimicrobial and cytotoxic properties. *Arab Journal of Basic and Applied Sciences* 30(1):285–298. doi:10.1080/25765299.2023.2207957

Antido, J.W.A., and Climacosa, F.M.M. 2022. Enhanced isolation of *Streptomyces* from different soil habitats in Calamba City, Laguna, Philippines using a modified integrated approach. *International Journal of Microbiology* 2022: 2598963. doi: 10.1155/2022/2598963

Arul Jose, P., Sivakala, K.K. and Jebakumar, S.R.D. 2013. Formulation and statistical optimization of culture medium for improved production of antimicrobial compound by *Streptomyces* sp. JAJ06. *International Journal of Microbiology* 2013:526260. doi:10.1155/2013/526260

Balasubramanian, S., Skaf, J., Holzgrabe, U., Bharti, R., Förstner, K.U., et al. 2018. A new bioactive compound from the marine sponge-derived *Streptomyces* sp. SBT348 inhibits Staphylococcal growth and biofilm formation. *Frontiers in Microbiology* 9:1473.

Baral, B., Akhgari, A., and Metsä-Ketelä, M. 2018. Activation of microbial secondary metabolic pathways: avenues and challenges. *Synthetic and Systems Biotechnology* 3:163–178.

Barka, E.A., Vatsa, P., Sanchez, L., et al. 2016. Taxonomy, physiology, and natural products of actinobacteria. *Microbiology and Molecular Biology Reviews* 80(1):1–43.

Baskaran, B., and Muthukumarasamy, A. 2017. Isolation, characterisation and enzymatic activity of Streptomyces sp. and its pH control during fermentation process. *IET Systematic Biology* 11(4):114–118. doi: 10.1049/iet-syb.2016.0048

Behroozian, S., Svensson, S.L., and Davies, J. 2016. Kisameet clay exhibits potent antibacterial activity against the ESKAPE pathogens. *MBio* 7:10–1128.

Bentley, S.D., Chater, K.F., Cerdeño-Tárraga, A-M., Challis, G.L., Thomson, N.R., et al. 2002. Complete genome sequence of the model actin- omycete *Streptomyces* coelicolor A3(2). *Nature* 417:141–147.

Bo, A.B., Kim, J.D., Kim, Y.S., Sin, H.T., Kim, H.J., and Khaitov, B., et al. 2019. Isolation, identification and characterization of *Streptomyces* metabolites as a potential bioherbicide. *PLoS One* 14(9): e0222933.

Cao, D.D., Trinh, T.T.V., Mai, H.D.T., Vu, V.N., Le, H.M. et al. 2019. Antimicrobial lavandulylated flavonoids from a sponge-derived *Streptomyces* sp. G248 in East Vietnam Sea. *Marine Drugs* 17:529.

Chakravarty, I., and Kundu, S. 2016. Improved production of daptomycin in an airlift bioreactor by morphologically modified and immobilized cells of *Streptomyces* roseosporus. *AMB Express* 6:101. doi: 10.1186/s13568-016-0274-0

Chakravorty, J., Ghosh, S., and Meyer-Rochow, V. 2011. Practices of entomophagy and entomotherapy by members of the Nyishi and Galo tribes, two ethnic groups of the state of Arunachal Pradesh (North-East India). *Journal of Ethnobiology and Ethnomedicine* 7:5. doi: 10.1186/1746-4269-7-5

Chanthasena, P., Hua, Y., Rosyidah, A., Pathom-Aree, W., Limphirat, W., and Nantapong, N. 2022. Isolation and identification of bioactive compounds from *Streptomyces* actinomycinicus PJ85 and their in vitro antimicrobial activities against methicillin-resistant Staphylococcus aureus. *Antibiotics (Basel)* 11(12):1797. doi: 10.3390/antibiotics11121797

Chater, K.F. 2016. Recent advances in understanding *Streptomyces*. *F1000Res* 5:2795. doi: 10.12688/f1000research.9534.1

Cheng, Y., Bi, X., Xu, Y., Liu, Y., Li, J., Du, G., Lv, X., and Liu, L. 2023. Artificial intelligence technologies in bioprocess: opportunities and challenges. *Bioresource Technology* 369:128451.

Chevrette, M.G., Carlson, C.M., Ortega, H.E., Thomas, C., Ananiev, G.E., et al. 2019. The antimicrobial potential of *Streptomyces* from insect microbiomes. *Nature Communications* 10(1):516. doi: 10.1038/s41467-019-08438-0.

Chiang, Y-M., Chang, S-L., Oakley, B.R., and Wang, C.C.C. 2011. Recent advances in awakening silent biosynthetic gene clusters and linking orphan clusters to natural products in microorganisms. *Current Opinion in Chemical Biology* 15:137–143.

Colombo, M.E., Kunova, A., Cortesi, P., Saracchi, M., and Pasquali, M. 2019. Critical assessment of *Streptomyces* spp. able to control toxigenic Fusaria in cereals: a literature and patent review. *International Journal of Molecular Sciences* 20:6119.

Cordell, G.A., Farnsworth, N.R., Beecher, C.W.W., Doel Soejarto, D., Kinghorn, A.D. et al. 1994. Novel strategies for the discovery of plant-derived anticancer agents. In: *Anticancer Drug Discovery and Development: Natural Products and New Molecular Models. Developments in Oncology*. Boston, MA: Springer.

Currie, C.R., Scott, J.A., Summerbell, R.C., and Malloch, D. 1999. Fungus-growing ants use antibiotic-producing bacteria to control garden parasites. *Nature* 398: 701–704.

Davies J. 2006. Where have all the antibiotics gone? *Canadian Journal of Infectious Diseases and Medical Microbiology* 17:287–290.

De Lima Procópio, R.E., da Silva, I.R., Martins, M.K., de Azevedo, J.L., and de Araújo, J.M. 2012. Antibiotics produced by *Streptomyces*. *The Brazilian Journal of Infectious Diseases* 16:466–471. doi: 10.1016/j.bjid.2012.08.014

Elias, F., Muddada, S., Muleta, D., and Tefera, B. 2022. Purification and characterization of bioactive metabolite from *Streptomyces* monomycini RVE129 derived from the rift valley soil of Hawassa, Ethiopia. *BioMed Research International* 2022:7141313.

Falkinham, J.O., Wall, T.E., Tanner, J.R., Tawaha, K., Alali, F.Q. et al. 2009 Proliferation of antibiotic-producing bacteria and concomitant antibiotic production as the basis for the antibiotic activity of Jordan's red soils. *Applied and Environmental Microbiology* 75:2735–2741.

Goodfellow, M., Kämpfer, P., and Busse, H. 2012. *Bergey's Manual of Systematic Bacteriology. 2nd Edition. Volume 5. The actinobacteria*. New York, NY, USA: Springer.

Gosse, J.T., Ghosh, S., Sproule, A., Overy, D., Cheeptham, N. et al. 2019. Whole genome sequencing and metabolomic study of cave *Streptomyces* isolates ICC1 and ICC4. *Frontiers in Microbiology* 10:1020.

Hamedi, J., Kafshnouchi, M., and Ranjbaran, M. 2019. A study on actinobacterial diversity of Hampoeil cave and screening of their biological activities. *Saudi Journal of Biological Sciences* 26:1587–1595.

Harir, M., Bendif, H., Bellahcene, M., & Fortas, Z. and Rebecca, P. 2018. Streptomyces secondary metabolites. *IntechOpen*. doi: 10.5772/intechopen.79890

Huang, X., Kong, F., Zhou, S., Huang, D., Zheng, J. et al. 2019. *Streptomyces* tirandamycinicus sp. nov., a novel marine sponge-derived actinobacterium with antibacterial potential against Streptococcus agalactiae. *Frontiers in Microbiology* 10:482.

Ju, Y., Son, K.H., Jin, C., Hwang, B.S., Park, D.J. and Kim, C.J., 2018. Statistical optimization of culture medium for improved production of antimicrobial compound by *Streptomyces* rimosus AG-P1441. *Food Science and Biotechnology* 27:581–590.

Kelly, K.L. and Judd, D.B., 1965. ISCC-NBS color-name charts illustrated with centroid colors. Washington, DC, USA: US Government Printing Office.

Khan, S.T. 2011. *Streptomyces* associated with a marine sponge Haliclona sp.; biosynthetic genes for secondary metabolites and products. *Environmental Microbiology Black Sci Pub*. 13:391–403.

Khushboo, Y., Singhvi, N., Gupta, V., Dhaka, N., and Dubey, K.K. 2023. Draft genome sequence of *Streptomyces* sp. KD18, isolated from industrial soil. *3 Biotech* 13(1):34.

Kumar, P., and Dubey, K.K. 2016. Modulation of fatty acid metabolism and tricarboxylic acid cycle to enhance the lipstatin production through medium engineering in *Streptomyces* toxytricini. *Bioresource Technology* 213:64–68.

Latha, S., Sivaranjani, G. and Dhanasekaran, D., 2017. Response surface methodology: A non-conventional statistical tool to maximize the throughput of *Streptomyces* species biomass and their bioactive metabolites. *Critical Reviews in Microbiology* 43(5):567–582.

Liu C, Han C, Jiang S, Zhao X, Tian Y et al. 2018. *Streptomyces* lasii sp. nov., a novel actinomycete with antifungal activity isolated from the head of an ant (Lasius flavus). *Current Microbiology* 75:353–358.

Liu, M., Abdel-Mageed, W.M., Ren, B., He, W., Huang, P. et al. 2014. Endo- phytic *Streptomyces* sp. Y3111 from traditional Chinese medicine produced antitubercular pluramycins. *Applied Microbiology and Biotechnology* 98:1077–1085.

López-García, M.T., Rioseras, B., Yagüe, P., Álvarez, J.R., and Manteca, Á. 2014. Cell immobilization of *Streptomyces* coelicolor: effect on differentiation and actinorhodin production. *International Microbiology* 17:75–80

Maciejewska, M., Adam, D., Martinet, L., Naome, A., Calusinska, M. et al. 2016. A phenotypic and genotypic analysis of the antimicrobial potential of cultivable *Streptomyces* isolated from cave moonmilk deposits. *Frontiers in Microbiology* 7:1455.

Maiti, P.K., Das, S., Sahoo, P., and Mandal, S. 2020. *Streptomyces* sp SM01 isolated from Indian soil produces a novel antibiotic picolinamycin effective against multi drug resistant bacterial strains. *Scientific Reports* 10(1):1–12. doi:10.1038/s41598-020-66984-w

Martens, E., and Demain, A.L. 2011. Platensimycin and platencin: promising antibiotics for future application in human medicine. *The Journal of Antibiotics* 64:705–710.

Mistiaen V. 2002. Time, and the great healer. The Guardian, 2 November 2002. https://www.theguardian.com/education/2002/nov/02/research.highereducation

Müller, W.E.G., Batel, R., Schröder, H.C., and Müller, I.M. 2004. Traditional and modern biomedical Prospecting: part I – the history: sustainable exploitation of biodiversity (sponges and invertebrates) in the Adriatic Sea in Rovinj (Croatia). *Evidence-Based Complementary and Alternative Medicine* 1:71–82.

Nakaew, N., Lumyong, S., Sloan, W.T., and Sungthong, R. 2019. Bioactivities and genome insights of a thermotolerant antibiotics-producing Strep-tomyces sp. TM32 reveal its potentials for novel drug discovery. *Microbiologyopen* 8:e842.

Nalini, M.S., and Prakash, H.S. 2017. Diversity and bioprospecting of actinomycete endophytes from the medicinal plants. *Letters in Applied Microbiology* 64:261–270.

Ni, H.J., Lv, S.Y., Sheng, Y.T., Wang, H., Chu, X.H. and Zhang, H.W., 2021. Optimization of fermentation conditions and medium compositions for the production of chrysomycin a by a marine-derived strain *Streptomyces* sp. 891. *Preparative Biochemistry & Biotechnology* 51(10):998–1003.

Oberhofer, M., Hess, J., Leutgeb, M., Gössnitzer, F., Rattei, T., et al. 2019. Exploring actinobacteria associated with rhizosphere and endo- sphere of the native alpine medicinal plant Leontopodium nivale subspecies alpinum. *Frontiers in Microbiology* 10:2531.

Okada, B.K., and Seyedsayamdost, M.R. 2017. Antibiotic dialogues: induction of silent biosynthetic gene clusters by exogenous small molecules. *FEMS Microbiology Reviews* 41:19–33.

Olmos, E., Mehmood, N., Haj Husein, L., Goergen, J-L., Fick, M., and Delaunay, S. 2013. Effects of bioreactor hydrodynamics on the physiology of *Streptomyces*. *Bioprocess and Biosystems Engineering* 36:259–272

Omura, S., Ikeda, H., Ishikawa, J., et al. 2001. Genome sequence of an industrial microorganism *Streptomyces* avermitilis: deducing the ability of producing secondary metabolites. *Proceedings of the National Academy of Sciences* 98:12215–12220.

Oskay, M. 2009. Comparison of Streptomyces diversity between agricultural and nonagricultural soils by using various culture media. *Scientific Research and Essays* 4(10):997–1005.

Patzer, S.I., and Volkmar, B. 2010. Gene cluster involved in the biosynthesis of griseobactin, a catechol-peptide siderophore of *Streptomyces* sp. ATCC 700974. *Journal of Bacteriology* 192:426–435.

Pereira, T., Nikodinovic, J., Nakazono, C., Dennis, G.R., Barrow, K.D., and Chuck, J.A. 2008. Community structure and antibiotic production of *Streptomyces* nodosus bioreactors cultured in liquid environments. *Microbial Biotechnology* 1:373–381.

Poulsen, M., Cafaro, M.J., Erhardt, D.P., Little, A.E.F., Gerardo, N.M., et al. 2010. Variation in Pseudonocardia antibiotic defence helps govern parasite-induced morbidity in Acromyrmex leaf-cutting ants. *Environmental Microbiology Reports* 2:534–540.

Poulsen, M., Oh, D.-C., Clardy, J., and Currie, C.R. 2011. Chemical analyses of wasp-associated *Streptomyces* bacteria reveal a prolific potential for natural products discovery. *PLoS One* 6:e16763.

Pronzato, R., and Manconi, R. 2008. Mediterranean commercial sponges: over 5000 years of natural history and cultural heritage. *Marine Ecology* 29:146–166.

Qin, S., Li, J., Chen, H-H., Zhao, G-Z., Zhu, W-Y. et al. 2009. Isolation, diversity, and antimicrobial activity of rare actinobacteria from medicinal plants of tropical rain forests in Xishuangbanna, China. Applied and Environmental Microbiology 75:6176.

Qin, Z., Munnoch, J.T., Devine, R., Holmes, N.A., Seipke, R.F. et al. 2017. Formicamycins, antibacterial polyketides produced by *Streptomyces* formicae isolated from African Tetraponera plant-ants. *Chemical Science* 8:3218–3227.

Quinn, G.A., Banat, A.M., Abdelhameed, A.M., and Banat, I.M. 2020. Streptomyces from traditional medicine: sources of new innovations in antibiotic discovery. *Journal of Medical Microbiology* 69(8):1040–1048. doi: 10.1099/jmm.0.001232

Rangseekaew, P., and Pathom-Aree, W. 2019. Cave actinobacteria as producers of bioactive metabolites. *Frontiers in Microbiology* 10:387.

Rathod, B.B., Korasapati, R., Sripadi, P., and Reddy Shetty, P. 2018. Novel actinomycin group compound from newly isolated *Streptomyces* sp. RAB12: isolation, characterization, and evaluation of antimicrobial potential. *Applied Microbiology and Biotechnology* 102:1241–1250.

Reen, F.J., Romano, S., Dobson, A.D.W., and O'Gara, F. 2015. The sound of silence: activating silent biosynthetic gene clusters in marine microorganisms. *Mar Drugs*13:4754–4783.

Ribeiro da Cunha, B., Fonseca, L.P., and Calado, C.R.C. 2019. Antibiotic discovery: where have we come from, where do we go? *Antibiotics* 8:45.

Romano, S., Jackson, S.A., Patry, S., Dobson, A.D.W. 2018. Extending the 'One Strain Many Compounds' (OSMAC) principle to marine microorganisms. *Marine Drugs* 16:244.

Saadoun, I., Alawawdeh, M., Jaradat, Z., Ababneh, Q., Al-Joubori, B. M., & Elsheikh, E. A. E. 2023. Characterization of diesel-degrading, hydrolytic enzymes-producing Streptomyces spp. isolated from fuel-oil polluted soils. *Arab Journal of Basic and Applied Sciences* 30(1):248–255. doi:10.1080/25765299.2023.2196110

Sanchez, J., Yague, P., Manteca, A. 2012. New insights in *Streptomyces* fermentations. *Fermentation Technology* 1(2):e105. doi: 10.4172/2167-7972.1000e105

Sarmiento-Vizcaíno, A., Espadas, J., Martín, J., Braña, A.F., Reyes, F., et al. 2018. Atmospheric precipitations, hailstone and rainwater, as a novel source of *Streptomyces* producing bioactive natural products. *Frontiers in Microbiology* 9:773.

Schatz, A., Bugle, E., and Waksman, S.A. 1944. Streptomycin, a substance exhib- iting antibiotic activity against Gram-positive and Gram-negative bacteria. *Proceedings of the Society for Experimental Biology and Medicine* 55:66–69.

Seabrooks, L., Hu, L. 2017. Insects: an underrepresented resource for the discovery of biologically active natural products. *Acta Pharmaceutica Sinica B* 7:409–426.

Shaik, M., Girija Sankar, G., Iswarya, M., and Rajitha, P. 2017. Isolation and characterization of bioactive metabolites producing marine Strepto- myces parvulus strain sankarensis-A10. *Journal, Genetic Engineering & Biotechnology* 15:87–94.

Shirling, E.B., Gottlieb, D. 1996. Methods for characterization of *Streptomyces* species. *International Journal of Systematic Bacteriology* 16(3):313–340.

Shishlyannikova, T.A., Kuzmin, A.V., Fedorova, G.A., Shishlyan nikov, S.M., Lipko, I.A., et al. 2017. Ionofore antibiotic polynactin produced by *Streptomyces* sp. 156A isolated from Lake Baikal. *Natural Product Research* 31:639–644.

Shrestha, B., Nath, D.K., Maharjan, A., Poudel, A., Pradhan, R.N., Aryal, S. 2021. Isolation and characterization of potential antibiotic-producing actinomycetes from water and soil sediments of different regions of Nepal. *International Journal of Microbiology* 2021:5586165. doi: 10.1155/2021/5586165

Singh, A., Singhal, B. 2022. Role of Machine Learning in Bioprocess Engineering: Current Perspectives and Future Directions. In *Design and Applications of Nature Inspired Optimization. Women in Engineering and Science*, ed. Singh, D., Garg, V., Deep, K. Springer, Cham. doi:10.1007/978-3-031-17929-7_3

Singh, V., Haque, S., Niwas, R., Srivastava, A., Pasupuleti, M., Tripathi, C.K. 2017. Strategies for fermentation medium optimization: An in-depth review. *Frontiers in Microbiology* 7:2087. doi: 10.3389/fmicb.2016.02087

Sivalingam, P., Hong, K., Pote, J., Prabakar, K. 2019. Extreme environment *Streptomyces*: Potential sources for new antibacterial and anticancer drug leads? *International Journal of Microbiology* 2019:5283948. doi: 10.1155/2019/5283948

Srivastava, A., Singh, V., Haque, S., Pandey, S., Mishra, M., Jawed, A., Shukla, P.K., Singh, P.K. and Tripathi, C.K.M., 2018. Response surface methodology-genetic algorithm based medium optimization, purification, and characterization of cholesterol oxidase from *Streptomyces* rimosus. *Scientific Reports* 8(1):10913. doi:10.1038/s41598-018-29241-9

Svensson, S.L., Behroozian, S., Xu, W., Surette, M.G., Li, L., et al. 2017. Kisameet glacial clay: an unexpected source of bacterial diversity. *MBio* 8:e00590-17.

Tenebro, C.P., Trono, D.J.V.L., Vicera, C.V.B. et al. 2021. Multiple strain analysis of Streptomyces species from Philippine marine sediments reveals intraspecies heterogeneity in antibiotic activities. *Scientific Reports* 11:17544. doi:10.1038/s41598-021-96886-4

Terkina, I.A., Parfenova, V.V., and Ahn, T.S. 2006. Antagonistic activity of actinomycetes of Lake Baikal. *Applied Biochemistry and Microbiology* 42:173–176.

Terra, L., Dyson, P.J., Hitchings, M.D., Thomas, L., Abdelhameed, A., et al. 2018. A novel alkaliphilic *Streptomyces* inhibits ESKAPE pathogens. *Frontiers in Microbiology* 9:2458.

Vijayakumar, V.R., Dhanasekaran, D. 2022. Isolation of *Streptomyces* from Soil Sample. In *Practical Handbook on Agricultural Microbiology. Springer Protocols Handbooks, ed.* Amaresan, N., Patel, P., Amin, D. Humana, New York, NY. doi:10.1007/978-1-0716-1724-3_16

Wainwright, M. 1991. Streptomycin: discovery and resultant controversy. *History and Philosophy of the Life Sciences* 13:97–124.

Waksman, S.A., Schatz, A., and Reynolds, D.M. 2010. Production of antibiotic substances by actinomycetes. *Annals of the New York Academy of Sciences* 1213:112–124. doi: 10.1111/j.1749-6632.2010.05861.x

Waksman, S.A., and Schatz, A. 1946. Soil enrichment and development of antagonistic microorganisms. *Journal of Bacteriology* 51:305–316.

Wang, J., Soisson, S.M., Young, K., Shoop, W., Kodali, S., et al. 2006. Platensimycin is a selective FabF inhibitor with potent antibiotic proper- ties. *Nature* 441:358–361.

Wardecki, T., Brotz, E., De Ford, C., von Loewenich, F.D., Rebets, Y., et al. 2015. Endophytic *Streptomyces* in the traditional medicinal plant Arnica montana L.: secondary metabolites and biological activity. *Antonie Van Leeuwenhoek* 108:391–402.

Yoon, S. H. et al. 2017. Introducing EzBioCloud: a taxonomically united database of 16S rRNA gene sequences and whole-genome assemblies. *International Journal of Systematic and Evolutionary Microbiology* 67:1613–1617.

Yu, Z., Fu, H., and Wang, J. 2024. Modeling and optimization of the culture medium for efficient 4′-N-demethyl-vicenistatin production by *Streptomyces* parvus using response surface methodology and artificial-neural-network-genetic-algorithm. *Fermentation* 10(3):154.

Zhang, D., Bao, Y., Ma, Z., Zhou, J., Chen, H., Lu, Y., Zhu, L. and Chen, X. 2023. Optimization of fermentation medium and conditions for enhancing valinomycin production by *Streptomyces* sp. ZJUT-IFE-354. *Preparative Biochemistry & Biotechnology* 53(2):157–166.

Zhao, H., Yang, A., Zhang, N., Li, S., Yuan, T., et al. 2020. Insecticidal endostemonines A–J produced by endophytic *Streptomyces* from Stemona sessilifolia. *Journal of Agricultural and Food Chemistry* 68:1588–1595.

Zhou, T., Reji, R., Kairon, R.S., and Chiam, K.H. 2023. A review of algorithmic approaches for cell culture media optimization. *Frontiers in Bioengineering and Biotechnology* 11:1195294. doi:10.3389/fbioe.2023.1195294

Zotchev, S.B. 2024. Unlocking the potential of bacterial endophytes from medicinal plants for drug discovery. *Microbial Biotechnology* 17(2):e14382. doi: 10.1111/1751-7915.14382

Chapter 11

Genetic engineering of streptomycetes to enhance the production of secondary metabolites

Naveen Kumar
Kurukshetra University, Haryana, India

Aarzoo
University College Dublin, Ireland

Nisha Sharma
Baba Haridas College of Pharmacy, Nazafgarh, Delhi, India

11.1 INTRODUCTION

A genus of bacteria called streptomycetes is well known for its capacity to generate a broad range of secondary metabolites (SMs), such as antibiotics and other bioactive substances. Genetic engineering of streptomycetes can be employed to improve the production of these SMs for various applications, such as drug discovery and development. The antibacterial activity of soil *Streptomyces* against specific microbes has been used to characterize a large number of novel antibiotics since the 1940s, when the capacity of *Streptomyces* to produce antibiotics was first reported. The majority of the antibiotic classes that are currently on the market were found in and derived from *Streptomyces* species that were identified between 1940 and 1962. Despite two decades of success using standard biochemical screening procedures, the rediscovery rate of known species and compounds has continuously increased and reached 99% in the late 1980s, even though cultured soil microorganisms make up less than 0.1% of all soil microorganisms. According to Lee et al. (2020a), no new classes of antibiotics have been authorized since that time. By breaking down plant biomass and chelating minerals in soil, actinobacteria have a major impact on agricultural production and mineral recycling in ecosystems, in addition to their significance in the medical area. Actinobacteria are reported to make up over 20% of the soil microbial flora in the spring, over 30% in the fall, and just 13% in the winter, along with other soil-dwelling organisms like pathogens and commensals (Barka et al., 2016).

 DOI: 10.1201/9781032694634-11

More than two-thirds of clinically effective antibiotics are specifically generated by *Streptomyces*. The secondary metabolism of *Streptomyces* is of tremendous importance to combat the emergence of multidrug-resistant bacteria and the identification of other useful chemicals and substances. Using NGS and in silico genome mining tools has made it easier to find new SMs in *Streptomyces*. A single *Streptomyces* genome should have between 20 and 50 smBGCs. This is a lot more than the number of SMs that have already been found (Blin et al., 2019). More than 200 full genomes have been recovered so far. It is notable that the NCBI taxonomy database has descriptions of over 1000 species in the genus *Streptomyces*, which is one of the largest genera. Tens of thousands of smBGCs are still unidentified as a result, and *Streptomyces* species continue to be the primary source for the discovery of novel SMs. However, the actual finding of the compounds is not certain, even though the genome of *Streptomyces* guarantees the existence of numerous unique bioactive chemicals (Lee et al., 2020b). The smBGCs are silent under normal circumstances since the SMs are typically not necessary for life. However, under certain circumstances, such as osmotic pressure, nutritional constraints, and interspecies competition, the smBGCs become active for their own objectives. As a result, intricate regulation is applied to the smBGCs in order to manufacture SMs in response to their requirements. To make the most of *Streptomyces'* industrial and medical potential, it is important to understand its complex regulation, especially that which controls the secondary metabolism.

11.2 FEATURES OF *STREPTOMYCES* GENOMES

Streptomyces have big linear chromosomes with a high G+C concentration, in contrast to other bacteria. The beginning of replication (oriC) of a linear chromosome is usually in the middle. The ends of the chromosome are home to TIRs, which are linked to the terminal proteins by covalent bonds. It's interesting to note that TIR lengths and sequences differ greatly throughout species, and there is no relationship between the number of TIR iterations and genome size. The high level of chromosomal instability in the *Streptomyces* genome, which frequently results in spontaneous deletions and changes, particularly at the ends of the chromosome, is its most distinctive trait. For instance, in laboratory culture settings, approximately 0.5% of the growing spores of *S. lividans* have significant deletions that eliminate up to 25% of the genome (~2 Mbp). As a result, the "core" portion of the chromosome contains vital genes involved in cell maintenance. On the other hand, genes that are conditionally adaptable, particularly those associated with secondary metabolism, are typically found in the chromosome's "arm" sections. The smBGCs exhibit a significant degree of variety due to their chromosomal plasticity, which may have been brought about by horizontal gene transfer with other bacteria or *Streptomyces* species or by common gene duplications (Lee et al., 2020b).

11.3 GENE COMPONENTS FOR SYNTHETIC BIOLOGY IN *STREPTOMYCES*

The vast ability of *Streptomyces* to develop novel SMs has been highlighted by genome information on several strains of the fungus. Nonetheless, in standard laboratory culture conditions, the majority of smBGCs in *Streptomyces* remain inactive. For example, only 3–5 SMs have been found, even though *S. coelicolor, S. griseus*, and *S. avermitilis* all have genomes that code for more than 30 smBGCs. Despite the fact that a lot of work has been done to activate silent smBGCs in *Streptomyces* strains, such as cocultivation techniques and the OSMAC strategy, these efforts have only partially succeeded in activating silent BGCs. Various techniques have been employed to achieve targeted smBGCs' particular activation, including substitute promoters, overexpressing or repressing regulatory genes, heterologous expression in distinct hosts, and remodeling targeted smBGCs. As a result, as shown in Figure 11.1, the number of genetic components that are now available for *Streptomyces* species has significantly expanded, including promoters, RBS, and terminators (Baral et al., 2018; Lee et al., 2019).

11.4 MOLECULAR TOOLS FOR STREPTOMYCETES ENGINEERING

Recently, there has been a notable advancement in computational science, along with parallel developments in molecular tools to address the difficulties in engineering streptomycetes. The λ Red system, which comes from *E. coli*, was used to get around the problems that come with using standard restriction- and ligation-based cloning methods to change genes. This enabled *Streptomyces* strains to disrupt genes, substitute genes, and restructure entire gene clusters more quickly and efficiently. The *S. cerevisiae* I-SceI homing endonuclease was used as an alternative to this traditional method for DNA double-strand break–based genome editing. This will help the genetic engineering work in streptomycetes even more. This meganuclease is less likely to produce off-targets when employed for gene editing because it has a longer recognition site (18 bp) than conventional restriction enzymes. The I-SceI recognition gene and the appropriate resistance marker are introduced to the recipient streptomycetes' genomes in a single crossover event by using the first plasmid in a two-plasmid system. The synthetic codon-optimized I-SceI, driven by either the constitutive ermE* or thiostrepton-inducible tipA promoter, is present in the second plasmid that induces double crossover. Marker-free gene deletions are possible using

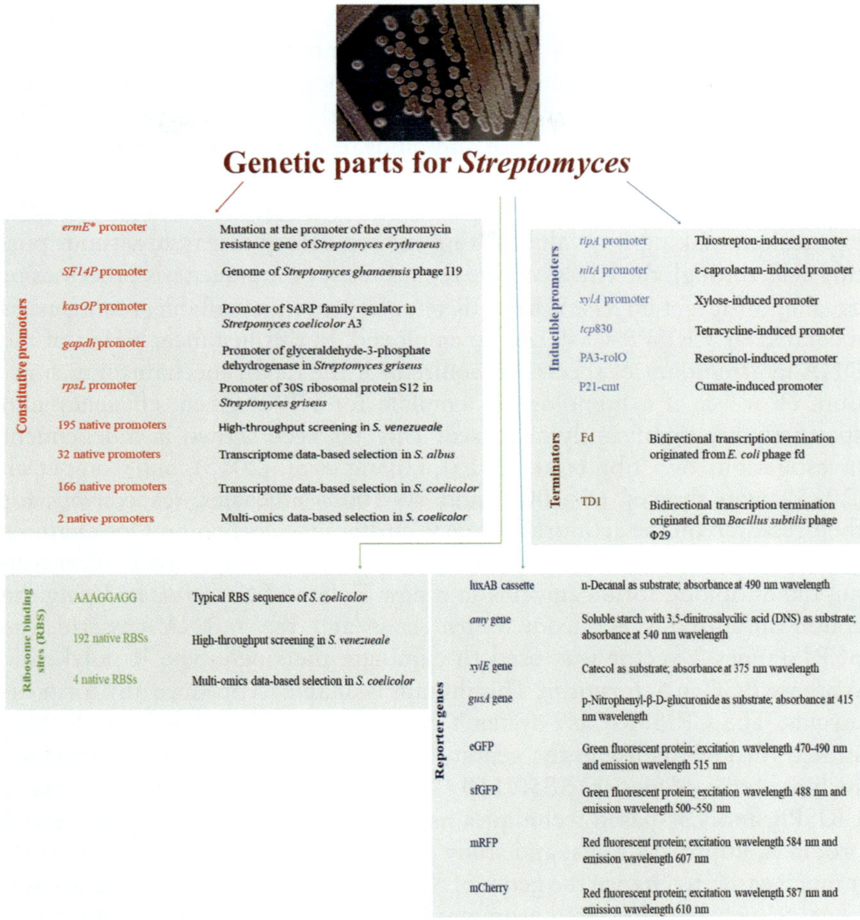

Figure 11.1 Genetic parts of *Streptomyces*.

pSG5–derived plasmids that use temperature-sensitive replicons. When the I-SceI technique was employed to replace a gene in *Streptomyces* sp. Tu 6071, it was demonstrated that I-SceI–mediated HR was 25 times more efficient than spontaneously occurring HR (Zhang et al., 2016; Fernández-Martínez and Bibb, 2014).

Recent breakthroughs in research have led to the identification of an even more effective and versatile system for genome engineering in streptomycetes: the type II CRISPR–associated protein (Cas9) system from *Streptococcus pyogenes*. The CRISPR-Cas9 method can be used for induction, reversible regulation of gene expression, and multiplex deletion of individual genes and gene clusters. The optimized system is based on two components: the endonuclease Cas9, also known as catalytically inactive dCas9, which interacts with the sgRNA to search the genome for PAM sequences, and the appropriate 20-bp recognition site. The sgRNA is a synthetic RNA made up of a crRNA and tracrRNA complex that is important for guiding Cas9 to change the targeted genomic sequence. Although the inactive dCas9 will attach to the target DNA without modifying the sequence, the active Cas9 detects the target DNA sequence and binds to and cleaves the precise location in the genome.

Here, dCas9 offers a clone-easy method for reversible control over gene expression, making it a valuable tool for regulatory research without putting cells through the stress of gene replacement or mutagenesis processes or causing off-target effects. When HR templates are unavailable, mutagenesis mediated by CRISPR-Cas9 can be employed. In this instance, NHEJ of the DNA in *Streptomyces* species is required for the repair mechanism to function. However, if a homologous template for HR is given, efficiency and specificity are significantly increased. This has been shown in independent investigations by Cobb et al. (2015), Huang et al. (2015), and Tong et al. (2015), who showed 60–100% and 70–100% efficacies, respectively, for their research on the actinorhodin gene cluster in *S. coelicolor*. One example is the use of CRISPR-Cas9 technology to locate and analyze the BGC encoding the antibiotic formicamycins in a new isolate of *S. formicae*. Using the codon-optimized Cas9 from *S. pyogenes* and the sgRNA cassette, the pCRISPomes-2 vector was used to eliminate the single-type II polyketide synthase BGC in *S. formicae*. This mutant is unable to produce any formicamycins. The CRISPR-Cas9 system's overall adaptability has been demonstrated through its use in the genome editing of uncommon actinomycetes, such as *Actinoplanes* sp. SE50/110 (Wolf et al., 2016). Recent advances in CRISPR and Cas-based techniques have made it faster to find new natural products, improve strains, and study genomes. These techniques also made it much easier to change the genes of *Streptomyces* sp. Utilizing genetic components such as synthesized promoters, ribosome-binding sites, terminators, and reporter genes further enhanced the arsenal for *Streptomyces* engineering (Lee et al., 2019) (Table 11.1).

11.5 SELECTION OF THE METHOD OF GENETIC MODIFICATION

There are several approaches to metabolic engineering that can be used to improve the SM synthesis pathway's efficiency. These strategies all focus on various elements of a common metabolic pathway. Precursor engineering, structural gene engineering, regulatory network engineering, and genome shuffles are a few examples. Over the past ten years, *streptomyces* genes have evolved from genetic modification to metabolic modification. Because complicated pathways, rather than single isolated genes, are used to generate secondary metabolites, metabolic engineering is required. As a result, a thorough examination of the pathways is necessary, after which it may be

Table 11.1 A combined list of methods utilized in *Streptomyces* engineering

Category	*Method*	*Feature*	*References*
Genetics parts	Reporter genes	Genes *luxAB*, *amy*, *xylE*, and *gusA* as well as *eGFP*, *sfGFP*, *mRFP*, and *mCherry*	Lee et al., 2019
	Terminator	Fd and TD1	
	RBSs	AAAGGAGG and diverse native or synthetic RBSs	
	Synthetic promoters	Constitutive *ermE*, SF14P, *kasOP*, *gapdh*, *rpsL* promoters as well as inducible *tipA nitA* and *xylA* promoters	
Genetic manipulation strategy	Target SM production by expression of the BGC vector	Expression of integrative (pSET152, pCAP01, pESAC) and replicative (pSKC2 and pUWL201) vectors	Lee et al., 2019
	Transfer of the BGC–encoded vector to the heterologous host for expression	Conjugation and protoplast transformation	
	Ligation/assembly of the BGC to the vector	Sticky/blunt end ligation, Gibson cloning, and recombination in different hosts	
Bioinformatics	ClustSCAN	Semi-automatic annotation of modular BGCs and in silico prediction of new chemical structures	Krysenko, 2023
	NP searcher	Automated genome mining for natural products and quick screening for potentially valuable chemicals	
	GNP/PRISM	Identifying groups of biosynthetic genes, guessing non-ribosomal peptides and type I and II polyketides with genetic coding, and bio- and cheminformatic replication of compounds that happen naturally	

necessary to target many genes within the pathway. The by-products of the catabolism of carbon molecules during primary metabolism are frequently precursors for the formation of SMs, which are essential. Precursors are therefore present where primary and SMs converge. The precursor must be easily accessible to boost the synthesis of; however, in some situations, this isn't the case. Therefore, the engineering strategy entails genetically modifying certain enzymes to boost the availability of specific precursors via various routes (Sharma et al., 2021).

One such pathway is the glycolytic (Embden-Meyerhof) pathway, which is involved in the conversion of glucose to pyruvate. The clavulanic acid manufacturing process was conducted using the precursor engineering approach. Glyceraldehyde-3-phosphate dehydrogenases are encoded by the *gap1* and *gap2* genes, which were disrupted in order to accomplish this. These contribute to the transformation of 1,3-BPG into D-G3P. G3P is one of the starting points for producing clavulanic acid. The genes were tampered with using ligation and excision techniques in several plasmids. The generation of clavulanic acid increased by a factor of two as a result of the buildup of the G3P precursor (Figure 11.2).

The polyketide antibiotic actinorhodin, which is generated by *S. coelicolor*, is another instance of how improved precursor availability boosts product production. By putting an extra copy on a plasmid, Ryu and associates attempted overexpression. The genes *accA2*, *accB*, and *ace* encode distinct subunits of the ACC enzyme (Fatma et al., 2020). These genes are used as precursors to produce actinorhodin. The results showed an improvement in the carbon flow to the actinorhodin precursor enzyme malonyl-CoA, resulting in a six-fold increase in product output. Metabolic engineering

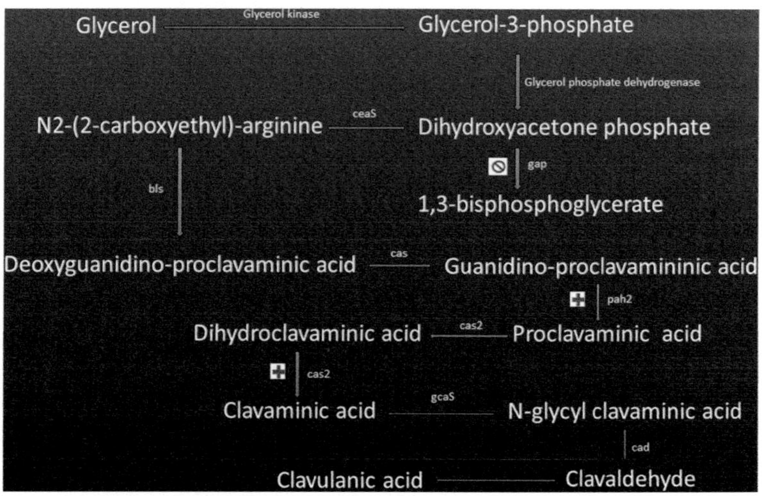

Figure 11.2 Enhancing the generation of clavulanic acid through metabolic engineering of different genes in the glycolytic pathway.

typically finds structural genes involved in the metabolic processes that produce the SMs to be appealing targets. In these instances, the gene dose might be enhanced; alternatively, the gene might be altered, deleted, or declared inactive. Theoretically, increasing the dosage of a specific gene would result in an increase in the target protein and, consequently, in the synthesis of SMs. The synthesis of clavulanic acid rose by up to five times when cas2 was introduced into the *S. clavuligerus* chromosome (López-Agudelo et al., 2021).

Utilizing recombinant plasmids created for the study, this was accomplished. Different plasmids for chromosomal integration and overexpression were created, and these plasmids were then transformed into *S. clavuligerus*. Moreover, it has recently been shown that the duplication of *pah2*, which codes for a proclavaminate amidino hydrolase, can enhance the synthesis of clavulanic acid by introducing a single extra copy of *pah2* into the chromosome through the use of an integrative plasmid. The *sanO* gene, which codes for a NRPS, was duplicated in order to boost the gene dose and hence the synthesis of Nikkomycin X, an antifungal drug produced by *S. ansochromogenes*. This was accomplished by creating a recombinant strain with the *sanO*-containing pWO6 plasmid. This resulted in a double rise. By deleting or inactivating genes that code for processes that change a metabolite into something else, metabolite synthesis can be enhanced. Furthermore, it may entail the elimination of genes whose by-products hinder or deteriorate the intended result. The insertion of a kanamycin-resistant tag into the coding region of the *nysF* gene and subsequent in-frame deletion result in a 60% increase in nystatin production, indicating a potential negative role for the gene in nystatin biosynthesis. On chromosomes, genes involved in secondary metabolic biosynthesis are frequently arranged in clusters along with regulatory genes unique to each route. In the expression of gene cluster elements, pathway-specific regulators can act as either positive (activators) or negative (repressors) factors. Distinct numbers of pathway-specific positive regulatory genes are found in distinct clusters. The most well-known example is *S. coelicolor*, which generates a number of antibiotics, including actinorhodin, a calcium-dependent antibiotic; undecylprodigiosin; and methylenomycin.

Since all of these previous results have been taken into account, it is clear that the best way to increase the production of secondary metabolite pathways is to either overexpress pathway-specific positive regulators or turn off pathway repressors. Regulatory sequences on DNA typically regulate the upregulation of genes. Such regulators exist in *Streptomyces*, and they cause the transcription of genes involved in biosynthesis. It has been demonstrated that several proteins, like SARP and AfsR-p from *S. peucetius*, improve the production of doxorubicin. Also, putting too much AfsR-p on a plasmid raises the levels of actinorhodin, clavulanic acid, streptomycin, and pikromycin in *S. lividans, S. clavuligerus, S. griseus*, and *S. venezuelae*, in that order (Rastegari et al., 2019). DNA regulatory sequences also govern the downregulation of genes; in this instance, the regulatory sequences would be pathway-specific repressors. Overproduction of SMs may result from their inactivation. This is the case of chromomycin, whose synthesis is elevated by plasmid-mediated inactivation of the pathway-specific transcriptional repressor cmmRII in *S. griseus* subsp. *griseus*. An additional pleiotropic gene, *nsdA*, has been found in *S. coelicolor*. Antibiotic production is slowed down by it, and when it's interrupted, the production of methylenomycin, actinorhodin, and calcium-dependent antibiotics goes up. *ActII-orf4* mRNA levels, a pathway-specific regulator, are higher in the *nsdA* mutant. DNA fragmentation and ligation are steps in the genome-shuffle process that may result in advantageous mutations. One novel technique that has been reported to quickly increase the production of SMs is genome shuffling. This approach has been used most recently to produce ε-PL in five different *Streptomyces* species. Using UV light to make hybrids through genome switching and interspecific hybridization, which is when DNA from two species in the same genus is mixed, led to three times more ε-PL production (Li et al., 2013).

11.6 CRISPR/CAS-BASED GENOME EDITING

Genome engineering has been completely transformed by CRISPR/Cas system-based genome editing tools. Genome editing has been extensively used using these technologies across all kingdoms, including bacteria, plants, and animals. In contrast to Cre and I-SceI nuclease-based genome editing, CRISPR/Cas uses transcribed synthetic guide RNA, or just crRNA, to direct Cas proteins to any site on the genome instead of requiring the pre-integration of a unique enzyme recognition sequence into the genome. The most commonly utilized Cas enzyme in *Streptomyces* among the several Cas endonucleases that are known to exist is Cas9, originating from *S. pyogenes* and classified as Class 2 type II. The Class 2 type V bacteria *Francisella novicida*, which is the source of Cpf1 (also called Cas12a), has recently been employed to construct streptomycetes. The CRISPR/Cas-assisted method provides a number of benefits over previous genome editing techniques, including increased efficiency, simplicity of use, and shorter operation times.

Furthermore, a number of CRISPR/Cas-derived technologies have recently been developed, including the BEs for targeted base mutagenesis based on dCas9 or Cas9n (a nickase version of Cas9 with the mutation of D10A), and the CRISPRi–mediated gene repression tool based on dCas9 (a nuclease-deficient Cas9 with two mutations of D10A and H840A) or ddCpf1 (a nuclease-deficient Cpf1 with the mutation of E1006A) and the BEs. They are potent additions to *Streptomyces*' arsenal of CRISPR/Cas created genome editing tools. In 2015, CRISPR/Cas9-based genome editing tools were developed successively by four research groups. These tools allowed for effective and quick genetic manipulation, such as point mutations in

Streptomyces and the deletion of single or multiple genes or gene clusters. In order to precisely delete different DNA sizes (ranging from 20 bp to 31.4 kb), including single genes, double genes simultaneously, and single antibiotic BGCs, Cobb et al. (2015) developed two sets of CRISPR/Cas9 genome editing systems, pCRISPomyces-1 and pCRISPomyces-2. They then used HDR on three different *Streptomyces* species. This resulted in an efficiency of 21–100%. Using this CRISPR/Cas9 technique, they later identified diverse metabolites in five native *Streptomyces* hosts, including a novel pentangular type II polyketide in *S. viridochromogenes*, by performing promoter knock-in for the activation of silent BGCs of various classes. In *S. coelicolor*, researchers created pKCcas9dO, a CRISPR/Cas9-based editing system that is comparable. Using this approach, genetic engineering was carried out at several levels with great efficiency ranging from 60–100%. Individual genes (*actII-orf4, redD*, and *glnR*) as well as single antibiotic BGCs (measuring 21.3, 31.6, and 82.8 kb, respectively) were deleted (Huang et al., 2015). Simultaneously, Tong et al. (2015) developed the pCRISPR-Cas9 editing method, which allows for the accurate deletion of single and multiple genes in *S. coelicolor* by high-dose RNA interference. Additionally, scientists developed the editing plasmid pCRISPR-Cas9-ScaligD and restored the NHEJ repair pathway by introducing the *S. carneus* ligD gene. When homologous DNA templates were not available, this technique achieved highly efficient inactivation of the target genes. These four genomes editing systems pCRISPR-Cas9, pKCcas9dO, pCRISPomyces-1, and pCRISPomyces-2 are all derived from plasmids that carry the temperature-sensitive replicon pSG5. The CRISPR/Cas9 editing system pWHU2653 was created by Zeng et al. (2015), using a plasmid containing the segregationally unstable replicon pIJ101. Furthermore, the D314A mutant of cytosine deaminase CodA, or counter-selection marker CodA(sm), was developed. 5-FC is converted to 5-FU, a hazardous chemical. The insertion of CodA (sm) made plasmid curing simpler. Plasmid curative efficiency can be increased to 95%, which is greater than the 39% of plasmid editing produced from pIJ101 that lacks CodA (sm).

Recently, two CRISPR/Cas9 genome editing tools from pWHU2653 were modified. Using the thiostrepton-inducible promoter tipAp instead of the Cas9 ermEp* promoter, Mo et al. (2019) generated the pMWCas9 editing system, which is based on pWHU2653 and has the potential to significantly boost the efficiency of DNA transformation. Notably, pMWCas9 has been successful in deleting highly repetitive DNA sequences, such as the eryAIII gene from erythromycin PKS in *Saccharopolyspora erythraea*. However, the target gene remained unaffected by the pCRISPR-Cas9 editing tool. This may have been caused by an unanticipated recombination of DNA originating from the pSG5 replicon.

Based on pWHU2653, Wang et al. (2020) created an improved dual-functional chromogenic-screening CRISPR/Cas9 tool (pQS-*gusA* and pQS-*idgS*) by substituting two reporter systems, GusA and IdgS, for the counter-selection marker CodA (sm). After chromogenic screening, these two reporter systems increased the effectiveness of plasmid curing and genome editing by 100% and substantially streamlined the procedure. Additionally, this editing approach was effectively used to manipulate the genetic makeup of *Verrucosispora* sp. MS100137, a rare actinomycete strain that grows slowly and is genetically resistant. With this genome editing technique, the deletion of the abyssomicin BGC (61 kb) and the carotenoid BGC (5.5 kb) was accomplished with an efficiency of up to 100%. Furthermore, a modified editing tool was created using the pCRISPR-Cas9 plasmid by combining the homology template (needed for HDR) and the *S. carneus*-derived ligD (needed for NHEJ) into a single plasmid, enabling the occurrence of any DNA repair pathway (Culp et al., 2019).

There are several issues with the CRISPR/Cas9-based genome editing technologies that were previously discussed. The PAM sequence is widely dispersed throughout *Streptomyces* genomes as a result of the high GC content (>70%) in the *Streptomyces* genome. However, the AT–rich DNA regions might not contain the PAM sequence. Moreover, the insertion of homologous DNA templates for DSB repair and the production of plasmids for the independent transcription of numerous sgRNAs each with its own promoter and terminator are difficult and time-consuming steps in the implementation of multiplex gene editing. Li et al., studied Cpf1 from *Francisella novicida* (FnCpf1), which recognizes T-rich PAM sequences (50–TTV-3), enabling *Streptomyces* genome editing in order to overcome these constraints (Li et al., 2018). One benefit of multiplex genome editing is that Cpf1 possesses RNase activity for pre-crRNA processing, and just one promoter is needed for the expression of numerous guide crRNAs. With *S. coelicolor*, it was possible to precisely delete single or double genes at the same time using HDR with a high degree of effectiveness (75–95%). In order to reconstruct the NHEJ pathway, we also added the codon-optimized *Mycobacterium smegmatis ligD* and *ku* genes. Random-sized DNA deletions may occur when NHEJ–assisted DSB repair inactivates target genes or gene clusters. Furthermore, it was noticed that in the seven *Streptomyces* species that were investigated, Cpf1 from *F. novicida* and Cas9 from *S. pyogenes* showed varying adaptability. The 5-oxomilbemycin A3/A4-producing strain *Streptomyces hygroscopicus* SIPI-KF, which cannot be edited by Cas9 due to its extreme toxicity, had a gene deleted thanks to the use of FnCpf1.

It has been shown that the CRIPSRi tool, which is based on the nuclease-defective Cas nuclease (dCas9 and ddCpf1), is a potent tool for metabolic engineering in bacteria and functional genome research. Three distinct CRISPRi tools for *Streptomyces* have been created recently. By substituting dCas9 for Cas9 in the genome editing system pCRISPR-Cas9, Tong et al. (2015) created an inducible CRISPRi that effectively repressed specific genes upon induction. The dCas9/sgRNA complex was expressed in this

CRISPRi system using the replicative plasmid pGM1190, which contains the temperature-sensitive replicant pSG5. The thiostrepton-inducible promoter (tipAp) controls the expression of the dCas9 gene. Based on dCas9 and ddCpf1, research team created two CRISPRi tools in *S. coelicolor*. The dCas9/sgRNAs or ddCpf1/crRNAs complex was expressed in these two systems using the integrative plasmid pSET152. Researchers were able to efficiently repress up to four genes at once using these two CRISPRi methods. The latter two approaches may offer two benefits over the replicative plasmid-based inducible CRISPRi tool (e.g., pGM1190). First, once they are incorporated into the genome, their repressive effects should be stable. Because the pSET152 transformation is more efficient than replicative plasmids, it can be used in a wider range of situations. Meanwhile, all *Streptomyces* strains with genome sequences now available have the φC31 attB site for pSET152 integration in their genomes (Zhao et al., 2020). It is significant to remember that creating numerous sgRNA expression cassettes with separate promoters and terminators is a time-consuming step in the dCas9-based system's simultaneous repression of several targets. On the other hand, because of ddCpf1's pre-crRNA processing capability, a single tailored CRISPR array with a single promoter is needed, which saves time and is convenient. Thus, for multiplexed gene repression, the ddCpf1-based CRISPRi system is superior to the dCas9-based systems.

Large chromosomal areas can be isolated via TAR-cloning, which eliminates the need to laboriously build a random clone library. The technique relies on a linear TAR-cloning vector with anchor-like, unique terminal targeting sequences that are homologous to the sequence of interest, as well as in vivo homologous recombination of a specific genome target. The process involves a cotransformation in which yeast is given both the vector and the genomic DNA. The flanking areas of the target sequences and the vector anchors are then homologously recombined. As a result, the target sequence is contained in circular YACs. DNA pieces up to 250 kb can be cloned using the usual TAR procedure; however, the success rate in producing clones with the right construct is only 1%–5%, which is extremely low. By inserting DSBs, usually by restriction enzymes, into the chromosomal DNA close to (>100 bp) the target area, the efficiency can be raised to 30%. Because the target genomic sequences are integrated into shorter DNA pieces, this facilitates more successful homologous recombination between the cloning vector anchor and the target genomic sequences.

In a more recent method, the DSBs were introduced using CRISPR/Cas9 to boost the specificity of TAR cloning. The benefit of this is that almost any desired sequence can be cloned into a YAC vector without the requirement for uncommon and specialized restriction enzymes that cut close to but not through the target sequence. Parts of the pristinamycin BGC from *S. pristinaespiralis* (Meng et al., 2017), and/or the prodigiosin BGC from *S. coelicolor* (Liu et al., 2017), were cloned using this technique.

11.7 REVERSE ENGINEERING APPROACHES

Since the public release of the genome sequences of the first model species, *S. coelicolor*, in 2002, numerous omics-guided methods have been employed to advance our comprehension of the global regulatory networks of *Streptomyces* involved in polyketide biosynthesis (Wang et al., 2024a). Even in well-characterized species, identifying global regulatory genes is far more challenging because *S. coelicolor* genome sequences contain over 300 annotated putative regulatory ORFs. This is despite the fact that most pathway-specific regulatory genes have been identified based on their typical location within the biosynthetic pathway gene cluster. A new technique called omics-guided reverse engineering has shown promise as a way to look at changes in gene expression that are connected to making too many polyketides in a number of commercial strains of *Streptomyces*. Using this strategy, we were able to compare the OIM that was randomly and recursively mutagenized with the wild-type strain at the molecular gene level. As a result, genes that were previously unknown for primary and/or secondary metabolism yet essential for the overproduction of secondary metabolites may be identified. The *S. coelicolor* DNA microarray technology was utilized to effectively treat the strains of *Saccharopolyspora erythraea* OIM that produced more erythromycin than tylosin and the strains of *S. fradiae* OIM that produced more tylosin. Similar genome-wide transcriptome research revealed that in *S. coelicolor* A3(2) and other genes that rely on AfsS, distinct nutritional stress responses regulate a pleiotropic antibiotic regulator known as AfsS (Yan and Xia, 2024; Wang et al., 2024b). Furthermore, a DNA microarray study on *S. peucetius* revealed that variations in culture conditions affect the expression patterns of *S. peucetius* genes involved in the production of doxorubicin. Furthermore, strategies based on genomics were used to activate cryptic pathways like stambomycins and 51-membered glycosylated macrolides. Significant connections to primary metabolism were also shown by chemically altering secondary metabolism, suggesting that small compounds may also increase secondary metabolite yields for biochemical characterization and discovery.

11.8 PERSPECTIVES FOR SECONDARY METABOLITE DISCOVERY IN *STREPTOMYCES*

Several methods for finding and producing SMs in *Streptomyces* were compiled in the current review. New techniques in engineering and the ability of NGS to fully sequence the genomes of target strains have shown that *Streptomyces* sp. may be a source of valuable chemicals that haven't been found yet. In the last few decades, genetic techniques based on the CRISPR/Cas system have been employed to increase the production of SMs and turn on quiet BGCs. However, additional work needs to be done to make the

CRISPR/Cas system better for designing streptomycetes. But, new "super-host" strains like *S. albus, S. coelicolor, S. lividans*, and *S. avermitilis* that allow the production of BGCs from different genes provide another way to get around the current problems. Super-host optimization, however, is not without its difficulties. For example, several metabolites were not synthesized to a significant degree. Several recent studies have also linked the optimization of conventional fermentation with the possibility of genetically modifying *Streptomyces*. In particular, looking into how nitrogen is supplied by amino acids and molecules made from amino acids, such as polyamines, gave us new ways to think about finding SMs. Different nitrogen sources, however, might reduce the formation of SMs at specific quantities and, in excess, can even be harmful to cells. For this reason, the supply of nutrients must be strictly regulated during fermentation. A lot of new bioinformatics tools, genetic engineering methods, and nutrient delivery methods have been used to make natural product discovery in *Streptomyces* sp. more successful. These tools have helped find more new compounds of interest.

11.9 CONCLUSION

The fields of genome and compound mining, omics, and molecular cloning techniques have experienced significant advancements and improvements. Systems metabolic engineering can harness the production capacity of streptomycetes. Ongoing systematic and worldwide investigations will continue to improve, resulting in faster and more informed decision-making for the redesign of specific strains to optimize the production of desired secondary metabolites and discover new ones. However, it is important to note that despite having access to all these tools, some repetitions of the design-build-test-learn cycle are still necessary, as demonstrated by previous instances of microbial metabolic engineering. This is particularly accurate when contemplating the complex control between primary and secondary metabolism in streptomycetes. *Streptomyces* wild-type producers naturally have a complex process for producing secondary metabolites, resulting in relatively low levels of these metabolites. Nevertheless, the manipulation of biosynthetic processes in both primary and secondary metabolism, together with their regulation, can be effectively utilized as targets for engineering purposes. Furthermore, the secondary metabolism of *Streptomyces* sp. relies on the provision of precursors from the primary metabolism, a process of great importance that can be further improved for biotechnological applications, alongside the utilization of current genetic modification techniques. Several techniques and instruments have been created in the past to make these steps easier. The main structural element of most constructions is an established vehicle, such as a vector (integrative or replicative system). It is possible to add more components to the construct, though. These components can be removed from preexisting sources or amplified in a PCR reaction.

Due to the high GC content of actinomycetes' DNA, the amplification and assembly of many pieces (such as the cloning of BGCs) may be difficult. Fast assembly techniques, combined with recently developed polymerases and tailored buffers, can solve this issue. It was shown, for instance, that the use of iCatch, DiPaC, AGOS, and a modified Gibson Assembly produced actinomycete engineering functional constructs. Engineered gene elements can be used in place of existing genetic components, which may not be effective in particular actinomycetes. A number of engineering studies that have been published recently have identified and effectively used additional new or modified genetic components (such as promoters, RBS engineering, and biosensor components) for the purpose of improving and fine-tuning the synthesis of actinomycete chemicals. Following the assembly of the necessary genetic components, the construct must be inserted into the actinomycete's genome. Usually, conjugation or protoplast transformation are utilized to achieve this. Replication and/or integration occur depending on which functions are available in the construct. For sequence and gene expression, for example, replicative and integrative plasmids are often utilized. For sequence and gene inactivation, including deletion, plasmids bearing homologous sections upstream and downstream of a target location are extensively exploited to aid in single- and double-crossing. Specifically, the conventional approach to producing deletion mutants is incredibly labor intensive. Actinomycetes may now quickly modify their genomes through the revolutionary CRISPR/Cas system. A number of variants, such as "pCRISPomyces, pKCcas9dO, pCRISPR-Cas9, CRISPR/dCas9, CRISPR/Cas9-CodA(sm), pCRISPomyces-2, and CATCH," were described in the interim. In fact, some *Streptomyces* strains and a few other actinomycetes were successfully engineered to have their genomes altered by the CRISPR/Cas systems.

ABBREVIATIONS

1,3-BPG	1,3-diphosphoglicerate
5' UTR	5' untranslated region
5-FC	5-fluorocytosine
5-FU	5-fluorouracil
ACC	acetyl-CoA carboxylase
BES	base editors
BGC	biosynthetic gene cluster
CAS	clavaminate synthase
CAS	CRISPR-associated protein
CDS	coding sequence
CRISPR	Clustered Regularly Interspaced Short Palindromic Repeats
CRISPRI	CRISPR interference
CRRNA	CRISPR RNA
G3P	glyceraldehyde-3-phosphate

HDR	homology-directed repair
HR	homologous recombination
NGS	next-generation sequencing
NHEJ	nonhomologous end joining
NRPS	non-ribosomal peptide synthetase
OIM	overproducing industrial mutant
ORFs	open reading frames
OSMAC	One Strain-Many Compounds
PAM	protospacer adjacent motif
PKS	polyketide synthase
RBS	ribosome binding site
SARP	Streptomyces antibiotic regulatory proteins
sgRNA	single guide RNA
smBGCs	secondary metabolite biosynthesis gene clusters
SMs	secondary metabolites
TAR	transformation-associated recombination
TIRs	terminal inverted repeat sequences
tracrRNA	transactivating crRNA
WT	wild type
YAC	yeast artificial chromosome

REFERENCES

Baral, B., Akhgari, A., & Metsä-Ketelä, M. 2018. Activation of microbial secondary metabolic pathways: Avenues and challenges. *Synthetic and Systems Biotechnology* 3:163–178.

Barka, E. A., Vatsa, P., Sanchez, L., Gaveau-Vaillant, N., Jacquard, C., Klenk, H. P., & van Wezel, G. P. 2016. Taxonomy, physiology, and natural products of Actinobacteria. *Microbiology and Molecular Biology Reviews* 80:1–43.

Blin, K., Shaw, S., Steinke, K., Villebro, R., Ziemert, N., Lee, S. Y., ... & Weber, T. 2019. antiSMASH 5.0: updates to the secondary metabolite genome mining pipeline. *Nucleic Acids Research* 47:W81–W87.

Cobb, R. E., Wang, Y., & Zhao, H. 2015. High-efficiency multiplex genome editing of Streptomyces species using an engineered CRISPR/Cas system. *ACS Synthetic Biology* 4:723–728.

Culp, E. J., Yim, G., Waglechner, N., Wang, W., Pawlowski, A. C., & Wright, G. D. 2019. Hidden antibiotics in actinomycetes can be identified by inactivation of gene clusters for common antibiotics. *Nature Biotechnology* 37:1149–1154.

Fatma, Z., Schultz, J. C., & Zhao, H. 2020. Recent advances in domesticating non-model microorganisms. *Biotechnology Progress* 36:e3008.

Fernández-Martínez, L. T., & Bibb, M. J. 2014. Use of the meganuclease I-SceI of Saccharomyces cerevisiae to select for gene deletions in actinomycetes. *Science Report* 4:7100.

Huang, H., Zheng, G., Jiang, W., Hu, H., & Lu, Y. 2015. One-step high-efficiency CRISPR/Cas9–mediated genome editing in Streptomyces. *Acta Biochimica et Biophysica Sinica* 47:231–243.

Krysenko, S. 2023. Impact of nitrogen-containing compounds on secondary metabolism in Streptomyces spp.—A source of metabolic engineering strategies. *SynBio* 1:204–225.

Lee, N., Hwang, S., Kim, J., Cho, S., Palsson, B., & Cho, B. K. 2020a. Mini review: Genome mining approaches for the identification of secondary metabolite biosynthetic gene clusters in Streptomyces. *Computational and Structural Biotechnology Journal* 18:1548–1556.

Lee, N., Hwang, S., Lee, Y., Cho, S., Palsson, B., & Cho, B. K. 2019. Synthetic biology tools for novel secondary metabolite discovery in Streptomyces. *Journal of Microbiology and Biotechnology* 29:667–686.

Lee, N., Kim, W., Chung, J., Lee, Y., Cho, S., Jang, K. S., ... & Cho, B. K. 2020b. Iron competition triggers antibiotic biosynthesis in Streptomyces coelicolor during coculture with Myxococcus xanthus. *The ISME Journal* 14:1111–1124.

Li, L., Wei, K., Zheng, G., Liu, X., Chen, S., Jiang, W., & Lu, Y. 2018. CRISPR-Cpf1-assisted multiplex genome editing and transcriptional repression in Streptomyces. *Applied and Environmental Microbiology* 84:e00827–18.

Li, S., Chen, X., Dong, C., Zhao, F., Tang, L., & Mao, Z. 2013. Combining genome shuffling and interspecific hybridization among Streptomyces improved ε-poly-L-lysine production. *Applied Biochemistry and Biotechnology* 169:338–350.

Liu, P., Zhu, H., Zheng, G., Jiang, W., & Lu, Y. 2017. Metabolic engineering of Streptomyces coelicolor for enhanced prodigiosins (RED) production. *Science China Life Sciences* 60:948–957.

López-Agudelo, V. A., Gómez-Ríos, D., & Ramirez-Malule, H. 2021. Clavulanic acid production by Streptomyces clavuligerus: Insights from systems biology, strain engineering, and downstream processing. *Antibiotics* 10:84.

Meng, J., Feng, R., Zheng, G., Ge, M., Mast, Y., Wohlleben, W., ... & Lu, Y. 2017. Improvement of pristinamycin I (PI) production in Streptomyces pristinaespiralis by metabolic engineering approaches. *Synthetic and Systems Biotechnology* 2:130–136.

Mo, J., Wang, S., Zhang, W., Li, C., Deng, Z., Zhang, L., & Qu, X. 2019. Efficient editing DNA regions with high sequence identity in actinomycetal genomes by a CRISPR-Cas9 system. *Synthetic and Systems Biotechnology* 4:86–91.

Rastegari, A. A., Yadav, A. N., & Yadav, N. 2019. Genetic manipulation of secondary metabolites producers. In *New and future developments in microbial biotechnology and bioengineering* (13–29) Elsevier.

Sharma, A., Gupta, G., Ahmad, T., Mansoor, S., & Kaur, B. 2021. Enzyme engineering: current trends and future perspectives. *Food Reviews International* 37:121–154.

Tong, Y., Charusanti, P., Zhang, L., Weber, T., & Lee, S. Y. 2015. CRISPR-Cas9 based engineering of actinomycetal genomes. *ACS Synthetic Biology* 4:1020–1029.

Wang, Q., Xie, F., Tong, Y., Habisch, R., Yang, B., Zhang, L., ... & Fu, C. 2020. Dual-function chromogenic screening-based CRISPR/Cas9 genome editing system for actinomycetes. *Applied Microbiology and Biotechnology* 104:225–239.

Wang, X., Chen, N., Cruz-Morales, P., Zhong, B., Zhang, Y., Wang, J., ... & Luo, X. 2024a. Elucidation of genes enhancing natural product biosynthesis through co-evolution analysis. *Nature Metabolism* 1–14.

Wang, Y., Yang, X., Yu, F., Deng, Z., Lin, S., & Zheng, J. 2024b. Structural and functional characterization of AfsR, an SARP family transcriptional activator of antibiotic biosynthesis in Streptomyces. *PLoS Biology* 22:e3002528.

Wolf, T., Gren, T., Thieme, E., Wibberg, D., Zemke, T., Pühler, A., & Kalinowski, J. 2016. Targeted genome editing in the rare actinomycete Actinoplanes sp. SE50/110 by using the CRISPR/Cas9 system. *Journal of Biotechnology* 231:122–128.

Yan, Y., & Xia, H. 2024. The roles of SARP family regulators involved in secondary metabolism in Streptomyces. *Frontiers in Microbiology* 15:1368809.

Zeng, H., Wen, S., Xu, W., He, Z., Zhai, G., Liu, Y., ... & Sun, Y. 2015. Highly efficient editing of the actinorhodin polyketide chain length factor gene in Streptomyces coelicolor M145 using CRISPR/Cas9-CodA (sm) combined system. *Applied Microbiology and Biotechnology* 99:10575–10585.

Zhang, M. M., Wang, Y., Ang, E. L., & Zhao, H. 2016. Engineering microbial hosts for production of bacterial natural products. *Natural Product Reports* 33:963–987.

Zhao, Y., Li, G., Chen, Y., & Lu, Y. 2020. Challenges and advances in genome editing technologies in Streptomyces. *Biomolecules* 10:734.

Chapter 12

A comprehensive study of the role of *Streptomyces* in the food industry and the contribution of biotechnological approaches

Preeti Chauhan and Palki Sahib Kaur
Department of Biotechnology, Chandigarh Colleges of Technology, CGC, Mohali, India

Tilak Raj
Department of Zoology, Pandit Neki Ram Sharma Government College, Rohtak, India

12.1 INTRODUCTION

Streptomyces species are exceedingly valuable for biotechnological uses. These are Gram-positive, filamentous, spore-forming bacteria belong to the Actinobacteria phylum and are grouped under the genus *Streptomyces*. *Streptomyces* can be found in a variety of habitats, such as symbionts, entophytes, mangroves, marine and terrestrial regions, and so on (Khadayat et al., 2020; Lee et al., 2020). There are more than 850 species of *Streptomyces* that have been investigated (Kemung et al., 2018; Aryal et al., 2020). Numerous metabolites with varied applications in biotechnology, agriculture, and medicine can be found naturally in *Streptomyces*. These *Streptomyces* metabolites are well known for their biological functions in morphogenesis, pathogen suppression, pest control, nutrition acquisition, and several medicinal uses (Khushboo et al., 2022).

Streptomyces species have complex morphological features and a complicated multicellular development that starts with hyphae development from spores and the formation of septa from uni-nucleated spores. They can also create aerial mycelia with numerous branching substrates (Chater, 2016). Among bacteria, *Streptomyces* are unique as they have relatively long and linear genomes of 8–10 Mb with very high G + C% (>70%) (Alam et al., 2022). The *Streptomyces* genome has many biosynthetic gene clusters (BGCs), which are the source of a wide variety of important bioactive compounds used in agriculture, medicine, and other industrial applications (Nicault et al., 2021).

Among many secondary metabolites produced by *Streptomyces* are several extremely important pharmaceuticals, such as antibiotics, anticancer,

 DOI: 10.1201/9781032694634-12

and anti-obesity medications (Khushboo et al., 2022). For decades, *Streptomyces* strains have been used to produce important chemicals on an industrial scale (Alam et al., 2022). One growing area of investigation is the application of *Streptomyces* in the food sector. Using the genus for biotechnological purposes presents a number of difficulties, and in order to get acceptable titers of the target chemicals, production strains frequently go through multiple rounds of strain improvement (Khushboo et al., 2022).

Previously, the producer strains improvement was frequently dependent on techniques like random mutagenesis (Wang et al., 2019; Khushboo et al., 2022). Although this approach is still effective, recent breakthroughs in synthetic biology and omics sciences have made it possible to optimize *Streptomyces* as cell factories by using a logical, data-driven approach (Ye et al., 2021).

The exceptionally intricate metabolism networks of bacteria are being revealed by the scientific community, owing to the vast data obtained from genomics as well as molecular profiling experiments and continuously developing tools for the various mathematical modeling (Barry et al., 2022). Upon investigating the metabolic capabilities of *Streptomyces* species, it was shown that numerous complex secondary metabolites are produced by sizable modular biosynthetic gene clusters (Lee et al., 2020). It has also helped in identifying single and multi-enzymatic complexes involved in biological reactions with medicinal relevance.

These multi-enzymatic complexes and enzymes have numerous uses in *in-vitro* and *in-vivo* bioconversion processes. The possibility of making extensive changes to these significant compounds to enhance their biotechnological characteristics has significantly increased with the introduction of these enzymes into host organisms (Zhu et al., 2021; Darsandhari et al., 2021). This chapter examines the most current and intriguing developments in the food sector that are greatly enhancing the capacity to fully utilize the genus *Streptomyces* as biotechnological potential.

12.2 BIOTECHNOLOGICAL IMPORTANCE OF *STREPTOMYCES*

Streptomyces produces both individual and spore chains on solid media during its life cycle. Following growth in liquid cultures, the spore chains and aerial hyphae are obtained. Due to unique phases of the life cycle and metabolic processes, *Streptomyces* can be utilized in a wide range of biotechnological applications as shown in Figure 12.1.

Streptomyces play a major role in biotechnology as it produces important secondary metabolites that are used in different industries like food, paper, pulp, leather, pharmacy, and so on (Sanchez et al., 2020). These metabolites are specific to each species and are necessary for the strains to live in a variety

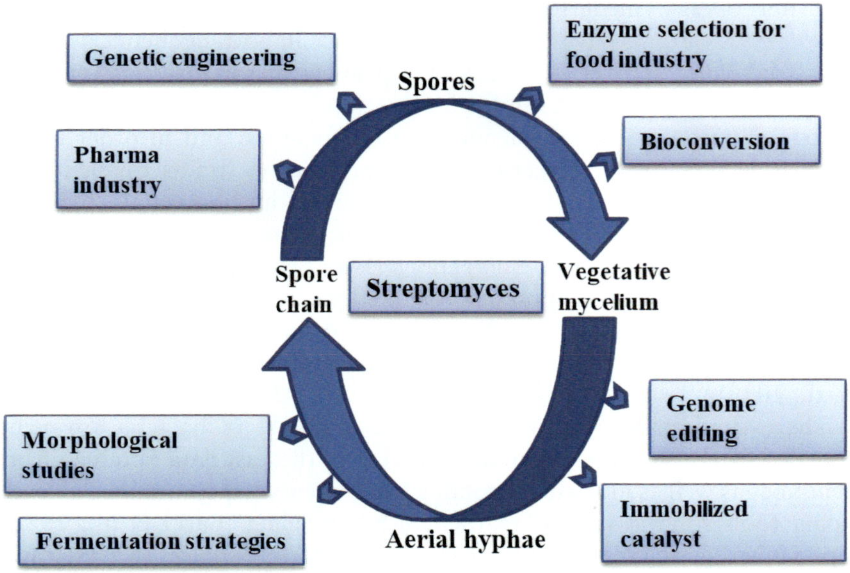

Figure 12.1 Various applications of *Streptomyces* based on their life cycle.

of environments, including terrestrial soils and marine sea beds, and to compete with other contacting microorganisms (Procopio et al., 2012).

These days, *Streptomyces* strains produce two-thirds of all bioactive compounds available in the market for use in medicine, veterinary care, agriculture, and so forth. These strains produce over five thousand different molecules, with a global market valued at around forty-six billion dollars in 2018 including lactams, aminoglycosides polyenes, macrolides, and tetracyclines (Sanchez et al., 2020). Some industrially important compounds used in the food industry are listed in Table 12.1.

The morphological differentiation is essential for the synthesis of secondary (2^0) metabolite production and is also required to get good yields (Li et al., 2020). The fermentation methods used by *streptomyces* to produce secondary metabolites pose several problems. Since these compounds are mostly produced in low concentrations during the stationary phase, industrial methods are typically very time consuming and expensive, and lead to low yields and productivity (Barbuto et al., 2021). The effective industrial process has traditionally included growth pH, medium composition, strain selection, transfer rate of oxygen in the bioreactor, temperature, and state of mycelium aggregation. Large-scale bioreactor fermentation (7-9L) is primarily used in the synthesis of secondary (2^0) metabolites by *Streptomyces* (El-Naggar et al., 2019). The majority of species in liquid-submerged cultures produce mycelium instead of spores (*Streptomyces clavuligerus*) to clump and form pellets of cellular aggregates.

Table 12.1 List of the important compounds produced by *Streptomyces* species having applications in food industry

Compounds	*Strain*	*Application*	*Reference*
Natamycin	*Streptomyces* natalensis	As preservative for food surfaces (cheese, yogurt, sausages, wine etc.)	Meena et al., 2021
		Antifoaming agents	
Napyradiomycin	*Streptomyces sp.* PM33	Antifouling	Sarmiento et al., 2022
Flavonol taxifolin			
Flavonoids			
Dihydroquercetin			
Undecylprodigiosin	*Streptomyces coelicolor*	Food grade colors	Sarmiento et al., 2022
Metacycloprodigiosin Prodigiosin	*Streptomyces sp.* JS520	Antimicrobial	
	Streptomyces sp. WMA-LM31	Antioxidant	
Actinorhodin	*Streptomyces coelicolor*	Food colorants	Venil et al., 2020
		Antimicrobial agents	
Carotenoids	*Streptomyces acidiscabies*	Food grade colors	Sarmiento et al., 2022
Phycocyanin	*Streptomyces* (AQBWWS1)		
Indigoids	*Streptomyces sp.* A1013Y		
TDTA (4, 8, 13-trihydroxy 6, 11 dione trihydrogranaticins A)			
Oxytetracyclin	*Streptomyces coelicolor*	Feed Supplement	Baghlaf et al., 1980
	Streptomyces rimosus (ATCC 10970)		Slemc et al., 2021
Tylosin	*Streptomyces fradiae*	Feed additives	Cuozzo et al., 2023
Hydroxycinnamic acid	*Streptomyces avermilitis* CECT 3339	Brewery	Bartolomé et al., 2003
Feruloyl esterase			
Chitinase	*Streptomyces violaceoruber* (pChi)	Baking process	Silano et al., 2019

The traditional methods of regulating these parameters have been supplemented in recent years with novel techniques such as mycelium-immobilized fermentation, metabolic pathway control, genetic modification of strains, antibiotic precursor supplementation, or metabolic engineering methods. According to this perspective, the morphological study using techniques such as scanning electron microscopy, transmission electron microscopy, and confocal-laser scanning fluorescence microscopy has also proven vital in both immobilized and submerged cell cultivations (Scaffaro et al., 2017).

12.3 FERMENTATION STRATEGIES FOR THE BIOTECHNOLOGICAL APPLICATIONS

In *Streptomyces* cultures, oxygen and food selection is very crucial; both must be carefully regulated to direct biosynthesis toward production of secondary metabolites rather than biomass. Elevated agitation speed or aeration fluxes have traditionally been employed to ensure an adequate oxygen transfer rate. Researches that integrate medium formulation with agitation speed modulation are regularly reported. For instance, agitation speed (ranging around 900 to 1300 rpm) utilized in *S. coelicolorA3(2) M145* affects the synthesis of red and blue pigments (Zhou et al., 2018). In order to achieve a highly reproducible process, 50% of dissolved oxygen (DO) is found necessary. The higher output values were reached by using non-stressful agitation conditions (900–1100 rpm) (Barbuto et al., 2021).

For the formation of secondary metabolites, increased percentages of dissolved oxygen in the fermentation vessel are necessary for pellet differentiation and *Streptomyces* mycelia formation. However, increasing the agitation speed to attain a high oxygen rate may have the opposite effect, for example, it may destroy strains of mycelium, which reduces the yield of secondary metabolites. Consequently, over the years, emphasis has been given to the investigation of novel impeller configurations in the vessel, as well as the introduction of some oxygen (O_2) vectors in the fermentation medium and also using some immobilized systems in place of liquid cultures.

12.4 *STREPTOMYCES* IN THE FOOD INDUSTRY

Food proteins are essential components that affect the physicochemical qualities of food; therefore for over 30 years, researchers have been using transglutaminase to crosslink food proteins and alter their functional properties (Vasic et al., 2023). Many studies discuss the use of transglutaminases in the food industry to crosslink proteins, such as those found in bread, cheese, yogurt, meat, and others.

The synthesis of nemadectin, a macrocyclic lactone that is utilized globally as microbial transglutaminases, belongs to the family of global

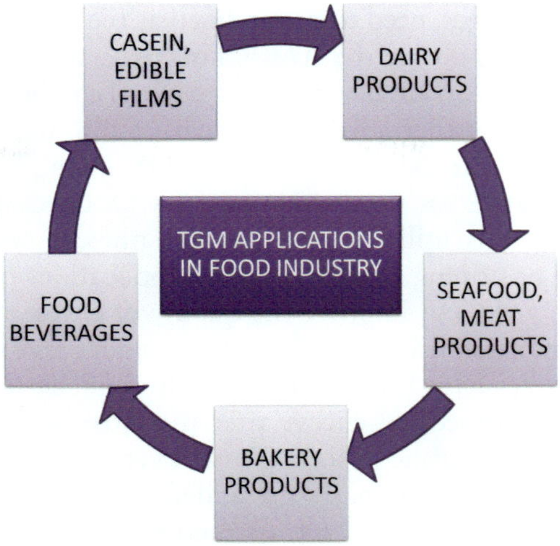

Figure 12.2 Applications of transglutaminases in the food industry.

transglutaminases that are characterized by different bacterial strains. They catalyze the cross-linking of the amino and acyl groups to produce an isopeptide bond. Owing to its wide range of enzymatic activity, transglutaminases have widespread application in the food sector. It is anticipated that the food processing industry will use enzymes at a net annual growth rate of 21.9% (Zhang et al., 2020). Figure 12.2 shows various applications of transglutaminases in the food industry.

Transglutaminases are used in the food sector to improve gel formation, foaming, viscosity, emulsification, and water-holding capacity. Enzymatic processing is encouraged by research on transglutaminases, particularly microbial transglutaminases, which offers a growing number of insights into the diverse applications of this enzyme in different industrial sectors. Additionally, it can be applied to create edible films. Transglutaminases catalyzed the crosslinks among its molecules as well as other proteins. This characteristic affects how protein capabilities including foaming, solubility, emulsifying ability, and gelation vary. Transglutaminases have a broad range specificity for substrate (Zhang et al., 2020).

12.5 DAIRY INDUSTRY

Enzymes like catalase, lactase, transglutaminases, laccase, and others are used in the dairy industry for treatment of skim milk yogurt for improvement in yogurt quality by making crosslinks with milk proteins by using *Streptomyces* sp. SB086 (Mokoonlall et al., 2016). Milk proteins like whey or casein are used as substrates to enhance the gelling, foaming, and emulsifying qualities of foods. Conversely, the whey proteins (globular), which contain disulfide linkages, need to be modified as they are inadequate transglutaminase substrates during the crosslinking process.

Certain reducing agents or elevations in pH level can be used to carry out such modifications. Furthermore, heat denaturation and the use of high hydrostatic pressure can be used to accomplish them. Transglutaminase-modified casein enables the production of dairy products with a good consistency structure, it is utilized in yogurts to prevent synergism and for texture softening or firming. Research on the applications of transglutaminases to boost the yield and improve the quality of cheese that is low in fat has also been conducted using a variety of techniques and protocols.

Subsequent studies have been documented for enhanced consistency and thermal stability in processed cheese after the introduction of transglutaminases in the manufacturing process (Fotschki et al., 2020; Hovjecki et al., 2021). Microbial transglutaminases are added to milk before fermentation to improve the rheological and ultra-structural characteristics of yogurt (Lang et al., 2020).

Transglutaminases are consistently being used in the baking industry to improve the quality of flour, which in turn enhances the texture and bread volume since transglutaminases use grain proteins as good substrates for cross-linking. For instance, it is commonly recognized that rice flour has important nutrients like fiber, proteins, and vitamins B and E. However, it is restricted to work with non-fermented baked goods.

One important use of transglutaminases is the restructuring of meat; there are various publications and research available examining the usage of transglutaminases in the improvement of meat products. The application of transglutaminases not only enhances the meat product's structure and texture but also adds cohesiveness without the need for heat processing or other chemicals like phosphates (Kaić et al., 2021; Kim et al., 2022). It also alters the heavy chain secondary (2^0) structure of myosin by adding more β-sheet and less α-helix, forming high-molecular-weight polymers. Furthermore, certain researches also demonstrate that the presence of transglutaminases might cause varying degrees of gelation (Tokay et al., 2021).

Transglutaminases also increase the nutritional value and quality of blood proteins and collagen by adding appropriate amino acids, such as lysine. Transglutaminases have made it possible to produce new meat products that employ inferior raw materials rather than premium meat products. The protein utilized as a raw material (soy powder or skim milk powder) and the nutritional value, texture, smell, or appearance of goods made with premium meat are not affected by transglutaminases (Vasic et al., 2023).

12.6 ENZYMES-BASED APPLICATION OF *STREPTOMYCES* IN THE FOOD INDUSTRY

Enzymes are becoming a more environment-friendly alternative to chemicals in many industrial production processes because they don't produce greenhouse emissions, they lower the need for energy in industries, and the waste products that are produced are environmentally friendly, biodegradable, and non-toxic. *Streptomyces* is a good source for industrially important enzyme production because they grow quickly, are easily available and affordable, and can be modified genetically to produce enzymes that function ideally under various industrial production conditions (Barbuto et al., 2021). *Streptomyces* can also withstand harsh production conditions and overcome the growing need for enzymes in various industries. The list of different enzymes produced by various strains of *Streptomyces* is given in Table 12.2.

Enzymes have found a broad range of uses in food processing, dairy, bakery, and research and development in animal feed, fruit juices, and beverages industries, making them indispensable in many industrial production processes as shown in Figure 12.3. Enzyme-based production techniques are now widely accepted in many industries, replacing chemical-based methods. One disadvantage of using enzymes in industry is that they can work poorly in some production circumstances, which can be rather harsh. The enzymatic activity is influenced by pressure, pH, presence of possible inhibitors, and temperature (Barbuto et al., 2021).

Proteases are among the most significant enzymes that are commercially used for meat tenderization, better oil recovery from seafood, decreased allergenicity, and improved palatability, digestibility, flavor, viscosity, and elasticity qualities. Based on their active site composition, proteases can be divided into two groups: metallo and serine proteases. Serine proteases are made up of serine residue that forms a catalytic site with histidine and aspartic acid in the active site. Alkaline protease is a key component in the production of value-added goods in the food sector. Alkaline proteases are commonly utilized to produce high-nutrition protein hydrolysates because of their capacity to break down proteins. To improve the nutritional content of food and feed mixtures, proteases are added. The production of cheese is the most significant use of proteases in the food business (Selamoglu et al., 2020).

Proteases play a catalytic role in the production of high-nutrient protein hydrolysate, which is employed in baby food products, and pharmaceutical dietary products (Singh et al., 2022). In the dairy industry, proteases are used to hydrolyze specific peptide bonds (Phe105–Met106 bond) during the cheese-making process, resulting in casein macro peptides of approximately 64 amino acid residues (Raveendran et al., 2018). Meat is tenderized by using proteases, which can hydrolyze muscle fiber proteins and connective components (Karkal et al., 2022). They are also used in the baking business to maintain bread's gluten strength, minimize dough consistency, and enhance bread's flavor and texture (Akbaş, 2023). These hydrolytic enzymes are used to hydrolyze gelatin and recover animal proteins, as well as to break down the turbidity complex that results from the protein in fruit juices (Shirvani et al., 2023).

It was reported that *Streptomyces* produces different types of proteases like serine proteases, metalloproteases, and amino-peptidases. Further serine

Table 12.2 List of the important enzymes produced by various strains of *Streptomyces*

Enzyme	*Strain*	*References*
Trypsin-like and chymotrypsin-like proteases	*Streptomyces griseus*, *Streptomyces fradie*	Read et al., 1984 Yamane et al., 1995
Serine protease	*Streptomyces hygroscopicus*, *Streptomyces spheroids*, *Streptomyces lividans*, *Streptomyces albogriseolus*	Zhang et al., 2008 Kitadokoro et al., 1993
Metalloprotease	*Streptomyces mobaraensis*, *Streptomyces lividans*, *Streptomyces cacaoi* *Streptomyces griseus*, *Streptomyces fradie*, *Streptomyces caespitosus*	Lichenstein et al., 1992 Zhang et al., 2008
Aminopeptidase	*Streptomyces griseus*, *Streptomyces lividans*, *Streptomyces mobaraensis*	Krieger et al., 1994 Umezawa et al., 2004
Alkaline protease	*Streptomyces flavogriseus* *Streptomyces fungicidicus* MML1614 *Streptomyces pulvereceus* MTCC 8374 *Streptomyces* sp. Al-Dhabi-82	Jayasree et al., 2010 Al-Dhabi et al., 2020
Transglutaminase	*Streptomyces mobaraensis* *Streptomyces paucisporogenes* ATCC *12596* *Streptomyces platensis* NRRL *2364* *Streptomyces hygroscopicus* *Streptomyces ladakanum* *Streptomyces libani* *Streptomyces nigrescens* *Streptomyces platensis* *Streptomyces sioyaensis*	Zhang et al., 2012 Ceresino et al. 2018
Cellulase	*Streptomyces strain* C188 *w* *Streptomyces argenteolus* AE58P *Streptomyces argenteolus* AE710A *Streptomyces argenteolus* AE82P *Streptomyces argenteolus* AP51A	Ventorino et al., 2016
Pectinase	*Streptomyces thermocarboxydus* *Streptomyces chartreusis* *Streptomyces* sp. GHBA10 *Streptomyces coelicoflavus* GIAL86	Salehghamari et al., 2019 Patel et al., 2021
Xylanase	*Streptomyces* sp. AOA40 *Streptomyces geysiriensis* *Streptomyces thermocoprophilus* TC13W *Streptomyces* sp. B6 *Streptomyces* sp. S1M3I	Poornima et al., 2020 Sinjaroonsak et al., 2020 Liu et al., 2020 Medouni-Haroune et al., 2023

(Continued)

Table 12.2 (Continued)

Enzyme	*Strain*	*References*
Amylase	*Streptomyces gulbargensis* *Streptomyces* sp. Al-Dhabi-46 *Streptomyces lopnurensis* KaM5 *Streptomyces rochei* *Streptomyces* MI-*1*	Al-Dhabi et al., 2020 Rathore et al., 2021 Al-Agamy et al., 2021 Ali et al., 2023
Lactase	*Streptomyces violaceus*	Sánchez et al., 1980
Lipase	*Streptomyces griseus* *Streptomyces clavuligerus* *Streptomyces coelicolor* *Streptomyces rimosus*	Abramić et al., 1999 Abba et al., 2021
Laccase	*Streptomyces* sp. SB086 *Streptomyces lavendulae* *Streptomyces* sp. HBD30	Jing, 2010 Hamoudi et al., 2021

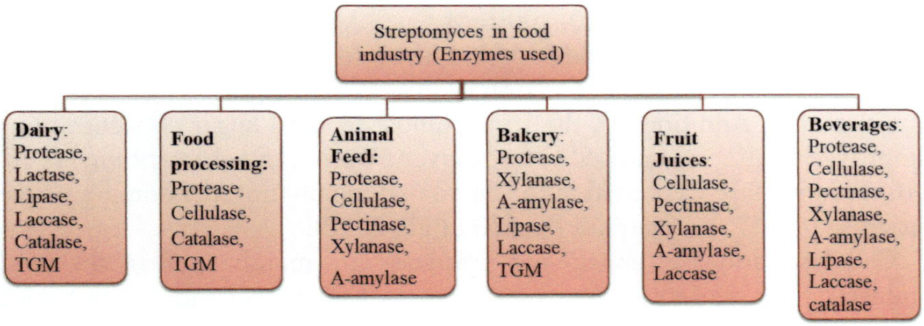

Figure 12.3 Applications of various enzymes in the food industry.

proteases can be classified into trypsin-like proteases, chymotrypsin-like proteases, elastase-like proteases, and amino acid–specific proteases on the basis of substrate specificity. Bentley et al. reported that *S. coeliocolor* genome possesses 8 metalloproteases, 21 aminopeptidases, and 27 serine proteases annotations (Bentley et al., 2002; Wi et al., 2006).

Enzymatic hydrolysis of cellulose, one of the most prevalent organic molecules that can be transformed into goods of considerable commercial importance, is carried out by cellulase enzymes. The use of cellulases in the food industry is still not well recognized (Ejaz et al., 2021). The food business and preservation can greatly benefit from the use of cellulases, which hold enormous potential for enhancing the global food supply for a better yield in food production (Ejaz et al., 2021).

12.7 FRUIT JUICE INDUSTRY AND BAKERY

AOA40 sp. of *Streptomyces* was identified from Mersin soil and is capable of producing both thermotolerant and halotolerant xylanase. With a chosen fermentation medium, different carbon sources were investigated for their ability to produce xylanase. Corn stover was shown to be a very effective carbon source. Afterward, xylanase is partially purified using gel filtration chromatography and ion-exchange chromatography as shown in Figure 12.4, to characterize the enzyme and use it to improve fruit juice and dough. Xylanase can be employed to hydrolyze the arabinoxylan in wheat during bread manufacture, rendering it soluble in water and extractable. This leads to improvements in the dough, volume, viscosity, softness, elasticity, and gluten structure (Both et al., 2021).

The ideal temperature for xylanase activity is 60°C, whereas the ideal pH in phosphate buffer is 6.0. Within the pH ranges of 4–9, the relative xylanase activity stayed between 59.93 and 54.43% of the activity at pH 6.0 (100.00%). At 70°C, the half-life of the xylanase activity was found at 172 minutes; at 80°C, it is around 75 minutes. The enzyme is significantly inhibited by concentrations of 10–100 mM Cu^{+2}, Mg^{+2}, EDTA, SDS, and Hg^{+2}. After partially purified xylanases are used to enzymatically treat grape (17.19%), apple (17.85%), and orange juice (18.36%), the fruit juices' clarity increased. Additionally, the fruit liquids' reduced sugar concentrations improved by 17, 16, and 19%, respectively. When refined *Streptomyces* sp. *AOA40* xylanase were used in the bread making process, the dough volume increased by 17.06% overall (Kumar et al., 2020).

Cellulase, amylase, lipase, and xylanases are mainly used in the food industry for a variety of purposes, such as fruit tenderization, fruit juice

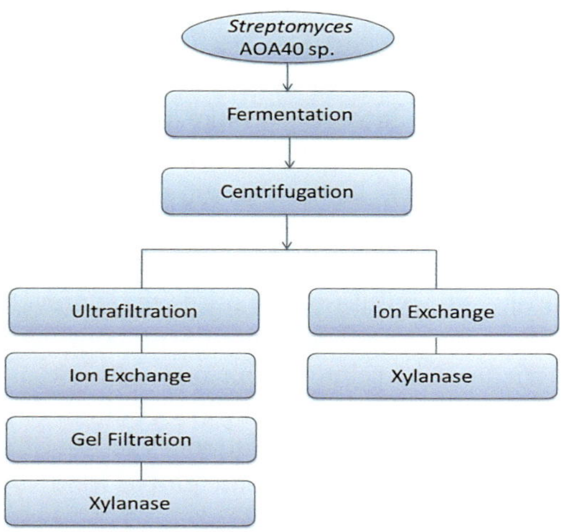

Figure 12.4 Extraction of xylanase with various techniques.

clarity, flavoring and essential oil extraction, and enhancing the overall yield of juices (Sondhi et al., 2021). A rising demand for fruit juices has been the outcome of growing public health consciousness. The standard methods of extracting apple juice are hampered by the presence of cellulose polysaccharides. Cellulases are added to fruits during processing to solubilize the cellulosic polysaccharides and reduce their strength, resulting in nearly total liquefaction.

Many species of *Streptomyces* have the ability to hydrolyze starch, and some of these species hydrolyze the raw starch granules, releasing maltose as the main result. This process is utilized in industry to convert raw starch into sugar for further fermentation. The four best cellulase-producing *Streptomyces* sp. are *S. argenteolus (*AP51A*), S. argenteolus (*AE82P*), S. argenteolus (*AE58P*), and S. argenteolus (*AE710*)* in liquid growth medium in the fruit juice industry (Ventorino et al., 2016). Cellulase is used to decrease the viscosity of nectar and puree from fruits like papaya, plum, mango, apricot, peach, and pear (Raveendran et al., 2018). Juice-based fibers can generate problems for the industry since they are denser and insoluble, which can jam production lines and result in significant losses (Kumar et al., 2019). Since fibers are naturally cellulosic, upon addition of cellulases there is reduction in fibers that facilitates easy juice filtration and obtains substantial reduction in viscosity and acidity by using a combined-enzyme preparation composed of xylanase and cellulase (Shariq et al., 2019). Citrus fruits are also known to taste and smell better when cellulases are added (Ejaz et al., 2021). Additionally, foods high in fiber can lower the risk of heart disease, diabetes, and some cancer types. They can also help maintain a healthy body weight (He et al., 2022).

12.8 ANIMAL FEED

Animal feed is enriched in polysaccharides like pectin, lignin, cellulose, hemicellulose, and others, which must be digestible by the animal. In the animal feed industry, alpha-amylase is introduced in preliminary treatment of animal feed processing to increase the palatability of the animal feed. For further improvement in digestibility of the animal feed, other enzymes like cellulases are also used because they can break down cellulose and hemicellulose into a more digestible form (Velazquez et al., 2021).

12.9 EDIBLE COLORS

There are two types of edible colors: natural and synthetic. The food sector has always used synthetic colorants extensively. Recent research has shown that none of the artificial coloring agents are particularly nutritious, and that the majority of them are hazardous to varying degrees, with some even having the potential to cause cancer. As consumers become more aware of how damaging synthetic colors can be, natural pigments are being highlighted more and more. Natural pigments have become economically viable due to their inherent qualities, safety, and prospective applications as additives (Venil et al., 2020). In the food industry, important indicators of edible colors are solubility, coloring, stability, and safety. The solubility of pigments depends on polarity and charges on the pigment molecules and the properties of the solvent as well.

Food-grade fermentative pigments, like phycocyanin and β-carotene, are now marketed progressively, and pigments with prospective applications in the food industry include indigoids, anthraquinones, and naphthoquinones. *Streptomyces coelicolor* are potent blue pigment producers and are used in a variety of ways. One of the by-products, actinorhodin, is an antibiotic made by *S. coelicolor* A3. Important compounds of the actinorhodin family (γ and λ Actinorhodin) have potential uses as food colorings and antimicrobial agents as well. Actinorhodin has no impact on Gram-negative bacteria like *E. coli* but can inhibit the majority of Gram-positive bacteria, like *Staphylococcus aureus*. It is a pH indicator that turns blue above pH 8.5 and red below (Zhang et al., 2008). Furthermore, prodigiosin family compounds, such as undecylprodigiosin and metacycloprodigiosin, have a wide range of uses, including food coloring, antimicrobials, and antioxidants (Sarmiento et al., 2022).

In *Streptomyces*, carotenoids (potent antioxidants) are produced constitutively, in response to light. It has been discovered that the *Streptomyces (*AQBWWS1*)* isolated from the sea sponge *Callyspongia diffusa* developed food-grade carotenoids pigments under fluorescent white light during fermentation (Dharmaraj et al., 2009). A novel blue pigment was first isolated from *Streptomyces* sp. A1013Y. The purified compound was recognized as TDTA (4, 8, 13 trihydroxy 6, 11 dione trihydrogranaticins A) with molecular weight 462Da. It has been shown that TDTA has excellent free radical scavenging characteristics and is a good source of natural pigment in the food sector (Zhu et al., 2020).

12.10 ENHANCING METABOLITE SYNTHESIS VIA REWRITING THE METABOLISM

The structural and functional relationships between the several genes within an operon controlling their production have been made possible by recent developments in "metabolites to gene" and "gene to metabolite" based techniques. In metabolic engineering, there are two ways of developing synthetic biology and genome editing technologies: the modification of silent gene clusters and the discovery of new bioactive compounds for enhanced production. These days research is focused on structure elucidation, biosynthetic pathways, gene cluster organization, and metabolic engineering for industrially important chemicals production obtained from *Streptomyces* as shown in Figure 12.5.

The discovery of cryptic pathways for secondary metabolites inside *Streptomyces* genomes has been made possible by the advancement of genomic techniques (Amara et al., 2018). It has been discovered that the *Streptomyces coelicolor* A3 genome has a considerable number of genes involved in the manufacture of secondary metabolites, which differ among species (Komaki et al., 2021). Around 62 gene clusters have been identified for *S. bingchenggensis*, as compared to 40 gene clusters observed in *S. albus*.

The production of industrially important secondary metabolites codes for gene clusters that are present together in operons. *Streptomyces* spp. has an intrinsic capacity to produce various industrially important compounds, by the abundance of gene clusters in their genomes (Lee et al., 2021). By using multiomics studies for better understanding of the metabolism of *Streptomyces* species, this genus can be optimized by rewiring its metabolism. Multiomics strategy combines metabolomics, transcriptomics, and 13C-labeled studies. During exponential growth, *S. coelicolor* accumulates triacylglycerides (TAGs), which are then destroyed during the stationary phase. The researchers developed a degradation strategy for TAG (ddTAG) which allowed them to manage the duration and strength of the TAG degradation by placing SCO6196 under the control of an inducible promoter. This resulted in rise in the titer. The same approach increased the titers among other polyketide-producing *Streptomyces* species (Wang et al., 2020). This is a good example of the metabolism rewiring via genetic engineering methods, controlled by molecular profile studies and can be applied to raise the concentrations of compound of interest.

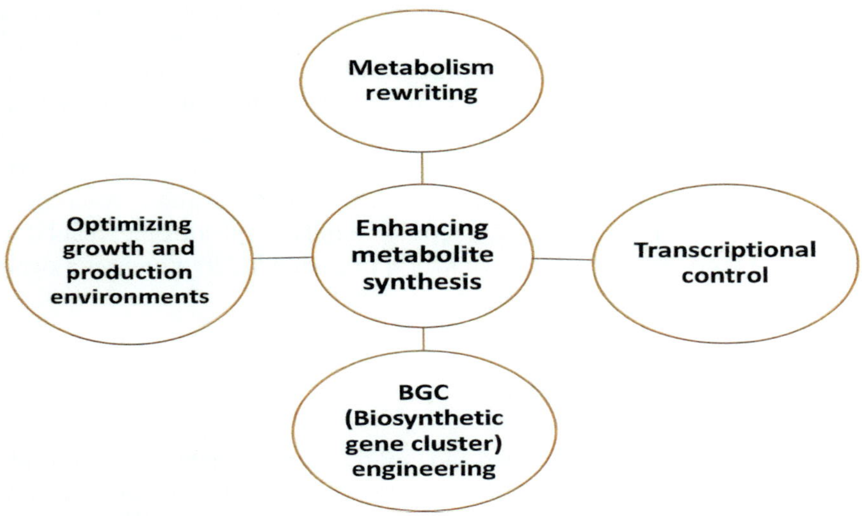

Figure 12.5 Strategies involving metabolite enhancements.

12.11 TRANSCRIPTIONAL CONTROL

As demonstrated by genomic and transcriptome studies, *Streptomyces* has complex transcriptional regulatory systems that control a range of biological functions, including the metabolic shift from primary to secondary metabolism and dynamic morphological development in response to environmental stimuli. This transcriptional regulatory system comprises various transcriptional regulators, like sigma and transcription factors, which influence gene expression by binding with the promoter region (Lee et al., 2021). Consequently, the expression of genes is influenced by suppressing or promoting transcriptional regulator expression.

In *Streptomyces*, the production of important secondary metabolites is controlled by some ambiguous regulatory networks and regulated at the transcriptional level. The set of genes regulated by the transcriptional regulator can be identified using comparative transcriptome analysis by changing the expression of transcriptional regulators (Lee et al., 2021). The genes that produce the significant secondary metabolites are typically organized into clusters (Wei et al., 2018). These gene clusters perform important regulatory functions. Other additional regulators frequently govern these clusters of genes involved in biosynthesis are termed pleiotropic regulators (Wei et al., 2018). These transcriptional regulators are present outside the BGCs and have the ability to influence (positively or negatively) the production of important secondary metabolites. For example, the MarR transcriptional regulator MapR (SCO2398) has been shown to positively regulate the production of actinorhodin in *S. coelicolor* by use of transcriptional control (Xu et al., 2020).

Pleiotropic regulators are also known to inhibit the formation of important secondary metabolites. It has been revealed that the pleiotropic regulator (WblA) is involved in the later stages of morphogenetic expressions, regulating the metabolites production in *Streptomyces* species *CB03234* (Zhang et al., 2020). These regulators control morphological differentiation and primary metabolism in addition to secondary metabolites production. As a result of which their removal may cause unexpected alterations that are frequently harmful to the cell. For example, *S. ghanaensis* contains the pleiotropic regulator AdpA that influences both accumulation and chromosome replication (Makitrynskyy et al., 2021). Similarly pleiotropic regulator WblA has been demonstrated to control sporulation in *Streptomyces* as well as suppress the production of secondary metabolites.

12.12 BGC ENGINEERING

One possible approach is to design and modify the BGCs of bioactive secondary metabolites to improve their production. Bottom-up synthetic biology approaches include assembly, synthesis, and cluster redesign. The application of BGCs remains technically problematic because of their repeating nature, structural complexity, and typically huge sizes. This is particularly true for BGCs that encode non-ribosomal peptide synthetases (NRPS) or big, multi-modular polyketides (Dong et al., 2021; Chang et al., 2022). Main strategies include the introduction of constitutive promoters and insertion of mini cluster in the chromosome without modifying the original cluster to increase the bioactive metabolite titer (Chang et al., 2022).

12.13 OPTIMIZING GROWTH AND PRODUCTION ENVIRONMENTS

Streptomyces strains frequently interact with plants and animals in their natural habitat, where they reside in intricate microbial communities (Connolly et al., 2022). Interspecies interactions frequently lead to the production of secondary metabolites, a phenomenon that can be used for biotechnological applications. It is possible to produce a variety of secondary (2^0) metabolites by the cocultivation of interacting species, however optimizing the growth conditions might be difficult (Boruta et al., 2021). Alternately, adding fragments of other organisms' cell walls or extracts from their cells might imitate interspecies interactions and potentially boost the synthesis of secondary metabolites. By adding biomass from *Fusarium oxysporum* and *Saccharomyces cerevisiae* to the culture broth increased production of some metabolites was noted (Song et al., 2020).

Addition of heat-killed *Staphylococcus aureus* to the medium increased metabolite synthesis from *Streptomyces* sp. *AC117* by fivefold (Shakeri et al., 2021). Furthermore, the presence of foreign signaling molecules released by animals and plants can activate secondary metabolism in *Streptomyces* species (Van et al., 2022). It is common practice to optimize growth conditions in order to maximize the titers of bioactive chemicals produced by *Streptomyces*. In certain instances, notable enhancements can be achieved by varying the carbon supply controlling the availability of phosphate and nitrogen and raising or lowering the oxygen-transfer rate (Barbuto et al., 2021).

Inorganic microparticle supplementation has the potential to increase secondary metabolite biosynthesis as well. Microparticle-enhanced cultivation (MPEC) is a morphological engineering approach used to increase cultivation intensity. The introduction of some talc particles to recombinant *S. albus* cultures has been reported to accelerate morphogenesis or the up regulation of the morphology control process during both the initial and production phase. This increases the intracellular availability of CoA and ultimately leads to a three-fold increase in secondary metabolites production (El-Housseiny et al., 2021). Upon comparison with control experiment without the addition of microparticles, 50% higher levels of poly-l-lysine (2.51 ± 0.08 g L−1) was obtained upon the addition of 6 μm talc microparticles to culture *of Streptomyces* sp. *M-Z18*. More recent studies also focus on optimizing the fermentation parameters like temperature conditions (24 to 30°C), 50% dissolved oxygen percentage, and non-stressful agitation from 900 to 1300 rpm (Barbuto et al., 2021).

12.14 CONCLUSION

Streptomyces is a very important genera in biotechnology. It has the capacity to generate a wide range of natural compounds as part of its primary metabolism. Recent developments in synthetic biology and systems biology have made it possible to comprehend the intricate biology of this genus much better, creating intriguing new opportunities for harnessing the biotechnological potential of *Streptomyces*. The biotechnological potential of *Streptomyces* species expanded with the greater advancements in various fields like genetic engineering, pharmaceuticals, bioinformatics, molecular medicine, and others. Although *Streptomyces* is widely used in many areas, including agriculture, medicine, and the environment, very few studies have investigated the technical aspects of these biotechnological applications in food processing that aim to increase the quality of food. Recent research in the field has established the basis for designing the next generation of *Streptomyces* by reprogramming transcriptional regulation, BGC engineering, and optimizing growth and production environments to extract a good quantity of secondary metabolite. The present chapter described significant developments in various methods and technologies related to the biotechnological applications of *Streptomyces* in the production of vital food-grade enzymes such as xylanase, cellulase, laccase, amylase, lipase, and pectinase in the food industry, with the nutrient enrichment and antibiotic-supplemented food to enhance shelf life and quality of food. Further research can focus more on innovative techniques to better understand the potential that *Streptomyces* portrays to improve the quality of food.

CONFLICT OF INTEREST

Nil

ACKNOWLEDGMENT

The authors acknowledge the Chandigarh group of colleges, Landran, Mohali for providing necessary help and support.

REFERENCES

Abba, P. O. 2021. *Screening and production of lipase from Streptomyces sp.*

Abramić, M., Leščić, I., Korica, T., Vitale, L., Saenger, W., and J. Pigac. 1999. Purification and properties of extracellular lipase from *Streptomyces* rimosus. *Enzyme and Microbial Technology* 25(6):522–9.

Akbaş, P. 2023. *Uses of proteases obtained from microorganisms in the food industry*. 81(3):289.

Al-Agamy, M. H., Alhuzani, M. R., Kelany, M. S., and M. M. Hamed. 2021. *Production and partial characterization of α-amylase enzyme from marine actinomycetes*. BioMed Research International.

Alam, K., Mazumder, A., Sikdar, S., Zhao, Y.M., Hao, J., Song, C., Wang, Y., Sarkar, R., Islam, S., Zhang, Y., and A. Li. 2022. Streptomyces: the biofactory of secondary metabolites. *Frontiers in Microbiology* 29(13):968053.

Al-Dhabi, N.A., Esmail, G.A., Ghilan, A.K., Arasu, M.V., Duraipandiyan, V., and K. Ponmurugan. 2020. Characterization and fermentation optimization of novel thermo stable alkaline protease from *Streptomyces* sp. Al-Dhabi-82 from the Saudi Arabian environment for eco-friendly and industrial applications. *Journal of King Saud University-Science* 32(1):1258–64.

Ali, I., Sultan, S., Tahir Mahmood, R., Tariq, M., Shamim, Z., Mushtaq, A., and M. Asiri. 2023. Production and characterization of α-amylase from indigenously isolated *Streptomyces* sp. *BioResources* 18(1):6.

Amara, A., Takano, E., and R. Breitling. 2018. Development and validation of an updated computational model of *Streptomyces* coelicolor primary and secondary metabolism. *BMC Genomics* 19:1–9.

Aryal, S., Neupane, L., Adhikari, R., Regmi, B., Koirala, N., and D. R. Joshi. 2020. Novel *Streptomyces* sp. reported in 2018: a meta-analysis. *AntiInfect Agents* 19: 2–13.

Baghlaf, A. O., Abou-Zeid, A. Z., El-Dewany, A. I., Eissa, A. E., Fouad, M., and M. Yassein. 1980. *Production of oxytetracycline by Streptomyces rimosus 12,907 as an animal feed supplement*. 135(5):427–34.

Barbuto, F. S., Cammarota, M., Schiraldi, C., and O. F. Restaino. 2021. Streptomycetes as platform for biotechnological production processes of drugs. *Applied Microbiology and Biotechnology* 105:551–68.

Barry, C. P., Gillane, R., Talbo, G. H., Plan, M., Palfreyman, R., Haber-Stuk, A. K., Power, J., Nielsen, L. K., and E. Marcellin. 2022. Multi-omic characterisation of *Streptomyces* hygroscopicus NRRL 30439: detailed assessment of its secondary metabolic potential. *Molecular Omics* 18(3):226–36.

Bartolome, B., Gómez, C. C., Sancho, A. I., Dıez, N., Ferreira, P., Soliveri, J., and J. L. Copa. 2003. Growth and release of hydroxycinnamic acids from brewer's spent grain by *Streptomyces* avermitilis CECT 3339. *Enzyme and Microbial Technology* 32(1):140–4.

Bentley, S., Chater, K., Cerdeño-Tárraga, A.M. et al. 2002. Complete genome sequence of the model actinomycete Streptomyces coelicolor A3(2). *Nature* 417:141–47.

Boruta, T., Ścigaczewska, A., and M. Bizukojć. 2021. "Microbial wars" in a stirred tank bioreactor: Investigating the co-cultures of *Streptomyces* rimosus and Aspergillus terreus, filamentous microorganisms equipped with a rich arsenal of secondary metabolites. *Frontiers in Bioengineering and Biotechnology* 9:713639.

Both, J., Biduski, B., Gómez, M., Bertolin, T. E., Friedrich, M. T., and L. C. Gutkoski. 2021. Micronized whole wheat flour and xylanase application: dough properties and bread quality. *Journal of Food Science and Technology* 58:3902–12.

Ceresino, E. B., de Melo, R. R., Kuktaite, R., Hedenqvist, M. S., Zucchi, T. D., Johansson, E., and H. H. Sato. 2018. Transglutaminase from newly isolated *Streptomyces* sp. CBMAI 1617: Production optimization, characterization and evaluation in wheat protein and dough systems. *Food Chemistry* 241:403–10.

Chang-Hun, Ji. 2022. Top-down synthetic biology approach for titer improvement of clinically important antibiotic daptomycin in *Streptomyces* roseosporus. *Metabolic Engineering* 69:40–49.
Chater, K. F. 2016. Recent advances in understanding *Streptomyces*. *F1000Res* 5:2795.
Connolly, J. A., Harcombe, W. R., Smanski, M. J., Kinkel, L. L., Takano, E., and R. Breitling. 2022. Harnessing intercellular signals to engineer the soil microbiome. *Natural Product Reports* 39(2):311–24.
Cuozzo, S., de LeBlanc, A. D., LeBlanc, J. G., Hoffmann, N., and G. R. Tortella. 2023. *Streptomyces* genus as a source of probiotics and its potential for its use in health. *Microbiological Research* 266:127248.
Darsandhari, S., Dhakal, D., Shrestha, B., Lee, S., Jung, N., Jung, H. J., and J. K. Sohng. 2021. Biosynthesis of bioactive tamarixetin in recombinant Escherichia coli. *Biotechnology and Applied Biochemistry* 68(3):531–7.
Deveoğlu, O., and R. Karadağ. 2019. A review on the flavonoids—a dye source. *International Journal of Advances in Engineering and Pure Sciences* 3:188–200.
Dharmaraj, S., Ashokkumar, B., and K. Dhevendaran. 2009. Food-Grade Pigments from *Streptomyces* Sp. isolated from the marine sponge Callyspongia diffusa. *Food Research International* 42(4):487–92.
Dong, Li. 2021. Reconstitution of a mini-gene cluster combined with ribosome engineering led to effective enhancement of salinomycin production in *Streptomyces* albus. *Microbial Biotechnology* 14:2356–68.
Ejaz, U., Sohail, M., and A. Ghanemi. 2021. Cellulases: from bioactivity to a variety of industrial applications. *Biomimetics* 6(3):44.
El-Housseiny, G. S., Ibrahim, A. A., Yassien, M. A., and K. M. Aboshanab. 2021. Production and statistical optimization of Paromomycin by *Streptomyces* rimosus NRRL 2455 in solid state fermentation. *BMC Microbiology* 21:1–13.
El-Naggar, N. E., Moawad, H., El-Shweihy, N. M., El-Ewasy, S. M., Elsehemy, I. A., and N. A. Abdelwahed. 2019. Process development for scale-up production of a therapeutic L-asparaginase by *Streptomyces* brollosae NEAE-115 from shake flasks to bioreactor. *Scientific Reports* 9(1):13571.
Fotschki, J., Wróblewska, B., Fotschki, B., Kalicki, B., Rigby, N., and A. Mackie. 2020. Microbial transglutaminase alters the immunogenic potential and cross-reactivity of horse and cow milk proteins. *Journal of Dairy Science* 103(3):2153–66.
Hamoudi, H., Giacobbe, S., Strazzulli, A., Pezzella, C., Sannia, G., Moracci, M., Piscitelli, A., and M. Kecha. 2021. A novel *Streptomyces* strain isolated by functional bioprospecting for laccases. *Process Biochemistry* 111:315–24.
He, Y., Wang, B., Wen, L., Wang, F., Yu, H., Chen, D., Su, X., and C. Zhang. 2022. Effects of dietary fiber on human health. *Food Science and Human Wellness* 11(1):1–10.
Hovjecki, M., Miloradovic, Z., Mirkovic, N., Radulovic, A., Pudja, P., and J. Miocinovic. 2021. Rheological and textural properties of goat's milk set-type yogurt as affected by heat treatment, transglutaminase addition and storage. *Journal of the Science of Food and Agriculture* 101(14):5898–906.
Jayasree, D., Kumari, T. D., Kishor, P. B., Lakshmi, M. V., and M. L. Narasu. 2010. Optimization of production protocol of alkaline protease by *Streptomyces* pulvereceus. arXiv preprint arXiv:1002.0048.
Jing, D. 2010. Improving the simultaneous production of laccase and lignin peroxidase from *Streptomyces* lavendulae by medium optimization. *Bioresource Technology* 101(19):7592–7.

Kaić, A., Janječić, Z., Žgur, S., Šikić, M., and K. Potočnik. 2021. Physicochemical and sensory attributes of intact and restructured chicken breast meat supplemented with transglutaminase. *Animals* 11(9):2641.

Karkal, S. S., Venmarath, A., Velappan, S. P., and T. G. Kudre. 2022. Enzymes in meat, fish, and poultry product processing and preservation-II. *In Novel Food Grade Enzymes: Applications in Food Processing and Preservation Industries* 22:193–216.

Kemung, H. M., Tan, L. T., Khan, T. M., Chan, K. G., Pusparajah, P., and B. H. Goh. 2018. *Streptomyces* as a prominent resource of future anti-MRSA drugs. *Frontiers in Microbiology* 9:2221.

Khadayat, K., Sherpa, D. D., Malla, K. P., Shrestha, S., Rana N., and B. P. Marasini. 2020. Molecular identification and antimicrobial potential of *Streptomyces* species from nepalese soil. *International Journal of Microbiology* 2020:8817467.

Khushboo, K. P., Dubey, K. K., Usmani, Z., Sharma, M., and V. K. Gupta. 2022. Biotechnological and industrial applications of *Streptomyces* metabolites. *Biofuels, Bioproducts and Biorefining* 16(1):244–64.

Kim, H., Lee, M. Y., Lee, J., Jo, Y. J., and M. J. Choi. 2022. Effects of glucono-δ-lactone and transglutaminase on the physicochemical and textural properties of plant-based meat patty. *Foods* 11(21):3337.

Kitadokoro, K., Nakamura, E., Tamaki, M., Horii, T., Okamoto, H., Shin, M., Sato, T., Fujiwara, T., Tsuzuki, H., Yoshida, N., and H. Teraoka. 1993. Purification, characterization and molecular cloning of an acidic amino acid-specific proteinase from *Streptomyces* fradiae ATCC 14544. *Biochimica et Biophysica Acta (BBA)-Protein Structure and Molecular Enzymology* 1163(2):149–57.

Komaki, H., and Tamura. T. 2021. Differences at species level and in repertoires of secondary metabolite biosynthetic gene clusters among *Streptomyces* coelicolor A3 (2) and type strains of S. coelicolor and its taxonomic neighbors. *Applied Microbiology* 1(3):573–85.

Krieger, T. J., Bartfeld, D., Jenish, D. L., and D. Hadary. 1994. Purification and characterization of a novel tripeptidyl aminopeptidase from *Streptomyces* lividans 66. *FEBS Letters* 352(3):385–8.

Kumar, M., Kumar, P., Das, P., Solanki, R., and M. K. Kapur. 2020. Potential applications of extracellular enzymes from *Streptomyces* spp. in various industries. *Archives of Microbiology* 202:1597–615.

Kumar, P., Kundu, A., Kumar, M., Solanki, R., and M. K. Kapur. 2019. Exploitation of potential bioactive compounds from two soil derived actinomycetes, *Streptomyces* sp. strain 196 and RI. 24. *Microbiological Research* 229:126312.

Lang, G. H., Kringel, D. H., dos Santos, A. T., Ferreira, C. D., Dias, Á. R., Da Rosa Zavareze, E., and M. de Oliveira. 2020. Cake of brown, black and red rice: Influence of transglutaminase on technological properties, in vitro starch digestibility and phenolic compounds. *Food Chemistry* 318:126480.

Lee, N., Hwang, S., Kim, J., Cho, S., Palsson, B., and B. K. Cho. 2020. Mini review: Genome mining approaches for the identification of secondary metabolite biosynthetic gene clusters in *Streptomyces*. *Computational and Structural Biotechnology Journal* 18:1548–56.

Lee, N., Hwang, S., Kim, W., Lee, Y., Kim, J. H., Cho, S., Kim, H. U., Yoon, Y. J., Oh, M-K., Palsson, B. O., and B. K. Cho. 2021. Systems and synthetic biology to elucidate secondary metabolite biosynthetic gene clusters encoded in *Streptomyces* genomes. *Natural Product Reports* 38:1330–61.

Li, Y., Kong, D., Fu, Y., Sussman, M. R., and H. Wu. 2020. The effect of developmental and environmental factors on secondary metabolites in medicinal plants. *Plant Physiology and Biochemistry* 148:80–9.

Lichenstein, H. S., Busse, L. A., Smith, G. A., Narhi, L. O., McGinley, M. O., Rohde, M. F., Katzowitz, J. L., and M. M. Zukowski. 1992. *Cloning and characterization of a gene encoding extracellular metalloprotease from Streptomyces lividans Gene* 111(1):125–30.

Liu, L., Xu, M., Cao, Y., Wang, H., Shao, J., Xu, M., Zhang, Y., Wang, Y., Zhang, W., Meng, X., and W. Liu. 2020. Biochemical characterization of xylanases from *Streptomyces* sp. B6 and their application in the xylooligosaccharide production from viscose fiber production waste. *Journal of Agricultural and Food Chemistry* 68(10):3184–94.

Makitrynskyy, R., Tsypik, O., and A. Bechthold. 2021. Genetic engineering of *Streptomyces* ghanaensis ATCC14672 for improved production of moenomycins. *Microorganisms* 10 (1):30.

Medouni-Haroune, L., Medouni-Adrar, S., Houfani, A. A., Bouiche, C., Azzouz, Z., Roussos, S., Desseaux, V., Madani, K., and M. Kecha. 2023. Statistical optimization and partial characterization of xylanases produced by *Streptomyces* sp. S1M3I using olive pomace as a fermentation substrate. *Applied Biochemistry and Biotechnology* 196 (4):1–9.

Meena, M., Prajapati, P., Ravichandran, C., and R. Sehrawat. 2021. Natamycin: a natural preservative for food applications—a review. *Food Science and Biotechnology* 5:1–6.

Mokoonlall, A., Sykora, L., Pfannstiel, J., Nobel, S., Weiss, J., and J. Hinrichs. 2016. A feasibility study on the application of a laccase-mediator system in stirred yogurt at the pilot scale. *Food Hydrocolloids* 60:119–27.

Nicault, M., Zaiter, A., Dumarcay, S., Chaimbault, P., Gelhaye, E., and P. Leblond. 2021. Elicitation of antimicrobial active compounds by *Streptomyces*-fungus co-cultures. *Microorganisms* 9:178.

Patel, P. H., Panchal, K. S., Patel, B. N., and H. P. Patel. 2021. Production and partial purification of pectinase from *Streptomyces* chartreusis. *Crop Research* 56(1and2):67–73.

Poornima, S., Divya, P., Karmegam, N., Karthik, V., and R. Subbaiya. 2020. Aqueous two-phase partitioning and characterization of xylanase produced by *Streptomyces* geysiriensis from low cost lignocellulosic substrates. *Journal of Bioscience and Bioengineering* 130(6):571–6.

Procopio, R. E., Silva, I. R., Martins, M. K., Azevedo J. L., and J. M. Araújo. 2012. Antibiotics produced by Streptomyces. *Brazilian Journal of Infectious Diseases* 16:466–71.

Rathore, D. S., and S. P. Singh. 2021. Kinetics of growth and co-production of amylase and protease in novel marine actinomycete, *Streptomyces* lopnurensis KaM5. *Folia Microbiologica* 66(3):303–16.

Raveendran, S., Parameswaran, B., Ummalyma, S. B., Abraham, A., Mathew, A.K., Madhavan, A., Rebello, S., and Pandey, A.. 2018. Applications of microbial enzymes in food industry. *Food Technology and Biotechnology* 56(1):16–30.

Read, R. J., Brayer, G. D., Jurasek, L., and M. N. James. 1984. Critical evaluation of comparative model building of *Streptomyces* griseus trypsin. *Biochemistry* 23(26):6570–75.

Salehghamari, E., Nasrollahzadeh, Z., Tahmaseb, M., and M. A. Amoozegar. 2019. Pectinase enzyme from *Streptomyces* coelicoflavus GIAL86 isolated from Meyghan Salt Lake, Arak, Iran. *International Journal of Aquatic Biology* 7(2):106–11.

Sanchez, De La, Nieta, R., Antoraz, S., Alzate, J. F., Santamaría, R. I., and M. Díaz. 2020. Antibiotic production and antibiotic resistance: the two sides of AbrB1/B2, a two-component system of *Streptomyces* coelicolor. *Frontiers in Microbiology* 11:587750.

Sarmiento, T. A. A., Silva, L., Sanchez-Suarez, J., and L. Diaz. 2022. Streptomyces-derived bioactive pigments: Ecofriendly source of bioactive compounds. *Coatings* 12(12):1858.

Scaffaro, R., Lopresti, F., Sutera, A., Botta, L., Fontana, R. M., and G. Gallo. 2017. Plasma modified PLA electrospun membranes for actinorhodin production intensification in *Streptomyces* coelicolor immobilized-cell cultivations. *Colloids and Surfaces B: Biointerfaces* 157:233–41.

Selamoglu, Z. 2020. Use of enzymes in dairy industry: a review of current progress. *Archives of Razi Institute* 75(1):131.

Shakeri, M. H., Shahnavaz, B., Makhdoumi, A., and M. Iranshahy. 2021. Evaluating the effect of various bacterial consortia on antibacterial activity of marine *Streptomyces* sp. AC117. *Biocontrol Science and Technology* 31(11):1248–66.

Shariq, M., and M. Sohail. 2019. Citrus limetta peels: a promising substrate for the production of multienzyme preparation from a yeast consortium. *Bioresources and Bioprocessing* 6(1):1–15.

Shirvani, A., Mirzaaghaei, M., and S. A. Goli. 2023. Application of natural fining agents to clarify fruit juices. *Comprehensive Reviews in Food Science and Food Safety* 22(6):4190–216.

Silano, V. B., Baviera, J. M., Bolognesi, C., Brüschweiler, B. J., Cocconcelli, P. S., Crebelli, R., Gott, D. M., Grob, K., Lampi, E., and A. Mortensen. 2019. Safety evaluation of the food enzyme chitinase from *Streptomyces* violaceoruber (strain pC hi). *EFSA Journal* 17(7):e05767.

Singh, T., Sharma, S., and Y. Tiwari. 2022. Microbial proteases: a significant tool for industrial applications. *InMicrobial Products* 16:11–24.

Sinjaroonsak, S., Chaiyaso, T., and Kittikun, A.H. 2020. Optimization of cellulase and xylanase productions by *Streptomyces* thermocoprophilus TC13W using low cost pretreated oil palm empty fruit bunch. *Waste and Biomass Valorization* 11:3925–36.

Slemc, L., Pikl, Š., Petković, H., and M. Avbelj. 2021. Molecular biology methods in *Streptomyces* rimosus, a producer of oxytetracycline. *InAntimicrobial Therapies: Methods and Protocols* 12:303–30).

Sondhi, S., Kaur, P. S., Sura, H., Juglani, M., and D. Sharma. 2021. Amylase based clarification of apple, orange and grape juice. *International Journal of Contemporary Research* 3(2):187–90.

Song, Z., Ma, Z., Bechthold, A., and X. Yu. 2020. Effects of addition of elicitors on rimocidin biosynthesis in *Streptomyces* rimosus M527. *Applied Microbiology and Biotechnology* 104:4445–55.

Tokay, F. G., Alp, A. C., and P. Yerlikaya. 2021. Production and shelf life of restructured fish meat binded by microbial transglutaminase. *LWT* 152:112369.

Umezawa, Y., Yokoyama, K., Kikuchi, Y., Date, M., Ito, K., Yoshimoto, T., and H. Matsui. 2004. Novel prolyl tri/tetra-peptidyl aminopeptidase from *Streptomyces* mobaraensis: substrate specificity and enzyme gene cloning. *Journal of Biochemistry* 136(3):293–300.

van Bergeijk, D. A., Elsayed, S. S., Du, C., Santiago, I. N., Roseboom, A. M., Zhang, L., Carrión, V. J., Spaink, H. P., and G. P. van Wezel. 2022. The ubiquitous catechol moiety elicits siderophore and angucycline production in Streptomyces. *Communications Chemistry* 5(1):14.

Vasic, K., Knez, Ž., and M. Leitgeb. 2023. Transglutaminase in foods and biotechnology. *International Journal of Molecular Sciences* 24(15):12402.

Velazquez-De, B. S., Hernández-Domínguez, E. M., Villa-Garcia, M., Diaz-Godinez, G., Mandujano-Gonzalez, V., and Álvarez-Cervantes, J. 2021. Alvarez-Cervantes. Exogenous enzymes as zootechnical additives in animal feed: a review. *Catalysts* 11(7):851.

Venil, C. K., Dufossé, L., and P. Renuka. 2020. Bacterial pigments: sustainable compounds with market potential for pharma and food industry. *Frontiers in Sustainable Food Systems* 4:100.

Ventorino, V., Ionata, E., Birolo, L., Montella, S., Marcolongo, L., De Chiaro, A., Espresso, F., Faraco, V., and O. Pepe. 2016. Lignocellulose-adapted endo-cellulase producing *Streptomyces* strains for bioconversion of cellulose-based materials. *Frontiers in Microbiology* 7:217082.

Wang, W., Li, S., Li, Z., Zhang, J., Fan, K., Tan, G., Ai, G., Lam, S.M., Shui, G., Yang, Z., and H. Lu. 2020. Harnessing the intracellular triacylglycerols for titer improvement of polyketides in Streptomyces. *Nature Biotechnology* 38(1):76–83.

Wang, X., Yin, S., Bai, J., Liu, Y., Fan, K., Wang, H., Yuan, F., Zhao, B., Li, Z., and W. Wang. 2019. Heterologous production of chlortetracycline in an industrial grade *Streptomyces* rimosus host. *Applied Microbiology and Biotechnology* 103:.6645–55.

Wei, J., He, L., and G. Niu. 2018. Regulation of antibiotic biosynthesis in actinomycetes: perspectives and challenges. *Synthetic and Systems Biotechnology* 3(4):229–35.

Wi, K. D., Gyun, K., Seop, K., Kyu, L. B., Taik, R., and L. K. Joon. 2006. Proteases and protease inhibitors produced in streptomycetes and their roles in morphological differentiation. *Journal of Microbiology and Biotechnology* 16(1):5–14.

Xu, Z., and Y. Li. 2020. A MarR-family transcriptional factor MapR positively regulates actinorhodin production in *Streptomyces* coelicolor. *FEMS Microbiology Letters* 367(18):140.

Yamane, T., Iwasaki, A., Suzuki, A., Ashida, T., Kawata, Y., and F. Sakiyama. 1995. Crystal Structure of *Streptomyces* erythraeus Trypsin at 1.9 Å Resolution. *The Journal of Biochemistry* 118(5):882–94.

Ye, L., Ye, R., Hu, F., and G. Wang. 2021. Combination of atmospheric and room temperature plasma (ARTP) mutagenesis, genome shuffling and dimethyl sulfoxide (DMSO) feeding to improve FK506 production in *Streptomyces* tsukubaensis. *Biotechnology Letters* 43(9):1809–20.

Zhang, F., Gao, D., Lin, J., Zhu, M., Zhuang, Z., Duan, Y., and X. Zhu. 2020. Construction of inducible genetic switch for the global regulator WblA to sustain both overproduction of tiancimycins and on-demand sporulation in *Streptomyces* sp. CB03234. *ACS Synthetic Biology* 9(6):1460–7.

Zhang, H., Zhan, J., Su, K., and Y. Zhang. 2008. A kind of potential food additive produced by *Streptomyces* coelicolor: characteristics of blue pigment and identification of a novel compound, λ-actinorhodin. *Food Chemistry* 95:186–92.

Zhang, L., Zhang, L., Han, X., Du, M., Zhang, Y., Feng, Z., Yi, H., and Y. Zhang. 2012. Enhancement of transglutaminase production in *Streptomyces* mobaraensis as achieved by treatment with excessive MgCl 2. *Applied Microbiology and Biotechnology* 93:2335–43.

Zhou, Y., Han, L. R., He, H. W., Sang, B., Yu, D. L., Feng, J. T., and X. Zhang. 2018. Effects of agitation, aeration and temperature on production of a novel glycoprotein GP-1 by *Streptomyces* kanasenisi ZX01 and scale-up based on volumetric oxygen transfer coefficient. *Molecules* 23(1):125.

Zhu, J., Shen, C., Zhao, W., Liu, X., Liu, J., and B. Yu. 2021. Regio-and stereoselective hydroxylation of testosterone by cytochrome P450 from *Streptomyces* griseus ATCC 13273. *Biocatalysis and Biotransformation* 39(2):130–7.

Zhu, Y., Shang, X., Yang, L., Zheng, S., Liu, K., and X. Li. 2020. Purification, identification and properties of a new blue pigment produced from *Streptomyces* sp. A1013Y. *Food Chemistry* 308:125600.

Index

Pages in *italics* refer to figures and pages in **bold** refer to tables.

For Product Safety Concerns and Information please contact our EU representative GPSR@taylorandfrancis.com Taylor & Francis Verlag GmbH, Kaufingerstraße 24, 80331 München, Germany